Ferroelastic Materials

Online at: https://doi.org/10.1088/978-0-7503-6089-0

Ferroelastic Materials

Edited by
Guillaume F Nataf
CNRS, University of Tours, INSA Centre Val de Loire, Tours, France

Blai Casals
Universitat de Barcelona, Barcelona, Spain

Ekhard K H Salje
University of Cambridge, Cambridge, United Kingdom

IOP Publishing, Bristol, UK

ISBN   978-0-7503-6089-0 (ebook)
ISBN   978-0-7503-6087-6 (print)
ISBN   978-0-7503-6090-6 (myPrint)
ISBN   978-0-7503-6088-3 (mobi)

DOI   10.1088/978-0-7503-6089-0

Version: 20251201

IOP ebooks

British Library Cataloguing-in-Publication Data: A catalogue record for this book is available from the British Library.

Published by IOP Publishing, wholly owned by The Institute of Physics, London

IOP Publishing, No.2 The Distillery, Glassfields, Avon Street, Bristol, BS2 0GR, UK

US Office: IOP Publishing, Inc., 190 North Independence Mall West, Suite 601, Philadelphia, PA 19106, USA

During the course of writing this book, Professor Ekhard K H Salje, a leading person in the field of ferroelasticity, and co-editor of this book, passed away. This book is dedicated to him.

Contents

6 Functional properties of ferroelastic domain walls   **6-1**

Guillaume F Nataf, Kumara Cordero-Edwards, Guangming Lu and Ekhard K H Salje

7 Ferroelastic domain walls at surfaces   **7-1**

Anna Morozovska, Eugène Eliseev and Nick Barrett

Editor biographies

Guillaume F Nataf

Guillaume F Nataf completed his joint PhD in 2016 at the University Paris-Saclay (France) and the University of Luxembourg. In 2017 he moved to the University of Cambridge as postdoc and obtained a Research Fellowship from the Royal Commission for the Exhibition of 1851. He also became a Junior Research Fellow of Wolfson College, Cambridge. In 2021 he became a CNRS Research Scientist at the GREMAN laboratory, Tours, France. A year later he was awarded an ERC Starting Grant to investigate thermal transport in ferroic materials. His research focuses on functional oxides (ferroelectrics and ferroelastics), with a particular interest in micro- and nano-structures (domains, domain walls).

Blai Casals

Blai Casals completed his PhD in Materials Science at the Universitat Autònoma de Barcelona (ICMAB-CSIC) in 2017. He then carried out postdoctoral research at the Institut Català de Nanociència i Nanotecnologia (ICN2) with the group of Professor Gustau Catalan, and at the University of Cambridge with the group of Professor Ekhard K H Salje, before joining the Department of Applied Physics at the University of Barcelona in 2022 as a tenure-track Assistant Professor. His research focuses on the dynamics of ferroic domain walls and avalanche phenomena, including magnetoacoustic interactions, ferroelastic and ferroelectric switching, and the mechanical and magnetic behavior of thin films.

Ekhard K H Salje

Ekhard K H Salje completed his PhD at Hannover University, Germany, in 1972 and became Head of the Institute for Crystallography and Petrology (Hannover University) in 1983. In 1985 he moved to the University of Cambridge where he became Professor of Mineral Physics (Department of Earth Sciences) in 1992. He was recognized by several academies: Leopoldina (1994), The Royal Society (1996), The Royal Society of Arts (1996), and Reial Acadèmia de Ciències i Arts de Barcelona (2011). He received numerous awards, including the Abraham-Gottlob-Werner Medal in Mineralogy (1994), the Agricola Medal for Applied Mineralogy (2006) of the German Mineralogical Society, the Schlumberger Medal of the Mineralogical Society of Great Britain (1998), and the Humboldt Prize of the Alexander von

Humboldt Foundation (2000). He is a Fellow of the Institute of Physics (1996) and the Geological Society (1996) and Honorary Fellow of Darwin (2001) and Clare Hall (2010) Colleges, in Cambridge. In 2004 he received the French Palmes Academiques and in 2006 the German Cross of the Order of Merit, first class. He has been honored by the Universities of Xi'an Jiaotong and of Würzburg. His research focused on mathematically correct and physically meaningful description of microstructures in minerals, in particular ferroelastics.

List of contributors

Jordi Baró
Departament de Física de la Matèria Condensada, Universitat de Barcelona, Barcelona, Spain

Nick Barrett
SPEC, CEA, CNRS, CEA Saclay, Université Paris-Saclay, Gif-sur-Yvette, France

Michael A Carpenter
Department of Earth Sciences, University of Cambridge, Cambridge, United Kingdom

Blai Casals
Departament de Física Aplicada, Universitat de Barcelona, Barcelona, Spain

Kumara Cordero-Edwards
Catalan Institute of Nanoscience and Nanotechnology (ICN2), Bellaterra, Barcelona, Spain

Eugène Eliseev
Frantsevich Institute for Problems in Materials Science, National Academy of Sciences of Ukraine, Kyiv, Ukraine

Jurij Koruza
Institute for Chemistry and Technology of Materials, Graz University of Technology, Austria

Etienne Lemaire
GREMAN UMR7347, CNRS, University of Tours, INSA Centre Val de Loire, Tours, France

Wei Li
School of Materials Science and Engineering and Tianjin Key Laboratory of Metal and Molecule-Based Material Chemistry, Nankai University, Tianjin, China

Guangming Lu
School of Environmental and Materials Engineering, Yantai University, Yantai, China

Anna Morozovska
Institute of Physics, National Academy of Sciences of Ukraine, Kyiv, Ukraine

Kevin Nadaud
GREMAN UMR7347, CNRS, University of Tours, INSA Centre Val de Loire, Tours, France

Guillaume F Nataf
GREMAN UMR7347, CNRS, University of Tours, INSA Centre Val de Loire, Tours, France

Petr Ondrejkovič
FZU—Institute of Physics of the Czech Academy of Sciences, Prague, Czech Republic

Mojca Otoničar
Electronic Ceramics Department, Jožef Stefan Institute, Ljubljana, Slovenia

Antoni Planes
Departament de Física de la Matèria Condensada, Universitat de Barcelona, Barcelona, Spain

Marcel Porta
Departament de Física Quàntica i Astrofísica, Universitat de Barcelona, Barcelona, Spain

Tadej Rojac
Electronic Ceramics Department, Jožef Stefan Institute, Ljubljana, Slovenia

Ekhard K H Salje
Department of Earth Sciences, University of Cambridge, Cambridge, United Kingdom

Wilfried Schranz
Faculty of Physics, University of Vienna, Wien, Austria

Andreas Tröster
Institute of Materials Chemistry, Technical University of Vienna, Vienna, Austria

Susan Trolier-McKinstry
Department of Materials Science and Engineering, The Pennsylvania State University, University Park, PA, USA

Dwight Viehland
Department of Materials Science and Engineering, Virginia Tech, VA, USA

Eduard Vives
Departament de Física de la Matèria Condensada, Universitat de Barcelona, Barcelona, Spain

IOP Publishing

Ferroelastic Materials

Guillaume F Nataf, Blai Casals and Ekhard K H Salje

Chapter 1

Introduction

Ekhard K H Salje and Guillaume F Nataf

Ferroicity encompasses the fundamental properties of ferroelectricity, ferromagnetism and ferroelasticity. Conceptually, they have different switching mechanisms where the defining hysteresis is electric, magnetic or mechanic, respectively. These external perturbations often combine, which is particularly important in the case of ferroelasticity because mechanical deformations are large in ferroelectrics and usually small, but not absent, in ferromagnets. Such coupling effects between the electrical polarization, the magnetic moment and the spontaneous deformations are the subject of much research and hold the key to understanding fundamental riddles such as the question 'is $BaTiO_3$ ferroelectric or ferroelastic?' The answer is 'both', and that either property would not be viable without the other.

It is probably the human fascination with curiosities that led to ferroelectricity being researched historically as early as in the seventeenth century, and then more widely in the early twentieth century when research on (anti-)ferromagnetism was initiated by Louis Néel in 1932. The field of ferroelasticity was founded much later with the first hysteresis study published in 1976 [1]. The term 'ferroelasticity' was first used by Aizu in 1970. The modern symmetry conditions for ferroelasticity are compiled well in the Bilbao Crystallographic Database [2]. The first textbook appeared in 1990 [3], with a second edition in 1993. The term 'multiferroic' often describes combined hysteresis effects, which are particularly attractive for the development of memory devices where magnetism and electric switching are combined.

The physical feature of mechanical switching relates to mobile interfaces. Often these interfaces are twin boundaries where the bulk crystal structure is slightly deformed with orientational anomalies such as domains with different shear strains and different orientations. Such twin walls (or domain walls) were known as early as 1669 [4], but systematic research of twin structures first started at the end of the nineteenth century. A huge amount of work went into the analysis of the geometrical pattern formations of twin structures [4, 5]. The key observation for ferroelastic materials was often overlooked, however. Importantly, the atomic arrangement

doi:10.1088/978-0-7503-6089-0ch1

inside the twin walls can differ greatly from the equivalent bulk structure [6]. This means that the physical properties of the wall are typically very different from the extrapolated bulk properties. The term 'emerging wall properties' was chosen for this phenomenon. The most common emerging property is that centrosymmetric bulk materials possess polar twin walls. Large dipole moments exist in walls but not in the adjacent domains [7].

Furthermore, the interatomic distances are often larger in the twin walls than in the bulk, so that atomic diffusion is greatly enhanced in the walls. This means that it is quite easy to dope such domain boundaries and, in the natural environment such as in natural minerals, they are always doped by environmentally available atoms/ions. Separating wall and bulk dopants has some of the greatest promise for unravelling the geological and palaeontological past of our planet. Artificial doping has been shown to produce typical wall effects ranging from superconductivity to photovoltaics, transistors, etc.

It is little wonder, therefore, that ferroelasticity advanced from being the Cinderella of multiferroic research to its forefront. Ferroelastics contain domain boundaries which have some of the most attractive properties—they produce atomically thin layers with extreme emerging properties such the that the device is the wall and not the bulk. The thickness of the device is thus just a few atoms thick. The ferroelastic domain boundaries are also mobile, such that they can be driven by external forces to exactly the sites where they are wanted. This enables them to be memory devices in memristors or neuromorphic computation, including the formation of magnetic vortices [8]. Their pattern formation is only now being explored. Proximity effects and the formation of walls inside walls can reduce the size of active device elements even further by the formation of Bloch walls such that the information density in ferroelastics becomes extremely high. Complex pattern formation, including domain glasses, has been envisaged and it appears that ferroelasticity as a collective phenomenon still has the potential to reveal a multitude of heretofore unexpected ferroelastic properties.

This book summarizes some of the most advanced studies of ferroelasticity and is suitable for researchers and students alike.

Chapter 2 provides an overview of the thermodynamics governing ferroelastic and antiferroelectric systems and formulates some open questions on these topics. Ferroelastic materials are defined by the reversible phase transition they undergo from a high-symmetry (high-temperature) paraelastic phase to a low-symmetry (low-temperature) ferroelastic phase, giving rise to spontaneous strain. If the two phases are group–subgroup related, this transition is described with a high accuracy by Landau-type theories, where coupling between order parameters and spontaneous strain is considered. Landau–Ginzburg theory can also be used to discuss the effects from inhomogeneous structures such as domains, domain walls or precursor structures. In contrast to the ferroelastic phase transitions that give rise to macroscopic tensors coupled to macroscopic fields, the criteria defining antiferroelectric phase transitions (that are also observed in ferroelastic materials) are still under discussion due to the absence of specific symmetry properties characterizing their existence. Recent approaches based on Landau-type theories tackle this issue.

Chapter 3 complements the overview of ferroelastic materials by providing another unifying perspective, consisting in the introduction of the formal concepts of ferroelastic species and layer groups, which are particularly useful to clarify the symmetry-imposed properties of ferroelastic domains and ferroelastic domain walls, respectively. At the same time, the concept of ferroelastic species also allows one to define and classify the ferrolastic materials themselves.

Chapter 4 focuses on the motion of ferroelastic domain walls, which has a profound influence on the material properties, such as the elastic stiffness, the relative permittivity, and the piezoelectric coefficients. Ferroelastic domain walls do not necessarily move smoothly through real solids, where 0D, 1D, 2D and 3D defects are omnipresent. Rather, they move through a complex energy profile with pinning centres that impede wall motion and induce hysteresis in the presence of oscillating fields. These movements are probed through a diverse range of (electrical) measurements: hysteresis loops, the Rayleigh law, third harmonic responses, and first order reversal curves. Structural characterizations, e.g. using x-ray diffraction and electron microscopy, complement these approaches.

Chapter 5 has another look at the motion of ferroelastic domain walls and phase boundaries by exploring the concept of avalanches. Avalanches are discrete events that occur when an external force is applied to a material. These events occur intermittently, and their magnitudes are statistically distributed according to a power law. In ferroelastic materials, avalanches are associated with changes in the order parameter of the phase transition, which manifests itself as variations in the spontaneous strain. They result from a competition between long-range elastic interactions related to the ferroelastic order and local disorder. Recording these avalanches remains an experimental challenge and requires experimental techniques capable of probing events on different spatial and temporal scales. The modeling of the avalanche statistics relies on stochastic models.

Chapter 6 provides an overview of ferroelastic domain walls. Ferroelastic domain walls have widths at the unit cell level and exhibit distinct mechanical properties. They host emerging properties, including (super-)conductivity, polarity and magnetic properties. They can combine to form different hierarchical structures where wall–wall interactions are important, and which can be controlled by an applied external field. Their ability to scatter phonons can be used to control reversibly thermal conductivity.

Chapter 7 focuses on the understanding and characterization of the surface discontinuity of ferroelastic domain walls. Both the specific domain wall properties and how the surface discontinuity can modulate them need to be understood, with the possible emergence of unique properties at the intersection of domain walls and the surface. Using the Landau–Ginzburg–Devonshire approach, elastic fields caused by the domain wall–surface junction in ferroelastics are analysed, and the appearance of improper ferroelectricity, carrier accumulation and vacancy segregation at the twin–surface intersection are considered. Electron-based techniques used to probe polarization at surfaces are reviewed, with a focus on the experimental demonstration of polarity in ferroelastic domain walls at the surface of bulk single crystal $CaTiO_3$.

Chapter 8 deals with martensitic transitions, which are a class of ferroelastic transitions taking place in metals and alloys. In particular, the focus is on those for which the symmetries of the high- and low-temperature phases show a group–subgroup relationship, which is the case for shape-memory materials. It is shown that these transitions often show an athermal character, which means that the transition dynamics is largely not influenced by thermal fluctuations. The consequence of this is that due to the existence of disorder, these transitions take place intermittently through a sequence of avalanches. Results show that avalanche size, duration and energy show power-law distributions to a very good approximation that corroborate the lack of characteristic scales and, thus, the fact that the transition process displays scale invariance. It is shown that avalanches require the existence of a critical amount of disorder, which is achieved through a self-organization process that occurs through cycling across the transition. At criticality, the critical exponents that characterize the avalanche distributions depend mainly on the symmetry change at the transition and on whether the driving parameter is a generalized force or its thermodynamically conjugate generalized displacement. A realistic, symmetry-based, Ginzburg–Landau model with disorder is proposed that reproduces martensitic microstructures and confirms avalanche criticality features.

Chapter 9 provides an overview of hybrid organic–inorganic perovskites, an emerging subclass of materials with perovskite architecture, which are chemically abundant and structurally diverse. They include halides, formates, azides, cyanides, hypophosphites, borohydrides, dicyanamides as well as dicynametallates. Compared to perovskite oxides, hybrid organic–inorganic perovskites show significantly more structural flexibility due to the existence of molecular components. This virtue allows hybrid organic–inorganic perovskites to show abundant phase transitions and many of them are ferroelastic. The ferroelastic phase transitions in hybrid organic–inorganic perovskites can be driven by complex mechanisms, which include not only conventional octahedral tilting and atomic displacements, but also order–disorder and rearrangement of molecular interactions.

Chapter 10 focuses on the applications of ferroelastic materials. Ferroelastics have been used for their shape-memory capabilities. Shape deformations occur under applied stress, and materials return to their previous shape upon thermal annealing. They have found unique applications in biomedical devices and implants, in aerospace applications as morphing wings and reconfigurable structures, and as stress driven actuators in engine and climate controls. The true ferroelastic switching nature has the potential for non-volatile memory and strain-tunable devices; however, in the past, it has proven difficult to drive these effects by applied strain. The coelastic materials, which include piezoelectric ceramics and single crystals, form the basis of electromechanical devices, acoustic sonar and medical transducers, and electrically driven actuators, just to give a few examples. Much research effort has been spent in developing high-performance piezoelectric materials, with high piezoelectric and electromechanical coupling coefficients. Efforts have focused on doping and compositional modifications, amongst other important variables. The transduction efficiency between electrical and mechanical energies can be $>90\%$ in the case of piezoelectric single crystals, along with low loss in their hysteresis loops.

The unique properties of these high-performance piezoelectric materials have been instrumental in their applications. More recently, the research focus has been on making heterostructure materials, using piezoelectric susbstrates and depositing other functional layers with other properties. These coelastic piezoelectric substrates can then be used to strain-tune ferroelastic, ferromagnetic and other important functional properties. This area of research is emerging, and may have important applications in future technologies.

Bibliography

[1] Salje E and Hoppmann G 1976 Direct observation of ferroelasticity in $Pb_3(PO_4)_2$–$Pb_3(VO_4)_2$ *Mater. Res. Bull.* **11** 1545

[2] Aroyo M I, Perez-Mato J M, Orobengoa D, Tasci E, De La Flor G and Kirov A 2011 Crystallography online: Bilbao Crystallographic Server *Bulg. Chem. Commun.* **43** 183

[3] Salje E K 1991 *Phase Transitions in Ferroelastic and Co-Elastic Crystals* (Cambridge: Cambridge University Press)

[4] Stenonis N 1669 *De Solido Intra Solidum Naturaliter Contento Dissertationis Prodromus* (Florence: Ex typographia sub signo Stellae)

[5] Nespolo M and Ferraris G 2000 Twinning by syngonic and metric merohedry analysis, classification and effects on the diffraction pattern *Z. Kristallogr.—Cryst. Mater.* **215** 77

[6] Janovec V and Přívratská J 2013 *Domain Structures* (Dordrecht: Springer) pp 484–543

[7] Van Aert S, Turner S, Delville R, Schryvers D, Van Tendeloo G and Salje E K H 2012 Direct observation of ferrielectricity at ferroelastic domain boundaries in $CaTiO_3$ by electron microscopy *Adv. Mater.* **24** 523

[8] Salje E K H 2021 Mild and wild ferroelectrics and their potential role in neuromorphic computation *APL Mater.* **9** 010903

IOP Publishing

Ferroelastic Materials

Guillaume F Nataf, Blai Casals and Ekhard K H Salje

Chapter 2

Ferroelastic phase transitions—well-known results and new perspectives

Wilfried Schranz, Andreas Tröster and Michael A Carpenter

2.1 Introduction

Phase transitions are of great importance in various scientific fields, including physics, chemistry, and materials science. They play a crucial role in understanding the behaviour of matter under different conditions and their property changes under various external conditions can be exploited in various ways. A structural phase transition at a critical temperature T_c (or pressure P_c, etc)—either driven by temperature, pressure, or another external field—is usually accompanied by a symmetry breaking, which can be quantified by an order parameter. The order parameter attains a non-zero value in the low symmetry phase (usually below T_c or above P_c) and is zero in the high-symmetry phase. It can be a scalar (e.g. density difference), a vector (e.g. polarization **P**), a pseudo-vector (e.g. magnetization **M** or rotation angle ϕ), or a strain tensor component ε_{ij}, etc. The character of the order parameter is determined by the so-called active irreducible representation (IR) [1], of the symmetry group of the high-symmetry phase. The active **IR** also determines the symmetry of the soft mode [2], a collective mode condensing at the transition.

In the case of a ferroelastic phase transition the order parameter is a strain (second rank tensor ε_{ij}, $i = 1, 2, 3$), if the strain fully breaks the symmetry of the paraelastic phase at the phase transition, or it is non-linearly coupled to a symmetry-breaking order parameter [3–5]. Strain is a unique quantity in the sense that it can couple to a variety of other variables such as polarization, polarization gradients, magnetic moments, etc. The resulting piezoelectric, flexo-electric or magnetoelastic, etc, properties make such materials important from the point of view of applications as, e.g. sensors, actuators, acousto-optic modulators and transducers, or for shape-memory applications, etc. Research on ferroelectrics and ferroelastics is increasingly dominated by the investigation of nano-scale structures [6–8]. Special interest is placed now on the emerging properties of domain walls [9], which has led to the new

doi:10.1088/978-0-7503-6089-0ch2
2-1

field called 'domain boundary engineering' [7, 10] with interesting opportunities for the development of novel device materials [11]. Ferroelastic materials play a special role here [12, 13] as they encompass a large fraction of materials exhibiting structural and other phase transitions. Ferroelastic phase transitions can be conveniently modelled by Landau's [1, 14] general and elegant theory of phase transitions of a mean-field type. Since at ferroelastic phase transitions the interactions mediated by strain fields are long-ranged, Landau theory often works quantitatively over a wide range of temperatures [23] providing valuable insights into thermodynamic properties such as specific heat, susceptibilities, etc. However, basic Landau theory only covers transitions between homogeneous phases. To account for the energy costs of inhomogeneities at lowest order, one adds invariants quadratic in the gradient of the order parameter (so-called Ginzburg terms [15]) to the Landau free energy density. The resulting Landau–Ginzburg free energies, which are also referred to as phase field models in materials science, allow the Landau framework to be extended to the study of domain walls and other inhomogeneous microstructures. Computational approaches to study ferroelastic phase transitions include large scale computer simulations of atomistic models [16, 17], density functional (DFT)-calculations [13, 18, 19], molecular dynamics (MD) simulations [20, 21], to name only a few.

From the experimental point of view, it is evident that studying the dynamic elastic response of ferroelastic phase transitions materials can play a vital role in understanding the physical phenomena near ferroelastic phase transitions. Because the elastic constants constitute tensorial susceptibilities, they can provide highly sensitive indicators of the strength and type of any strain coupling which may occur. A wide variety of patterns of evolution with temperature, pressure and applied field is possible and, in principle, should provide insights into the strength, mechanisms and dynamics of strain coupling for any particular material of interest. However, in contrast to dielectric spectroscopy where the frequency dependence of the electric field can be adjusted continuously through many orders of magnitude, elastic measurements need a number of different methods which work in relatively narrow frequency windows. Using torsion pendulum measurements (10^{-3}–1 Hz), dynamic mechanical analysis (10^{-1}–10^2 Hz), rod resonance (10^3–10^5 Hz), resonant ultrasound spectroscopy (10^5–10^6 Hz), pulse-echo ultrasonics (10^7–10^9 Hz) and Brillouin scattering (10^{10}–10^{11} Hz) 14 decades can be captured in principle. Although we are not aware of any material studied over such a broad frequency range, there are several examples where at least low frequency (10^{-1}–10^2 Hz) and high frequency (10^5–10^6 Hz) data are available, and we will discuss some of them in this chapter.

After some brief historical notes (section 2.2) and a short reminder of the definition of spontaneous strain in different domain states (section 2.3) we focus on a variety of selected examples displaying proper (section 2.4.1), pseudo-proper (section 2.4.2), and improper (section 2.4.1) ferroelastic (FEL), and co-elastic (section 2.4.4) phase transitions (PTs). Recently found relations [22] of ferroelasticity in antiferroelectric materials will be discussed briefly in section 2.5. Special attention is given (section 2.6) to the study of inhomogeneous structures at specific ferroelastic phase transitions including domains and domain walls (section 2.6.1) and precursor effects related to order parameter fluctuations (section 2.6.2). The chapter concludes with a short summary and a look to the future.

2.2 Some (almost) historical notes

The term 'ferroelasticity' was coined by Aizu [24] in 1969. A crystal is said to be *ferroelastic* if it has two or more orientation states (domain states (DSs)) in the absence of mechanical stress which can be switched from one to another by the application of a mechanical stress. Later, Aizu [25] introduced the term 'ferroics'. Phase transitions accompanied by a change of the point-group symmetry are called ferroic phase transitions. Aizu provided a unified symmetry classification of ferro-electrics, ferroelastics and ferromagnetics. He also considered [26, 27] the possibility of antiferroelasticity, in analogy with antiferroelectricity. At present, rigorous criteria defining antiferroelectric (AFE) phase transitions are still under discussion [28]. We will come back to this interesting topic and its relation to pure ferroelastics later.

Different group–subgroup pairs (species) were labelled as either full or partial, or non-ferroelectric, non-ferromagnetic and/or non-ferroelastic. A full ferroic material exhibits spontaneous order (such as polarization, magnetization, or strain) that can be reversed by a conjugated external field (electric field, magnetic field or mechanical stress, respectively). In a full ferroic crystal all the n domain states can be distinguished by distinct values of the components of a macroscopic tensorial property T_j ($j = 1, \ldots, n$). In a partial ferroic crystal only $m < n$ tensor components T_j are distinct [29]. Aizu derived and tabulated 773 possible ferroic species (now sometimes called Aizu species); 212 of them are nonmagnetic [30], with 88 of them being ferroelectric and 94 species ferroelastic.

Landau theory is a unique tool for the description of ferroelastic phase transitions and corresponding physical properties. A comprehensive work on purely ferroelastic phase transitions in the framework of Landau theory was performed by J C Tolédano and P Tolédano [31]. These authors have also carried out similar analyses for other ferroic phase transitions [32], concluding that most of the ferroelectric transitions are expected to be of the proper type (polarization is the primary order parameter), whereas most of the ferroelastic transitions should be of the improper type (primary order parameter is not a strain component). Another milestone was the calculation of elastic anomalies [23, 33] and spontaneous strain [3] for a large number of ferroelastic materials by Carpenter and Salje using Landau theory, following the work of Slonczewski and Thomas [34], Rehwald [35] and Luthi and Rehwald [36]. Pioneering work on the symmetry classification of structural phase transitions has also been done by Czech scientists Janovec, Dvorak and Petzelt [37].

2.3 Homogeneous strain

Let us consider a homogeneous anisotropic crystal. Its macroscopic properties are fully characterized by the point-group symmetry of the crystal. The Lagrangian (finite) strain tensor η_{ij} is defined as [38]

$$\eta_{ij} = \frac{1}{2}\left(\frac{\partial u_i}{\partial x_j} + \frac{\partial u_j}{\partial x_i} + \frac{\partial u_k}{\partial x_i}\frac{\partial u_k}{\partial x_j}\right), \tag{2.1}$$

where u_i $(i = 1, 2, 3)$ are lattice displacements. For small displacement gradients, equation (2.1) can be linearized as

$$\varepsilon_{ij} = \frac{1}{2}\left(\frac{\partial u_i}{\partial x_j} + \frac{\partial u_j}{\partial x_i}\right). \tag{2.2}$$

This linear approximation works usually well for temperature induced structural phase transitions [3] at ambient pressure, where strains are small. However, if high pressure is applied to a temperature induced phase transition or if the phase transition is induced by the application of a large hydrostatic pressure, the Lagrangian finite strain tensor is more appropriate [39].

In the presence of a structural phase transition with sufficiently small external fields, the strain can be written as the sum of a spontaneous part ε_{ij}^s and induced parts that may arise from the presence of, for example, mechanical stress σ_{ij}, electric field E_k or magnetic field H_k:

$$\varepsilon_{ij} = \varepsilon_{ij}^s + s_{ijkl}\sigma_{kl} + d_{ijk}E_k + Q_{ijk}H_k, \tag{2.3}$$

where s_{ijkl} is the elastic compliance tensor and d_{ijk} and Q_{ijk} are the piezoelectric and piezomagnetic coefficients, respectively.

For small mechanical stress in the absence of electric and magnetic fields, the strain is symmetric and given by equation (2.2). In the following, we will mainly use Voigt notation for strain, stress and elastic moduli, i.e. $(\varepsilon_{xx}, \varepsilon_{yy}, \varepsilon_{zz}, 2\varepsilon_{yz}, 2\varepsilon_{xz}, 2\varepsilon_{xy}) = (\varepsilon_1, \varepsilon_2, \varepsilon_3, \varepsilon_4, \varepsilon_5, \varepsilon_6)$, $(\sigma_{xx}, \sigma_{yy}, \sigma_{zz}, \sigma_{yz}, \sigma_{xz}, \sigma_{xy}) = (\sigma_1, \sigma_2, \sigma_3, \sigma_4, \sigma_5, \sigma_6)$ and $C_{ijkl}(i, j = 1, 2, 3) \rightarrow C_{\alpha\beta}(\alpha, \beta = 1, 2, ..., 6)$.

2.3.1 Domain states

In this section we summarize the main points that are necessary for understanding the concept of spontaneous strain [41] and its definition in different domains. The loss of one or more symmetry operations at a ferroic transition implies that the crystal can exist in a number of equivalent, equally stable DSs in the ferroic phase. Quite generally, the number n of possible DSs, resulting from a symmetry reduction of the parent space group G_0 to the low symmetry space group G—when G is a subgroup of G_0—is given as [43]

$$n = \frac{|G_0|}{|G|} = \frac{|F_0|}{|F|}z, \tag{2.4}$$

where $|G_0|$ and $|G|$ are the number of space group symmetry elements of G_0 and G, and F_0 and F are the corresponding point-group symmetry elements. z denotes the multiplication of the unit cell volume of the low symmetry phase with respect to the parent phase. Let us illustrate this on the example of PbTiO$_3$ (PTO). PTO exhibits a uniaxial ferroelectric and improper ferroelastic phase transition [44] from cubic ($Pm\bar{3}m$) to tetragonal ($P4mm$) structure (figure 2.1) without multiplication of the unit cell at $T_c \approx 765$ K.

The high and low symmetry phases consist of 48 and 8 symmetry elements, respectively, which leads to six different DSs (Figure 2.2). We will denote these domain state by 1_1, 1_2, 2_1, 2_2, 3_1, 3_2, where the three orientation states in terms of spontaneous polarization are written as $1_1 = (P_s, 0, 0)$, $2_1 = (0, P_s, 0)$ and $3_1 = (0, 0, P_s)$. The states 1_2, 2_2, 3_3 have opposite signs of polarization.

From a given DS, say $3_1 = (0, 0, P_3)$, to another domain state $3_2 = (0, 0, -P_3)$ the structure changes gradually, often forming a planar domain wall (DW), oriented with the DW normal **n**. In PbTiO$_3$ some researchers found [45, 46] that a DW ($3_1|3_2$) with **n** = [100] is not simply of the Ising type ($P_x(x) \rightarrow -P_3(-x)$), but forms at lower temperatures a DW of mixed Ising–Bloch–Néel type, i.e. $(0, 0, P_3) \rightarrow$

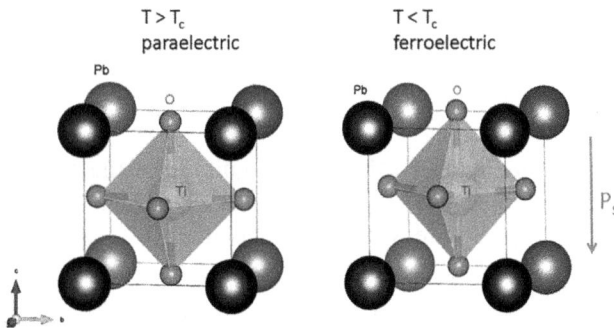

Figure 2.1. Lattice structure of paraelectric (left) and ferroelectric (right) PbTiO$_3$. A displacement of the Ti atom in the O octahedron leads to the spontaneous polarization of PbTiO$_3$, with the vector of polarization from the centre of the O octahedron to the Ti atom. The relative shift of the positive and negative ions from their centrosymmetric positions is ≈ 0.1 Å at room temperature. (Images produced using VESTA [40].)

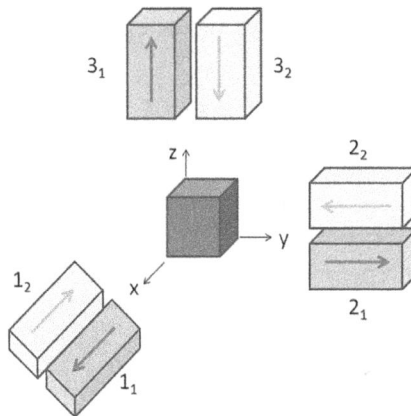

Figure 2.2. Six possible domain states of PbTiO$_3$, resulting from the symmetry reduction $Pm\bar{3}m \rightarrow P4mm$ at the ferroelectric/ferroelastic phase transition.

$(P_1(x), P_2(x), P_3(x)) \rightarrow (0, 0, -P_3)$. This change in the internal structure should occur as a result of a local 2D phase transition [45] in the DW at about 170 K below T_c. In a recent work [46], we have shown that this chiral phase transition from Ising to mixed Ising–Bloch–Néel type is probably triggered by a biquadratic coupling of the primary order parameter and its gradient originating from inhomogeneous electrostriction. However, at present this interesting chiral structure for pure PbTiO$_3$ is only based on theory and still awaits experimental verification.

Let us come back to the general case of a crystal which contains a number of domains. The application of an external stress σ_α, $(\alpha = 1, ..., 6)$ to such a crystal in a particular orientation state, e.g. S_i, leads to a change of the free energy $\Delta F = F(S_j) - F(S_i)$ as

$$\Delta F = F(S_j) - F(S_i) = -\left[\varepsilon_\alpha^s(S_j) - \varepsilon_\alpha^s(S_i)\right]\sigma_\alpha. \tag{2.5}$$

The sign of ΔF determines the necessary condition whether switching from S_i to S_j is possible. As seen from equation (2.5) it depends crucially on the magnitudes of the spontaneous strains in the respective DS S_i and S_j.

2.3.2 Spontaneous strain

It was argued a long time ago [41] that the magnitude of the spontaneous strain in equation (2.5) is not uniquely definable. According to Aizu [41], the spontaneous strain tensor in the domain state S_i can be calculated as

$$\varepsilon_\alpha^s(S_i) = \varepsilon_\alpha(S_i) - \frac{1}{N}\sum_{k=1}^{N}\varepsilon_\alpha(S_k). \tag{2.6}$$

This definition ensures that the spontaneous strain is independent of the choice of the coordinate system and vanishes over the whole temperature range in the prototypic (paraelastic) phase. However, even with this definition, the calculation of the spontaneous strain can be rather tricky when the strain is not the primary order parameter. Let us show this for the current example of PbTiO$_3$. Application of equation (2.6) leads to the magnitudes of the spontaneous strain components for domain state 3_1 (figure 2.2):

$$\varepsilon_1^s = \varepsilon_2^s = \frac{a - c}{3a_0}$$
$$\varepsilon_3^s = \frac{2(c - a)}{3a_0}. \tag{2.7}$$

Despite the elegance of this definition of the spontaneous strain tensor, in most cases the authors write

$$e_1^s = e_2^s = \frac{a - a_0}{a_0}$$
$$e_3^s = \frac{c - a_0}{a_0}. \tag{2.8}$$

However, irrespective of the definition, the calculation of the spontaneous strain components at any temperature below T_c requires knowledge of $a_0(T)$ extrapolated to the tetragonal low symmetry phase in addition to the tetragonal lattice parameters $a(T)$, $c(T)$. Quite often, the cubed root of the tetragonal volume $(a^2c)^{1/3} = a_0(T)$ is used at $T < T_c$ for this purpose. However, in this case the volume strain $\varepsilon_V = \dfrac{V - V_0}{V_0} = \dfrac{a^2c - a_0^3}{a_0^3} = 0$ in the tetragonal phase, in contrast to the experimental data (see figure 3(b) of [44]). Haun *et al.* [44] thus followed another procedure to obtain an improved estimate of the spontaneous strain. The authors extrapolated $a_0(T)$ from the cubic phase to few degrees below T_c. By calculating the ratio $\dfrac{\varepsilon_{33}^s}{\varepsilon_{11}^s} = \dfrac{c - a_0}{a - a_0} = -3.455$ and assuming it to be independent of temperature, they determined $a_0(T)$ for the whole tetragonal phase. Recently, we have applied finite strain theory combined with DFT calculations [47] to study the high pressure phase transition in PTO at $P_c \approx 12$ GPa. It turned out that a careful analysis yields a slightly different thermal baseline $a_0(T)$, with an enormous impact on the success of this theory.

Another prominent case is provided by the $Pm\bar{3}m$ to $I4/mcm$ phase transition in SrTiO$_3$, which is accompanied by a small but non-zero volume strain [42] reaching 10^{-4} at $T = 0$ K. This non-zero volume strain is crucial to obtain effects of hydrostatic pressure [48] on the phase transition properties and on the phase diagram [49] of SrTiO$_3$.

2.4 Landau theory of ferroelastic phase transitions

According to the Landau theory of phase transitions [1], the order parameter exhibits the transformation properties of a single irreducible representation (IR) of the parent-phase symmetry group. Spontaneous strain can arise at a ferroelastic phase transition for one of the following reasons. (i) The strain is the primary order parameter; or at least has the same symmetry as the primary order parameter. Such phase transitions are called *proper*, or *pseudo-proper* ferroelastic, respectively. (ii) The symmetry of the order parameter is different from that of the strain, but a spontaneous strain appears below T_c due to a non-linear coupling $Q^f \varepsilon$ between the primary order parameter (Q) and the strain (ε), where $f > 1$ is the so-called faintness index [1] of the spontaneous strain. In this case the term *improper* ferroelastic phase transition is commonly used.

Quite generally the Landau expansion of the free energy has the form [23]

$$F = F + \frac{a}{2}(T - T_c)Q^2 + \frac{b}{4}Q^4 + \cdots + \sum_{ijmf} \lambda_{ijmf}\, \varepsilon_{ij}^m Q^f + \frac{1}{2}\sum_{ijkl} C_{ijkl}^0 \varepsilon_{ij}\varepsilon_{kl}, \quad (2.9)$$

where Q is the primary order parameter (OP) that fully breaks the symmetry of the parent phase. Depending on the dimension of the active representation, Q may be a scalar, vectorial or tensorial quantity. For a proper ferroeastic transition, Q typically represents the symmetry-breaking strain ε_s, whereas for the case of a pseudo-proper phase transition Q is bilinearly coupled to the strain ε_s. For the case of an improper ferroelastic PT, $m = 1$ and Q is non-linearly ($f > 1$) coupled with some strain components. For most cases $f = 2$. We will demonstrate the different cases on some examples in the following sections.

2.4.1 Proper ferroelastic transitions

The simplest Landau potential which describes a second order proper ferroelastic transition can be written as

$$F(T, \, \varepsilon_s) = F(T, \, 0) + \frac{C_{2s}^0}{2}(T - T_c)\varepsilon_s^2 + \frac{C_{4s}^0}{4}\varepsilon_s^4, \qquad (2.10)$$

where we adopt the notation of [50]; ε_s is the primary order parameter, and C_{2s}^0 and C_{4s}^0 are combinations of second and fourth order elastic constants of the paraelastic phase.

Minimizing the free energy $\frac{\partial F}{\partial \varepsilon_s} = \sigma_s$, the temperature dependence of the spontaneous strain in the absence ($\sigma_s = 0$) of external stress reads

$$\varepsilon_s = 0 \text{ at } T > T_c$$

$$\varepsilon_s = \left(\frac{C_{2s}^0}{C_{4s}^0}(T_c - T)\right)^{1/2} \text{ at } T < T_c. \qquad (2.11)$$

The second derivative of F with respect to the strains determines the elastic constant as $C_{\alpha\beta} = \frac{\partial^2 F}{\partial \varepsilon_\alpha \partial \varepsilon_\beta}$, which for the present example yields

$$C_{2s} = C_{2s}^0(T - T_c) \text{ at } T > T_c$$

$$C_{2s} = 2C_{2s}^0(T_c - T) \text{ at } T < T_c. \qquad (2.12)$$

The softening of the transverse elastic constant and the resulting divergence of the compliance $S_{2s} = 1/C_{2s}$ is reminiscent of the Curie–Weiss law in proper ferroelectric materials. Materials known to undergo ferroelastic transitions have been listed by Folk *et al.* [51] and by Tolédano and Tolédano [31].

A beautiful example of a true proper ferroelastic transition is the case of LaNbO$_4$ [52]. These crystals undergo a phase transition from $I4_1/a$ (tetragonal) to $I2/a$ (monoclinic) symmetry at $T_c \approx 800$ K. The soft mode corresponds to a transverse acoustic mode propagating along the direction which is inclined by about 23° in the [110]-plane, which coincides with the orientation of the ferroelastic domain walls (in agreement with theoretical predictions [23]) in the low temperature phase. The spontaneous strains have components $\varepsilon_1 - \varepsilon_2$ and ε_6, implying that C_{2s} is a combination of elastic constants $C_{11} - C_{12}$, C_{66} and C_{16}. However, despite these complications, figure 2.3 shows the linear temperature dependence of the soft elastic constant C_{2s} (equation (2.12)) over a temperature range of about 100 K. The ratio of the slopes below and above T_c of 2.6:1 deviates somewhat from a pure Curie–Weiss behaviour (2:1), which indicates that higher order coupling terms between ε_s and

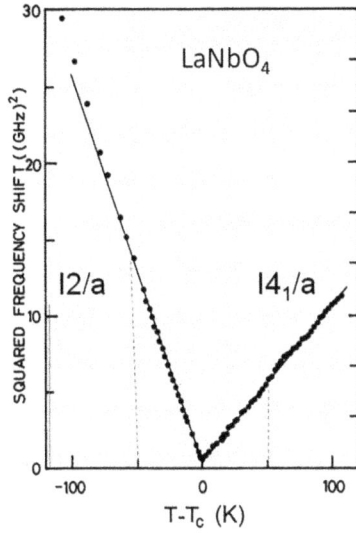

Figure 2.3. Temperature dependence of the squared soft mode frequency of the acoustic mode of LaNbO$_4$ measured by Brillouin scattering. The ratio of slopes (2.6:1) below and above T_c deviates slightly from the Curie–Weiss value (2:1). (Reproduced with permission from [52]. Copyright 1985 The Physical Society of Japan.)

other non-symmetry-breaking strain components must be taken into account (see below).

Probably the best example yet studied of elastic-constant variations due to proper ferroelastic behaviour is the pressure-induced tetragonal $(P4_12_12) \rightarrow$ orthorhombic $(P2_12_12_1)$ transition [53, 54] in paratellurite, TeO$_2$ at $P_c \approx 9$ kbar. An excellent review can be found in [23]. For this phase transition, the symmetry-breaking strain is $\varepsilon_s = \varepsilon_1 - \varepsilon_2$ with the corresponding elastic constant given as $C_{2s} = \frac{1}{2}(C_{11} - C_{12})$. Thus, the phase transition occurs at the pressure P_c, where $C_{11} = C_{22}$. In the orthorhombic phase $(P > P_c)$ the soft elastic constant is given by the combination [50] $C_{2s} = \frac{1}{4}(C_{11} + C_{22} - 2C_{12})$. Indeed, figure 2.4 shows the expected softening of the elastic constant C_{2s}. However, in this case the ferroelastic phase transition is induced by hydrostatic pressure and not by temperature.

Tolédano *et al.* [50] worked out the Landau theory for this phase transition in detail. We are using here a simplified version of the free energy expansion of [50] to show only the basic ideas behind the description of pressure-induced proper ferroelastic PTs. Similarly to the temperature induced phase transition (equation (2.10)), we first expand (as many authors do) the second order term directly in terms of pressure, bearing in mind that the coefficients no longer have the physical meaning of second, third and fourth order elastic constants, as in [50]:

$$F = F_0 + \frac{a(P_c - P)}{2}\varepsilon_s^2 + \frac{b}{4}\varepsilon_s^4. \tag{2.13}$$

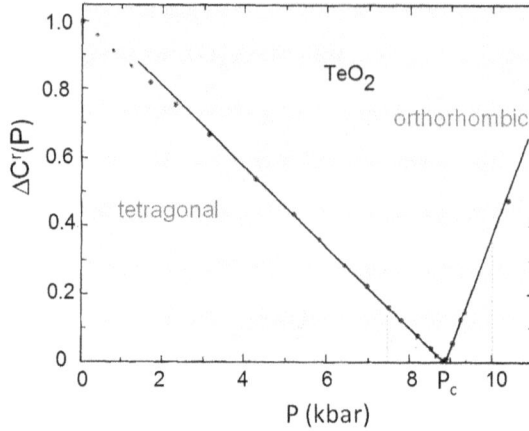

Figure 2.4. Pressure dependence of the reduced elastic constant $\Delta C^r(P) = \frac{(C_{11} - C_{12})_P}{(C_{11} - C_{12})_0}$ of TeO$_2$ determined from ultrasound velocity measurements. The ratio of the slopes above and below the transition is 3:1. Note that here the high-symmetry phase is below P_c in contrast to most temperature driven PTs, where the high symmetric phase is above T_c. (Reproduced with permission from [54]. Copyright 1975 Elsevier.)

After minimizing the free energy (2.13) with respect to ε_s we obtain

$$\varepsilon_s = 0 \quad \text{for} \quad P < P_c \quad \text{and} \tag{2.14a}$$

$$\varepsilon_s = \sqrt{\frac{a}{b}(P - P_c)} \quad \text{for} \quad P > P_c. \tag{2.14b}$$

The pressure dependence of the soft elastic constant calculated from equation (2.13) reads

$$C_s = a(P_c - P) \quad \text{for} \quad P < P_c \quad \text{and}$$
$$C_s = 2a(P - P_c) \quad \text{for} \quad P > P_c. \tag{2.15}$$

Equations (2.15) describe the linear pressure dependences of the experimental data in figure 2.4 excellently. However, the ratio of slopes (3:1) for C_s above and below P_c deviates from the Curie–Weiss prediction (2:1) considerably. To understand this behaviour—and to keep closer to the physical meaning of the expansion coefficients —we prefer to start from a Landau free energy written exclusively in terms of strains (and temperature) instead of mixing strain and pressure, taking into account also coupling terms between the symmetry-breaking strain ε_s and the non-symmetry-breaking strains ε of A_g-symmetry (fully symmetric). Then we can write

$$F = F_0 + \frac{C_{2s}^0}{2}\varepsilon_s^2 + \frac{C_{4s}^0}{4}\varepsilon_s^4 + \lambda\varepsilon_s^2\varepsilon + \frac{C^0}{2}\varepsilon^2. \tag{2.16}$$

To obtain the pressure dependent free energy we perform a partial Legendre transformation, replacing ε by P, i.e. $G(\varepsilon_s, \sigma = -P) = F(\varepsilon_s, \varepsilon) - \sigma\varepsilon(\sigma)$, leading to

$$G = \frac{a}{2}(P_c - P)\varepsilon_s^2 + \frac{b}{4}\varepsilon_s^4 - \frac{P^2}{2C^0}. \tag{2.17}$$

This has the same form as equation (2.16), but now we can express the expansion parameters a, b and P_c in terms of physically meaningful parameters such as elastic constants of second and fourth order and coupling coefficient(s):

$$a = \frac{2\lambda}{C^0} \tag{2.18a}$$

$$b = C_{4s}^0 - \frac{2\lambda^2}{C^0} \tag{2.18b}$$

$$P_c = \frac{C_{2s}^0 C^0}{2\lambda}. \tag{2.18c}$$

We observe the well-known fact that for sufficiently large coupling strength λ between symmetry-breaking and non-symmetry-breaking strains, b can become negative, in which case the phase transition is of first order. The magnitude of λ can be determined, e.g. from measurements of ε versus pressure (or versus temperature, if the phase transition is also driven by temperature). In fact, after minimizing equation (2.16) in the presence of P with respect to ε, we obtain

$$\varepsilon = -\frac{1}{C^0}(P + \lambda\varepsilon_s^2) \tag{2.19}$$

with ε_s given by equation (2.14a).

Using the experimental lattice parameters [55] for TeO$_2$, Carpenter and Salje [23] have indeed demonstrated the linear dependence of $\varepsilon_s^2 = (\varepsilon_1 - \varepsilon_2)^2$ and $\varepsilon = \varepsilon_1 + \varepsilon_2$ with pressure.

The effective elastic constants can be calculated from equation (2.16) as

$$C_s = a(P_c - P) \quad \text{for} \quad P < P_c \text{ and}$$
$$C_s = 2a\frac{C_{4s}^0}{(C_{4s}^0 - \frac{2\lambda^2}{C^0})}(P - P_c) \quad \text{for} \quad P > P_c. \tag{2.20}$$

Since $\frac{C_{4s}^0}{(C_{4s}^0 - \frac{2\lambda^2}{C^0})} > 1$ always, one can now easily understand the origin of the measured (see figure 2.4) change in slopes.

2.4.2 Pseudo-proper ferroelastic transitions

Here, the symmetry-breaking strain ε_s is not the primary (driving) order parameter, but it is linearly coupled to it, i.e. they both transform according to the same IR of the parent space group. Following the traditional approach [1, 23], the Landau free energy of a pseudo-proper ferroelastic phase transition of second order can be written in a very simplified version as

$$F = F_0 + \frac{a}{2}(T - T_c)Q^2 + \frac{b}{4}Q^4 + \gamma Q\varepsilon_s + \frac{1}{2}C_s^0\varepsilon_s^2. \tag{2.21}$$

Minimizing the free energy (2.21) with respect to Q and ε_s leads to

$$\varepsilon_s = 0 \quad \text{for} \quad T > T_c^* \quad \text{and} \tag{2.22a}$$

$$\varepsilon_s = -\frac{\gamma}{C^0}Q \quad \text{for} \quad T < T_c^* \tag{2.22b}$$

$$Q = 0 \quad \text{for} \quad T > T_c^* \quad \text{and} \tag{2.22c}$$

$$Q = \sqrt{\frac{a}{b}(T_c^* - T)} \quad \text{for} \quad T < T_c^* \tag{2.22d}$$

$$T_c^* = T_c + \frac{\gamma^2}{aC_s^0}. \tag{2.22e}$$

In the absence of strain coupling ($\gamma = 0$), the phase transition would occur at T_c, where the soft mode approaches zero. Coupling to the strain ε_s causes a renormalization of the transition temperature to $T_c^* = T_c + \frac{\gamma^2}{aC_s^0}$, where C_s reaches zero. Indeed, since the strain is not the driving OP, the elastic constant is calculated as

$$C_s = \frac{\partial^2 F}{\partial \varepsilon_s^2} - \left(\frac{\partial^2 F}{\partial \varepsilon_s \partial Q}\right)^2 \left(\frac{\partial^2 F}{\partial Q^2}\right)^{-1}_{\varepsilon_s = 0}, \tag{2.23}$$

which yields

$$C_s = C_s^0 - \frac{\gamma^2}{a(T - T_c)} = C_s^0 \frac{T - T_c^*}{T - T_c} \quad \text{for} \quad T > T_c^* \quad \text{and} \tag{2.24a}$$

$$C_s = C_s^0 \frac{2(T_c^* - T)}{2(T_c^* - T) + (T_c^* - T_c)} \quad \text{for} \quad T < T_c^*. \tag{2.24b}$$

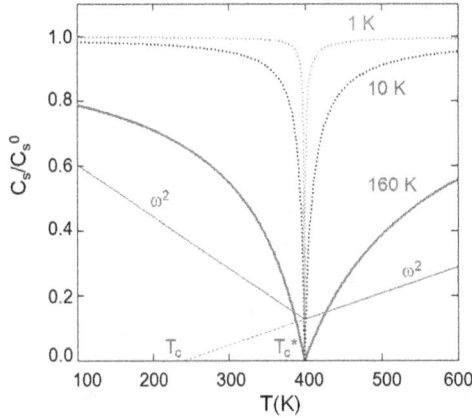

Figure 2.5. Characteristic temperature dependences of C_s according to equation (2.24a) for various values of $T_c^* - T_c = \frac{\gamma^2}{aC_s^0}$. With increasing OP-strain coupling, γ, the elastic anomaly approaches the linear (Curie–Weiss) behaviour of a proper ferroelastic case. The behaviour of the soft mode frequency $\omega^2(T)$ for the case of $T_c^* - T_c = 160$ K is also shown. It extrapolates to zero at $T_c < T_c^*$.

Figure 2.5 shows the typical behaviour of the soft elastic constant near a pseudo-proper ferroelastic phase transition for various values of the order parameter–strain coupling γ. With decreasing γ, it deviates increasingly from the shape of a pure proper ferroelastic PT.

The squared soft mode frequency ω^2 calculated from equation (2.21) reads

$$\omega^2 = \frac{1}{m}\left(\frac{\partial^2 F}{\partial Q^2}\right)_{\varepsilon_s = 0} = \frac{1}{m}[a(T - T_c) + 3(bQ^2)], \qquad (2.25)$$

where m is the effective mass of the soft mode. Inserting equation (2.22a) into (2.25) we obtain

$$\omega^2 = \frac{1}{m}[a(T - T_c)] = \frac{1}{m}\left[a(T - T_c^*) + a(T_c^* - T_c)\right] \quad \text{for} \quad T > T_c^* \text{ and} \quad (2.26a)$$

$$\omega^2 = \frac{1}{m}\left[2a(T_c^* - T) + a(T_c^* - T_c)\right] \quad \text{for} \quad T < T_c^*. \qquad (2.26b)$$

Equations (2.26a) imply that the soft mode frequency does not approach zero at T_c^*, i.e. at the temperature where the elastic constant softens, but stays finite at T_c^* and would approach zero at the extrapolated temperature T_c. Nevertheless, the Curie–Weiss slopes of 1:2 in the squared frequencies above and below T_c^* are still also recovered in this case.

A beautiful example in which all these features can be demonstrated is the pseudo-proper ferroelastic transition in LaP_5O_{14}. It undergoes a $Pmcm \rightarrow P2_1/c$ transition near $T_c^* \approx 400$ K. The primary order parameter belongs to an optic mode of B_{2g} symmetry, which softens when approaching the phase transition [56]. However, as expected for a pseudo-proper ferroelastic PT, with cooling the sample from the paraelastic phase, the soft mode does not approach zero at the extrapolated temperature $T_c \approx 237 \pm 10$ K, since due to the coupling with the strain ε_{13}, C_{55} reaches zero already at $T_c^* \approx 400$ K (see the red curve in figure 2.5, or figure 8 of [56]).

$LiCsSO_4$ is another example, exhibiting a pseudo-proper ferroelastic phase transition at about $T_c^* \approx 202$ K [57]. Here the symmetry-breaking strain is ε_6, implying that the transverse elastic constant C_{66} softens at $T_c^* = T_c + 15$ K. In fact, the experimental data of figure 1 in [57] are rather close to the blue curve ($T_c^* - T_c = 10$ K) of figure 2.5.

Another interesting case is $BiVO_4$ [58]. It exhibits the same symmetry change $I4_1/a \rightarrow I2/a$ as the isomorphous $LaNbO_4$. However—in contrast to $LaNbO_4$ (figure 2.3)—there is a soft optic mode in $BiVO_4$ which drives the transition and to which the strain is linearly coupled. This OP-strain coupling is rather strong, with $T_c^* = T_c + 163$ K [58].

Let us now consider a pseudo-proper FEL phase transition, driven by pressure instead of temperature. Then the free energy in simplified form can be written as

$$F = \frac{m\omega_0^2}{2}(T)Q^2 + \frac{\beta}{4}Q^4 + \gamma Q\varepsilon_s + \frac{C_s^0}{2}\varepsilon_s^2 + \lambda Q^2\varepsilon + \frac{C^0}{2}\varepsilon^2. \qquad (2.27)$$

In the presence of hydrostatic pressure $P = -\frac{\partial F}{\partial \varepsilon}$, the non-symmetry-breaking strain follows the relation

$$\varepsilon = -\frac{1}{C^0}(P + \lambda Q^2) \qquad (2.28)$$

and from $\frac{\partial F}{\partial \varepsilon_s} = 0$, we obtain

$$\varepsilon_s = -\frac{\gamma}{C_s^0}Q. \qquad (2.29)$$

Performing a Legendre transformation, the Gibbs free energy takes the simple form

$$G = \frac{a}{2}(P_c^* - P)Q^2 + \frac{b}{4}Q^4 - \frac{P^2}{C^0}, \qquad (2.30)$$

where the expansion parameters are given as

$$a = \frac{2\lambda}{C^0} \qquad (2.31a)$$

$$P_c^* = P_c - \frac{\gamma^2}{aC_s^0} \qquad (2.31b)$$

$$P_c = \frac{m\omega_0^2}{a} \qquad (2.31c)$$

$$b = \beta - \frac{2\lambda^2}{C^0}. \qquad (2.31d)$$

As demonstrated before, such an approach has the advantage that one can determine the Landau parameters from temperature dependent measurements, which is usually much easier than measuring a complete set of thermodynamic properties as a function of pressure. Minimizing equation (2.30) with respect to Q, we obtain the well-known relation for the primary OP:

$$Q^2 = \frac{a}{b}(P - P_c^*). \qquad (2.32)$$

The symmetry-breaking elastic constants take the form

$$C_s = C_s^0 - \frac{\gamma^2}{a(P_c - P)} = C_s^0 \frac{P_c^* - P}{P_c - P} \quad \text{for} \quad P < P_c^* \quad \text{and} \qquad (2.33a)$$

$$C_s = C_s^0 \frac{\dfrac{2a\beta}{b}(P - P_c^*)}{\dfrac{2a\beta}{b}(P - P_c^*) + a(P_c - P_c^*)} \quad \text{for} \quad P > P_c^* \qquad (2.33b)$$

and the non-symmetry-breaking elastic constants read

$$C = C^0 \quad \text{for} \quad P < P_c^* \quad \text{and} \qquad (2.34a)$$

$$C = C^0 - \frac{\dfrac{2\lambda^2 b}{\beta}}{1 + \dfrac{b^2}{2\beta}\dfrac{P_c - P_c^*}{P - P_c^*}} \quad \text{for} \quad P > P_c^*. \qquad (2.34b)$$

For the pressure dependence of the soft mode frequency, we obtain

$$\omega^2 = \frac{1}{m}[a(P_c - P)] \quad \text{for} \quad P < P_c^* \quad \text{and} \qquad (2.35a)$$

$$\omega^2 = \frac{1}{m}\left[\frac{2a\beta}{b}(P - P_c^*) + a(P_c - P_c^*)\right] \quad \text{for} \quad P > P_c^*. \tag{2.35b}$$

At this point we observe an interesting feature of the soft mode behaviour at pressure-induced proper, or pseudo-proper, ferroelastic phase transitions. Since $\beta/b = \frac{\lambda^2}{c^0}/b > 1$ always, a Curie–Weiss slope of 2:1 can never occur at pressure-induced ferroelastic phase transitions. The case 2:1 would only be approached if $\lambda \to 0$, but then $P_c = \frac{m\omega_0^2 c^0}{2\lambda} \to \infty$. This is in sharp contrast to temperature induced PTs, where $\lambda_1 = 0$ is possible and would lead to pure Curie–Weiss slopes.

An interesting example of a pseudo-proper ferroelastic phase transition driven by hydrostatic pressure is stishovite SiO_2. Very recently, Zhang *et al.* [59] succeeded in measuring the pressure dependencies of the elastic moduli C_{ij} of single-crystal stishovite (SiO_2) up to about 70 GPa using Brillouin light scattering. At room temperature, stishovite exhibits a pseudo-proper FEL phase transition [60] from a $P4_2/mnm$ rutile-type structure into $CaCl_2$-type post-stishovite (*Pnnm*) at $P_c^* \approx 50$–55 GPa. The ferroelastic transition in stishovite is of particular interest in geophysics. Due to the abundance of ≈ 25 vol% of this compound in basaltic subducting slabs, its elastic and anelastic properties contribute substantially to seismic signals of the Earth. The Landau free energy for the FEL phase transition was already derived in [23], and was later successfully used [61, 62] to calculate thermodynamic quantities, i.e. strains, elastic constants as well as soft mode frequency as functions of pressure.

Figure 2.6 shows the pressure dependencies of the elastic moduli of stishovite reproduced from the table of the supplemental material of [59].

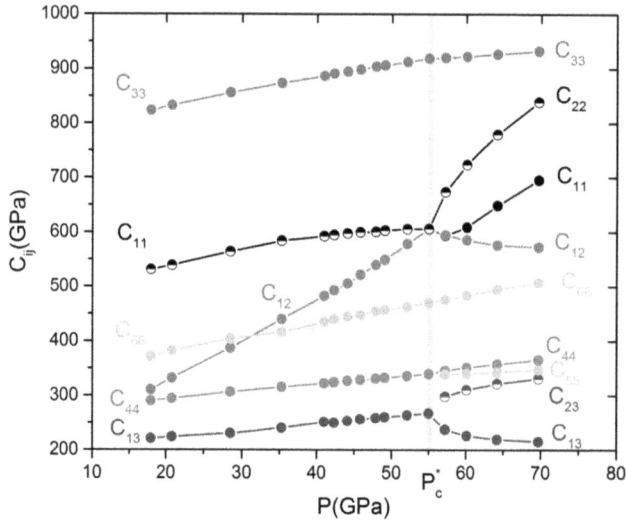

Figure 2.6. Pressure dependent elastic moduli of single-crystal stishovite around the pseudo-proper ferroelastic phase transition at $P_c^* \approx 55$ GPa from Brillouin light scattering experiments. The lines are guides for the eye. The data are from the table of supplemental material in [59].

As shown in figure 2 of [59], the data can be excellently fitted using the Landau expansion of [23, 62]. The non-symmetry-breaking longitudinal elastic constants (C_{ij} for $i, j \leqslant 3$) exhibit anomalies of the types described by equation (2.34a), i.e. they yield a negative or positive step-like anomaly, which is, however, not sharp but exhibits some rounding. Typically such a rounding effect increases with increasing magnitude of $P_c - P_c^*$ for a pseudo-proper PT. For stishovite (figure 2.7) $P_c - P_c^* \approx 55$ GPa turns out to be rather large. The limiting case $P_c - P_c^* = 0$ occurs for a proper FEL phase transition, or an improper one, implying for the longitudinal elastic constants showing a clear step (for second order PTs) or a cusp (for first order PTs).

The symmetry-breaking combination of elastic moduli in stishovite is $C_s = C_{11} - C_{12}$ for $P < P_c^*$ and $\frac{1}{2}(C_{11} + C_{22} - 2C_{12})$ for $P > P_c^*$. As figure 2.7 shows, it can also be excellently fitted using equation (2.33a). As predicted by equation (2.35a) the soft mode frequency approaches (from $P < P_c^*$) zero at $P_c = P_c^* + \frac{\gamma^2}{aC_s^0} \approx 110$ GPa. However, starting at pressures above P_c^*, the soft mode frequency should approach zero at $P_c^{**} = P_c^* - \frac{P_c - P_c^*}{2} \frac{\beta}{b}$. For a simple Curie–Weiss slope ($\beta/b = 1$), this would imply $P_c^{**} = 28$ GPa. But according to Carpenter et al. [33] $\beta/b = 1.135$, which yields $P_c^{**} = 31$ GPa, in good agreement with figure 2.7.

Here, we would like to stress again the advantage of starting with a temperature dependent free energy (2.27) as compared to approaches that directly expand the quadratic term in the Landau free energy as function of pressure. In the second case one loses some important information. Inspecting equations (2.31a) we see that,

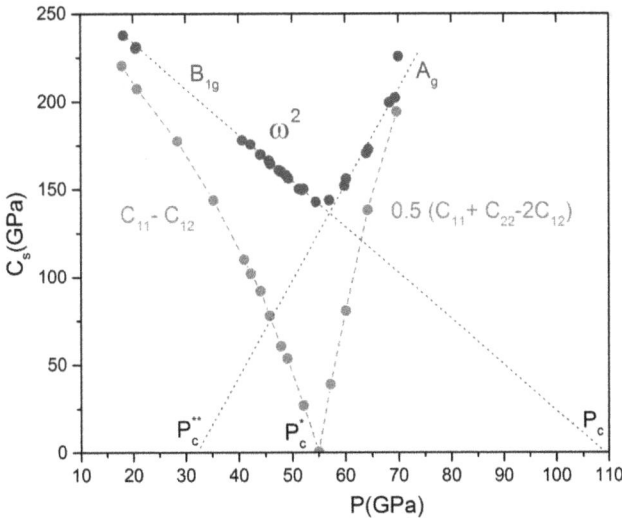

Figure 2.7. Pressure dependence of the symmetry-breaking elastic modulus of single-crystal stishovite around the pseudo-proper ferroelastic phase transition at $P_c^* \approx 55$ GPa and the square of the soft mode frequency of the B_{1g}-mode. Data are from the supplement material in [59].

$P_c = \frac{mC^0}{2\lambda}\omega_0^2(T)$. This has several implications. For example, if we assume (as usually in a Landau expansion) $\omega_0^2 = \alpha(T - T_c)$, we observe that the critical pressure increases linearly with temperature. Moreover, since P_c depends inversely on the coupling λ to the non-symmetry-breaking strains, P_c increases with decreasing λ_1. Since the magnitude of λ enters into the renormalized fourth order coefficients as $\beta = b + \frac{2\lambda^2}{c^0}$, which determine the Curie–Weiss slopes (1:$2\frac{\beta}{b}$), we observe that materials with lower deviations from the classical Curie–Weiss ratios (1:2) tend to have higher values of P_c. Compare, e.g. TeO_2 (figure 2.4; Curie–Weiss ratio 1:3, $P_c \approx 0.9$ GPa) with stishovite (figure 2.7; Curie–Weiss ratio 1:1.135, $P_c \approx 55$ GPa).

A highly interesting case of pseudo-proper ferroelastic behaviour is found in iron-based superconductors [63]. The end-member $BaFe_2As_2$ exhibits a structural phase transition from $I4/mcm$ (tetragonal) to an $Fmmm$ (orthorhombic) structure (figure 2.8) at $T_s = 140$ K [64, 65] and a stripe-type antiferromagnetic order at $T_N \approx T_s$.

Electron doping by substituting Fe^{2+} by Co^{2+} in $Ba(Fe_{1-x}Co_x)_2As_2$ reduces both [66] the structural transition temperature T_s and the antiferromagnetic one at T_N (figure 2.9). At $x \approx 0.03$ superconductivity emerges at $T_{sc} < T_N < T_s$ in these systems (figure 2.9).

To study this phase transition dynamic mechanical analysis (DMA) measurements of $BaFe_2As_2$ single crystals were performed in three-point bending geometry (figure 2.10) at dynamical force $F = F_S + F_D\cos(\omega t)$ with different static components F_S.

In this geometry the Young's modulus is determined by the relation $Y'_{[110]} \propto C_{66}\Delta/(C_{66} + \Delta)$, where Δ is a combination of non-critical longitudinal elastic moduli, which do not soften at T_s. Since $C_{66} \to 0$ when approaching the structural phase transition at T_s, $Y'_{[110]} \propto C_{66}(T)$ to a good approximation.

To account for the sequence of phase transitions in $BaFe_2As_2$ we write the free energy in the form [67, 68]

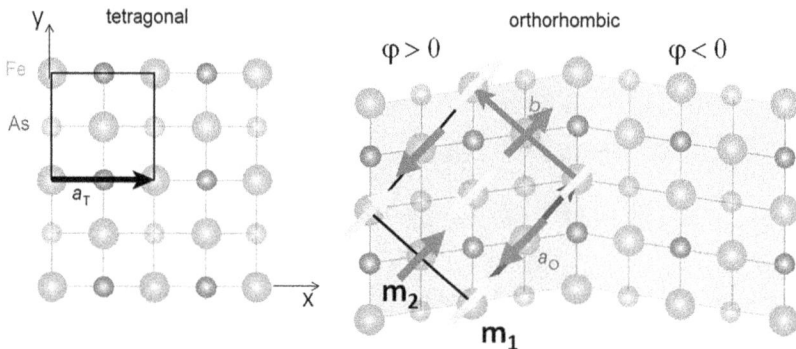

Figure 2.8. Sketch showing the relation between the tetragonal high-symmetry phase ($I4/mmm$) at $T > T_s$ and the two domain states ($\varphi > 0$ and $\varphi < 0$) of the orthorhombic low symmetry phase ($Fmmm$) of $BaFe_2As_2$. The collinear Néel order in the two Fe sublattices, marked by the magnetic moments m_1 and m_2 takes place at $T < T_N$.

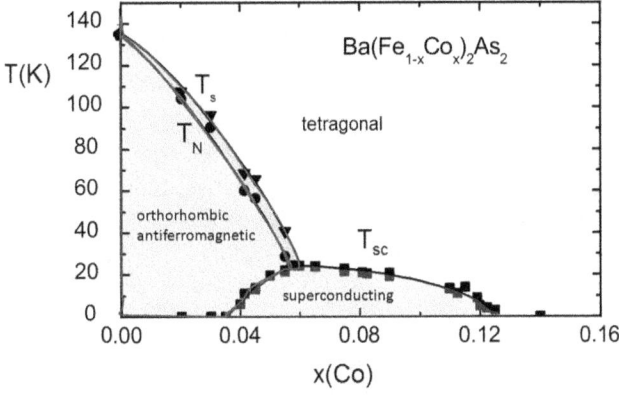

Figure 2.9. Phase diagram of Ba(Fe$_{1-x}$Co$_x$)$_2$As$_2$ showing the doping dependence of the structural (T_s), the antiferromagnetic (T_N) and the superconducting transition (T_{sc}). (Reproduced with permission from [66]. Copyright 2010 the American Physical Society.)

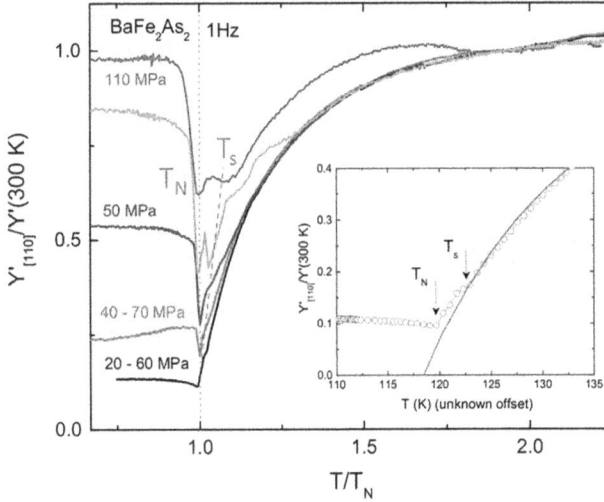

Figure 2.10. Real part of Young's modulus $Y'_{[110]} \propto C_{66}$ of BaFe$_2$As$_2$ obtained in dynamic three-point bending experiments in which the applied force $F = F_S + F_D \cos(\omega t)$ varies at a frequency of 1 Hz. Its behaviour above T_N is quite similar to the blue curve of figure 2.5. Below T_N the behaviour at low forces is dominated by domain wall motion. The inset shows the splitting of the structural (at T_s) and the magnetic (at T_N) phase transition in the pure sample.

$$F = F_0 + \frac{a_0}{2}(T - T_n^0)\varphi^2 + \frac{b}{4}\varphi^4 + \frac{C_{66}^0}{2}\varepsilon_6^2 + \frac{C_{4s}^0}{4}\varepsilon_6^4 + \gamma\varphi\varepsilon_6 +$$

$$+ \frac{A_0}{2}(T - T_N^0)Q_m^2 + \frac{B}{4}Q_m^4 + \lambda_{m\varphi}Q_m^2\varphi + \lambda_{m\varepsilon}Q_m^2\varepsilon_6 + \qquad (2.36)$$

$$+ \frac{\alpha_0}{2}(T - T_{sc})|\psi|^2 + \frac{\beta}{4}|\psi|^4 + \lambda_{sc}|\psi|^2\varepsilon_6 + \delta_m|\psi|^2 Q_m^2 + \delta_s|\psi|^2\varepsilon_6^2,$$

where we added the contribution of the superconducting transition to the free energy, and we have already inserted the value Q_m of the magnetic order parameter in the magnetic phase. Here T_n^0 is the temperature, where the nematic susceptibility χ_φ would diverge in the absence of strain coupling ($\gamma = 0$). T_N^0 is the Néel temperature (i.e. phase transition temperature of the magnetic phase, where $Q_m \neq 0$) and T_{sc} the superconducting phase transition temperature.

In the tetragonal phase, the temperature dependence of $C_{66}(T)$ can be excellently fitted by a relation similar to equation (2.24), i.e.

$$C_{66}(T) = C_{66}^0 - \frac{\gamma^2}{a_0(T - T_n^0)} = C_{66}^0 \frac{T - T_s}{T - T_n^0}, \qquad (2.37)$$

with $T_s - T_n^0 = \frac{\gamma^2}{a_0 C_{66}^0} \approx 30\text{--}40$ K. The non-linear softening—which is in contrast to a proper ferroelastic phase transition (figure 2.3)—as well as the clear difference between T_s (temperature of complete elastic softening) and T_n^0 (temperature of divergence of order parameter susceptibility) proves the pseudo-proper nature of the structural phase transition in BaFe$_2$As$_2$. It implies that the primary order parameter is not the strain ε_6, but another variable φ that couples linearly to ε_6. Since the appearance of φ breaks the four-fold rotational symmetry of the tetragonal phase, φ is called a 'nematic' OP. The microscopic origin (spin or orbital) of the nematic order parameter is still under debate [70, 71]. We will not enter the discussion here, because on the level of Landau theory it cannot be decided.

As already shown above, the shear elastic modulus C_{66} should approach zero at the structural PT. However, as shown in figure 2.10, C_{66} retains a positive residual value at T_s. To account for this, Böhmer and Meingast [72] added the contribution of external stress $-\sigma_6 \varepsilon_6$ to the free energy (2.36). Their model reproduced the main features of the data (figure 2.10) rather well, i.e. $C_{66}(\sigma_6 \neq 0)$ remains finite and an increasing up-shift of the structural phase transition with increasing σ_6 (whereas T_N remains constant) leads to an increasing splitting between T_s and T_N. The flat region of C_{66} at small forces below the phase transition is due to domain wall motion induced superelastic softening [69].

Due to the competition between the superconducting order parameter with the magnetic ($\delta_m > 0$) [73] and the structural ($\delta_s > 0$) [66] degrees of freedom, the superconducting transition is suppressed for pure BaFe$_2$As$_2$. Indeed, this competition was impressively demonstrated by high-resolution x-ray measurements [66] of Ba(Fe$_{1-x}$Co$_x$)$_2$As$_2$, i.e. the orthorhombic distortion approaches zero at $x \approx 0.063$, and superconductivity appears (figure 2.9).

Böhmer et al. [74] studied the phase diagram of electron and hole doped iron-based pnictides in detail, by measuring the shear modulus C_{66} for a wide range of concentrations and temperatures. Since it was argued several times that nematic fluctuations play a vital role for the glue of the superconducting pairing in iron-based superconductors, a study of the nematic susceptibility is of great importance. The authors made use of the fact that the shear elastic constant contains the nematic susceptibility, as can be seen by rewriting equation (2.37) as

$$(C_{66}^0 - C_{66}(T))^{-1} = \frac{a_0(T - T_n^0)}{\gamma^2} = \frac{1}{\gamma^2}\chi_\varphi^{-1}. \tag{2.38}$$

Interestingly enough, the measurements of $Ba(Fe_{1-x}Co_x)_2As_2$ show [72, 74] that the nematic susceptibility follows a mean-field-like temperature dependence, i.e. $\chi_\varphi \propto (T_n^0 - T)^{-1}$ up to $x = 0.09$. Additionally, $T_n^0(x)$ decreases strongly [74] (from $\approx$ 90 K to 30 K) with increasing x, but $T_s - T_n^0 = \frac{\gamma^2}{a_0 C_{66}^0} \approx$ 30–40 K is practically independent of doping in this range of concentration. The fact that the nematic susceptibility is enhanced over most of the superconducting dome (figure 2.9) underlines its possible role in promoting superconductivity in these materials.

These results show the potential of dynamic elastic measurements for obtaining valuable insights into the properties of quite different classes of materials.

2.4.3 Improper ferroelastic phase transitions

By definition, an improper ferroelastic transition is one in which the symmetry-breaking strain, ε_{sb} has symmetry which differs from that of the driving order parameter, q. The most common circumstance for this is in crystal structures such as perovskites where a soft optic mode related to a special point on the Brillouin zone (BZ) away from the gamma-point (e.g. at the BZ boundary) provides the transition mechanism. The lowest order coupling term is $\lambda\varepsilon_{sb}q^2$ and leads to characteristic patterns of variation for spontaneous strains and elastic constants through transitions which might be second order, tricritical or first order in character [23]. Whether q represents a mechanism that is displacive, electronic, charge ordering, etc, is not the primary consideration in this context, though the relaxation time of the order parameter in response to an induced strain will influence the variations in elastic moduli observed using different experimental techniques. The distinctive evolution of elastic moduli associated with improper ferroelastic transitions and the key issues relating their overall patterns of behaviour are most easily illustrated with specific examples, as represented here by $LaAlO_3$, $GeTe$, $PbZrO_3$, $SrZrO_3$, etc.

The cubic ($Pm\bar{3}m$)–rhombohedral ($R\bar{3}c$) transition at $\approx$ 820 K in $LaAlO_3$ is driven by an R-point soft optic mode described by an order parameter with symmetry properties of irreducible representation R_4^+. It displays variations in strain, heat capacity, soft mode frequencies and elastic moduli which fit with a classical second order transition described by the standard Landau expansion [75] in terms of a three component order parameter (q_1, q_2, q_3) with spontaneous value (q, 0, 0) and coupling terms of the type $\lambda_\alpha\varepsilon_\alpha q^2$.

Full expressions of the free energy and the corresponding elastic-constant variations are presented in [75]. For a second order transition, these predict a stepwise elastic softening of $C_{ik} - C_{ik}^0$ at and below T_c that is independent of temperature and has a magnitude which depends on terms with a form λ^2/b. Calculated variations of C_{ik} based on a complete calibration of all the coefficients for $LaAlO_3$ are compared with experimental data from Brillouin spectroscopy (in figure 6 of [75]). The classical result of discontinuous softening is observed, apart from slight additional precursor softening in narrow temperature intervals both above and below T_c.

These additional softening effects also highlight the fact that there are dynamical effects close to the transition point that are not accounted for by a simple Landau description. Evidence for dynamic relaxation processes in the vicinity of a transition point is often provided by a central peak (CP) in Raman, Brillouin and inelastic neutron scattering experiments. Analysis of the observed CP in LaAlO$_3$ yielded relaxation times in the picosecond range, with a maximum of $\approx 60 \times 10^{-12}s$ at ≈ 802 K. Details of the atomic motions responsible for dynamic effects on this time scale have not yet been resolved but a likely origin involves clusters of tilted octahedra flipping between four possible ferroelastic twin orientations immediately below T_c, where the energy barriers between orientations are expected to be low. Another contribution might come from phonon density fluctuations. Regardless of the exact mechanism, the additional (non-Landau) softening in the vicinity of T_c can be accounted for by coupling of acoustic phonons with the central peak mode(s).

Other driving mechanisms can give rise to the stepwise softening that is characteristic of an improper ferroelastic phase transition. For example, the cubic ($Fm\bar{3}m$)–rhombohedral ($R3m$) transition near 625 K in GeTe involves a combination of a soft optic mode with a Peierls distortion. GeTe is a material of intense topical interest due to its potential in the context of phase-change and nanowire memory devices, as a base for thermoelectric materials, and as a ferroelectric. The order parameter q has Γ_4^- symmetry and couples to the symmetry-breaking shear strain, ε_4, which has the symmetry of Γ_5^+, as $\lambda \varepsilon_4 q^2$. Values of ε_4 reach at least 3% and the transition is weakly first order. Resonant ultrasound spectroscopy (RUS) results [76] in the frequency range 100–1200 kHz show a steep elastic softening by up to $\approx$ 40% at 620 K during cooling and at 630 K during heating, followed by non-linear recovery below the transition point. The difference in form from the linear recovery in LaAlO$_3$ may be understood by observing that for a first order transition the free energy has to be expanded at least up to sixth order with respect to the order parameter components. High acoustic loss immediately below the transition is attributed to the mobility of domain walls, which freeze at about 200 K.

An important consideration in experimental determinations of elastic properties is always the timescale for relaxation of the order parameter in response to any strain induced in the measuring process. Elastic softening, as represented by terms of type $\lambda \varepsilon q^2$, will only occur if the time required for the order parameter to adjust to the strain is shorter than the timescale of the applied stress. If the relaxation is too slow, the elastic response is determined by higher order coupling terms such as $\lambda \varepsilon^2 q^2$ which are always allowed by symmetry. In principle, λ can be positive or negative, leading to stiffening or softening that will reflect the variation of $q^2(T)$ more directly.

An example is provided by the antiferroelectric ($Pm\bar{3}m - Pbam$) transition at $\approx$ 510 K in PbZrO$_3$ which is first order in character and also induces improper ferroelasticity. The variation of f^2 for resonances in the RUS spectra from a single crystal shows (figure 2.11(a)) discontinuous stiffening at the (first order) transition point and further non-linear stiffening upon further cooling [77]. The overall pattern

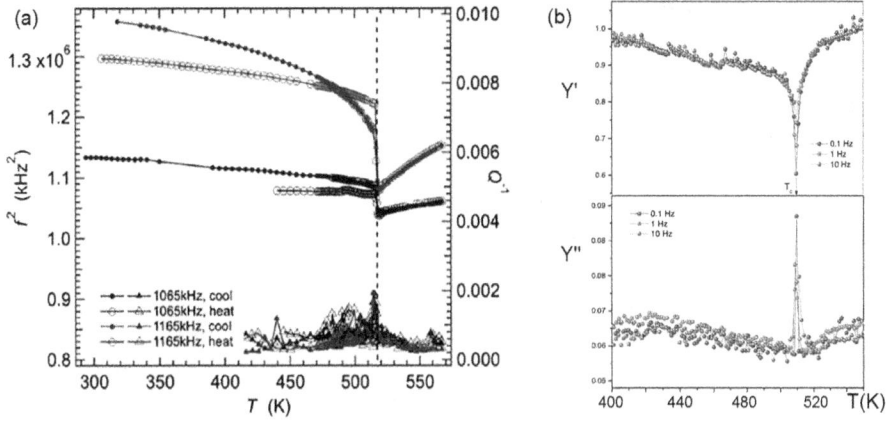

Figure 2.11. (a) Temperature variations of $f^2 \propto Y'$ of PbZrO$_3$ for two selected resonances at $f \approx 1$ MHz measured with RUS [77]. The evolution of f^2 and Q^{-1} for resonances with frequencies near 1165 and 1065 kHz at room temperature are representative of the variations of $\frac{1}{2}(C_{11} - C_{12})$ and C_{44}, respectively, in the stability field of the cubic structure and of some average of shear moduli in the stability field of the orthorhombic structure. Q^{-1} variations show a peak in loss at the transition point with evidence for local dynamical polar clusters. (b) Temperature dependencies of normalized Young's modulus Y' and Y'' of PbZrO$_3$ measured in the [100] c direction at various frequencies (0.1–10 Hz) with diamond DMA. ((a) Reproduced with permission from [77]. Copyright 2022 Elsevier. (b) Reproduced with permission from [78]. Copyright 2016 the American Physical Society.)

reflects the evolution of symmetry-breaking strains which, in turn, depend on the evolution of squared values of two coupled order parameters. In direct contrast with RUS results in the frequency range ≈ 0.1–1 MHz, results from DMA measurements made at 0.02–50 Hz show (figure 2.11(b)) classical softening due to coupling of the form [78] $\lambda \varepsilon q^2$. Thus the time scale for relaxation of the order parameter in response to an induced strain is between $\approx 10^{-2}$ s (softening, DMA) and $\approx 10^{-6}$ s (stiffening, RUS). Interestingly enough, a peak in Q^{-1} is observed in the vicinity of T_c. This is remarkable, since at RUS frequencies the relation $\omega \tau > 1$ should hold, which implies that no Landau–Khalatnikov (LK) damping due to order parameter relaxation (see section 2.6.2) is possible in this frequency range. We can thus assume that the variation of Q^{-1} close to T_c reflects the dynamics of polar clusters. We will come back to this in section 2.6.2. In contrast to the temperature range near T_c, there is no sign of sufficiently high values of Q^{-1} through a wide temperature interval below the transition point that would indicate mobility of ferroelastic domain walls on a timescale of $\approx 10^{-6}$ s. Under the conditions of the DMA experiment, the domain walls are mobile below the transition point and freeze at ≈ 300–400 K (see also section 2.6.1 below).

Ferroelastic behaviour can also be driven by magnetic ordering when the magnetic order parameter, m, couples to a symmetry-breaking shear strain ε_{sb}. Bilinear coupling of strain and a magnetic order parameter is only allowed in cases of piezomagnetic materials with specific magnetic symmetry. If the transition is ferroelastic, the lowest order coupling term permitted by symmetry would typically be improper [79], $\lambda\varepsilon_{sb}m^2$. Straightforward spin-lattice coupling of this type tends to be weak and in some cases the magnitude of ε_{sb} may be too small to be detectable by standard diffraction techniques. For example, the transition from paramagnetic to helical magnetic in Cu_2OSeO_3 at ≈ 60 K gives a change from cubic to rhombohedral symmetry, but no deviations from metrically cubic lattice geometry have been observed. The small degree of elastic stiffening which occurs below the transition point scales approximately with m^2, implying that the effective lowest order coupling terms have the form [80] $\lambda\varepsilon_{sb}^2 m^2$ rather than $\lambda\varepsilon_{sb}m^2$.

In contrast to Cu_2OSeO_3, ferrimagnetic transitions in $NdCo_2$ from a parent cubic structure to tetragonal and orthorhombic structures are accompanied by shear strains of up to ≈ 0.005. f^2 values from peaks in RUS spectra indicate that the shear modulus evolves as expected for improper ferroelastic behaviour in the sense that there is precursor softening in the stability field of the cubic structure, softening in the stability field of the tetragonal phase, and strong acoustic attenuation in the stability fields of both the tetragonal and orthorhombic phases [81]. However, the precursor softening occurs in a temperature interval of ≈ 400 K and can also be explained in terms of bilinear coupling of the symmetry-breaking strain with a driving order parameter, as would be expected ahead of a pseudo-proper ferroelastic transition.

2.4.4 Co-elastic phase transitions

A pervasive feature of almost all phase transitions, whatever the driving order parameter might be, is that they are accompanied by some degree of strain relaxation. The formal spontaneous strains need not include a symmetry-breaking shear strain, however. In other words, not all phase transitions are ferroelastic but all are expected to show elastic/anelastic anomalies which reveal the nature and strength of static/dynamic strain coupling. The term co-elastic has been used to identify transitions which are in that sense elastic without being ferroelastic [5]. Aside from the absence of ferroelastic twins, the patterns of behaviour are closely similar to those of improper ferroelastic transitions because the lowest order coupling of non-zero strains always has the form $\lambda\varepsilon q^2$.

A classic example of co-elastic behaviour is provided by the hexagonal ($P6_222$)–trigonal ($P3_221$) transition in quartz, SiO_2. It is weakly first order and the pattern of elastic-constant variations [82] through the transition point (figure 2.12) is closely similar to that expected for an improper ferroelastic transition with a Landau free energy expansion up to sixth order in components of the order parameter. As also in the case of improper ferroelastics, softening at temperatures of about 200 K above the transition point is not predicted by the standard Landau expansion and is due to dynamical precursor effects. They will be discussed in more detail in section 2.6.2.

The same discussion of relaxation times for the order parameter in response to an induced strain applies to co-elastic transitions as set out above for improper ferroelastic transitions. Let us illustrate this using two examples showing magnetic transitions. The first example of co-elastic behaviour in a magnetic system is

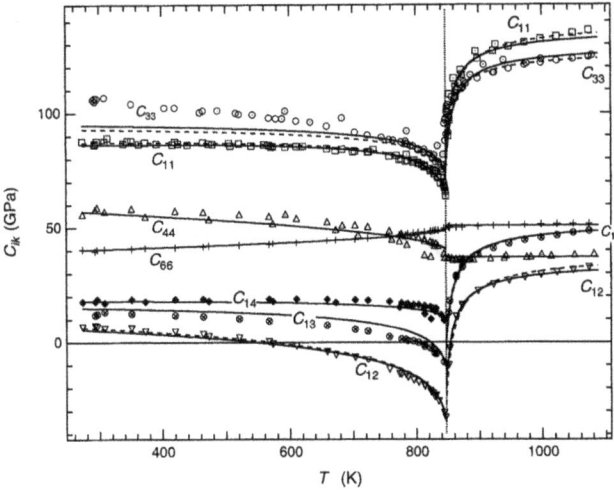

Figure 2.12. Variations of elastic constants through the hexagonal ($P6_2 22$)–trigonal ($P3_2 21$) transition in quartz. Symbols are experimental data from multiple sources in the literature. Solid curves in the stability field of the trigonal structure are solutions to a Landau expansion using a complete calibration of all the coefficients. Solid curves in the stability field of the hexagonal structure are calibrations of precursor softening described by $\Delta C_{ij} = A_{ij} |T - T_c|^{-\kappa}$. For the present case a value of $\kappa \approx 0.65$ was found. (Reproduced from [82].)

provided by the paramagnetic–antiferromagnetic transition in CoF_2 at ≈ 39 K [83]. The symmetry change is $P4_2/mnm1' \rightarrow P4'/mnm'$, for which the magnetic order parameter, Q_m, has the symmetry of irrep Γ_2^+. RUS data show that the pattern of discontinuous softening is followed by non-linear recovery below the transition point, resembling the classical consequence of coupling $\lambda \varepsilon q^2$, where the relaxation time of the order parameter is shorter than $\approx 10^{-6}$ s. The degree of softening is smaller than that shown by GeTe primarily because the magnitude of the strains is only up to 0.001 for CoF_2, as opposed to up to at least 0.03 for GeTe. An additional factor is that the values of Landau coefficients tend to be smaller for displacive transitions than for transitions in which a contribution to the excess entropy comes from a component of order/disorder. Evidence of dynamic effects ahead of the transition is provided both by the tail in ε_1 and precursor softening of the shear modulus.

By way of contrast to CoF_2, figure 2.13 shows the evolution of two resonances in RUS spectra from a single crystal of $YMnO_3$ through the paramagnetic–antiferromagnetic transition at about 70 K [84]. Rather than softening below the transition point, there is continuous stiffening which resembles the evolution of the square of the magnetic order parameter. Just as in the case of $PbZrO_3$ discussed above, the relaxation time of the order parameter must be longer than 10^{-6} s. The lack of any variation of Q^{-1} associated with the transition also confirms that there are no intrinsic acoustic loss mechanisms operating on the time scale of the resonance frequencies.

In summary, improper ferroelastic and co-elastic phase transitions display characteristic variations in elastic properties which reveal the nature and magnitude

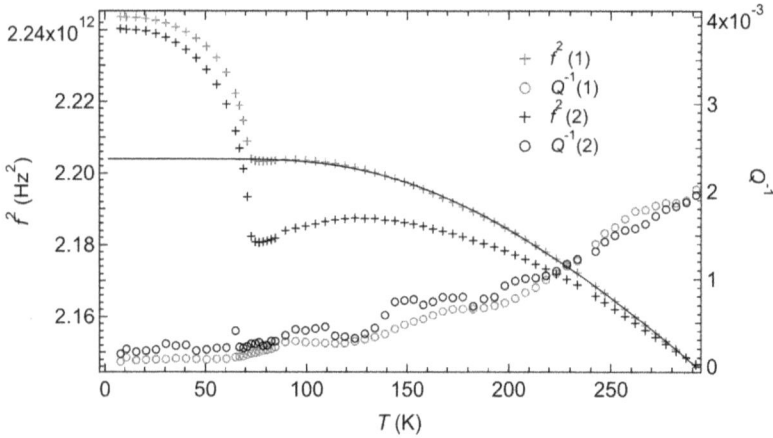

Figure 2.13. Variations of f^2 and Q^{-1} from two resonances with frequencies near 700 (black) and 1500 kHz (red) from RUS spectra through the paramagnetic–antiferromagnetic transition in a single crystal of YMnO$_3$. The solid line is a guide for the eye representing the expected evolution of f^2 for one of the peaks in the absence of any transition. Values of f^2 for the two resonances have been scaled along the y-axis so that they match at room temperature. (Reproduced from [84]. Copyright IOP Publishing. CC BY 3.0.)

of strain coupling with driving order parameters that can be due to any transition mechanism. Variations in anelastic properties are revealing not only of the presence of precursor effects, changes in relaxation time in the vicinity of transition points and the dynamics of ferroelastic domain wall motion, but also of the absolute relaxation times of the strain–order parameter coupling.

2.5 Antiferroelectrics and their relation to ferroelasticity

Antiferroelectrics are materials with a great potential for technical applications [85]. At variance with structural ferroic phase transitions which give rise to macroscopic tensors coupled to macroscopic fields, criteria defining AFE phase transitions are still under debate. Recently, Tolédano and Guennou [28] proposed specific symmetry criteria for AFE transitions. They showed that two symmetry-based conditions have to be fulfilled simultaneously in a phase transition to qualify as (paraelectric) PA–(antiferroelectric) AFE.

First condition: At the PA→AFE transition, a set of crystallographic sites undergoes a symmetry lowering such that their site symmetry group becomes polar and gives rise to a local polarization. The sites between the polar ones have to keep the inversion centres in order to restore the nonpolar macroscopic symmetry due to the anti-parallel local polarizations.

Second condition: The AFE space group has a symmorphic polar subgroup coinciding with the local symmetry of emerging polar sites (only simple symmetry operations such as rotations, reflections or inversion, no glide planes or screw axes).

If *both* conditions are fulfilled, the phase is referred to be AFE, i.e. characterized by the existence of an electric field induced ferroelectric (FE) phase associated with a double hysteresis loop and a characteristic anomaly of the dielectric permittivity.

Let us illustrate the application of the above two conditions on few quite well-known ferroic materials [28]. $SrTiO_3$ exhibits an antiferrodistortive and improper ferroelastic phase transition from the cubic $Pm\bar{3}m$ to the tetragonal $I4/mcm$ space group. The polar sites have symmetry $mm2$ and the polar subgroups of $I4/mcm$ are $I4$, $C2$, Cm, and $P1$ whose point groups differ from $mm2$. Thus, condition 2 is not fulfilled, and indeed no AFE properties have been detected for $SrTiO_3$. Another case is $PbZrO_3$ with a phase transition from $Pm\bar{3}m$ to $Pbam$. Crystallographic sites undergoing a lowering of their local symmetry to the polar point groups m, 2, 1 coincide with the symmorphic polar subgroups Pm, $P2$, and $P1$ of $Pbam$. $PbZrO_3$ can therefore be considered to be potentially AFE, and indeed double hysteresis loops and a field induced FE phase have been found [86] in this material.

Another interesting example, not mentioned in [28], is potassium thiocyanate (KSCN). This material is treated in more detail in section 2.6.2. It undergoes a phase transition [87, 88] from $I4/mcm$ to $Pbcm$ with doubling of the unit cell, at $T_c = 415$ K. In $Pbcm$ polar sites have symmetry 2 and m and the symmorphic subgroups of $Pbcm$ are Pm and $P2$, i.e. coinciding with the symmetry of the polar sites. Thus, both conditions are fulfilled, implying that KSCN is potentially AFE. Experiments under an electric field should be performed to check whether this material is indeed AFE.

Tolédano and Guennou [28] have shown that quite generally a PE → AFE transition can be explained by a Landau type free energy expansion in terms of an antiferroelectric order parameter η coupled to polarization P, which in a simplified version reads

$$\Phi(\eta, P, T) = \Phi_0(T) + \frac{a}{2}(T - T_c)\eta^2 + \frac{\beta}{4}\eta^4 + \frac{\gamma}{6}\eta^6 + \frac{1}{2\chi_0}P^2 + \frac{\delta}{2}\eta^n P^m - EP. \quad (2.39)$$

The authors have shown [28], that only couplings of the type $n = 1$, $m \geqslant 2$ or $n = 2$, $m = 2$ describe the property of AFE transitions with unstable polar phase at zero electric field. Interestingly, all experimental examples of AFE transitions found up to now correspond to a positive coupling coefficient $\delta > 0$, leading to a decrease of the dielectric permittivity below T_c. Also the above mentioned KSCN crystals show a decrease of dielectric permittivities in all three directions [89], which can be explained by a microscopic model [90]. Indeed, by expanding the free energy of the pseudospin model [91] of KSCN in terms of the AFE order parameter η_1 and the polarization P_x, one obtains a corresponding Landau free energy expansion, which contains a coupling term $\delta\eta_1^2 P_x^2$, with $\delta = \frac{2k_B}{T^3}T_{AF}^2 T_{FE}^2 > 0$, where T_{AF} and T_{FE} are the AFE and FE phase transition temperatures, respectively.

In a recent paper, Tolédano and Khalyavin [92] have shown that in real AFE systems, for example in $PbZrO_3$, $NaNbO_3$ or $PbHfO_3$, the free energy expansion can be more complicated. For example in $PbZrO_3$ two order parameters of different symmetries and many components couple to the polarization. However, despite the more complicated form of the free energy, the main reason for the AFE behaviour remains the biquadratic coupling between the OPs and the polarization.

Very recently, Tolédano and Li revealed [22] that most ferroelastic materials have polarizable crystallographic sites forming an antiferroelectric array of dipoles, which can be turned into polar phases upon application of high electric fields. In Landau theory it may be described by coupling terms between strain e_{ij} and polarization components P_i of the type $\delta e_{ij} P_i P_j$. An example presented in [22] is the orthorhombic (*Pmna*) to monoclinic ($P2_1/c$) ferroelastic phase transition of NdP_5O_{14}, which is accompanied by a spontaneous shear strain $e_{ij} = e_{xy}$. Including coupling terms of strain and polarization the free energy under applied electric field $\mathbf{E} = (E_x, E_y, 0)$ reads

$$\Phi = \Phi_0(T) + \frac{a}{2}(T - T_c)e_{xy}^2 + \frac{\beta}{4}e_{xy}^4 + \frac{1}{2\chi_{0x}}P_x^2$$
$$+ \frac{1}{2\chi_{0y}}P_y^2 + \delta e_{xy}P_x P_y - E_x P_x - E_y P_y. \tag{2.40}$$

The authors [22] have elaborated temperature and field dependencies of the dielectric permittivity, polarization and strain. However, since no attention has been paid to the changes of elastic susceptibilities (a tensor quantity of importance at ferroelastic phase transitions) under electric field and at different temperatures, it is planned to investigate this in the near future.

2.6 Inhomogeneous structures at ferroelastic phase transitions

It is well known that inhomogeneous structures near PTs can exist in a wide range of size and time scales. A comprehensive discussion would go beyond the scope of this chapter. Instead, we focus on some specific topics.

2.6.1 Domains and domain walls

An excellent review of domains, domain walls and their impact on susceptibilities is presented in the book by Tagantsev, Cross, and Fousek [93]. Here we will discuss ferroelastic domains and domain walls, a topic which is intimately related to the work of Ekhard Salje [5, 94–96].

A ferroelastic phase transition is accompanied by a number of different orientational DSs, which can be switched among each other by the application of an external stress. Starting, for example, with a crystal which contains an equal number of all possible orientational DSs, the average spontaneous strain is zero. If by application of an external stress some of the DSs are switched, the average strain is no longer zero, which leads to an effective elastic softening of the crystal. Let us illustrate this with the example of $SrTiO_3$. It undergoes an antiferrodistortive improper ferroelastic phase transition from $Pm\bar{3}m$ to $I4/mcm$ [2] at $T_c \approx 105$ K. The phase transition is due to static rotations $(0, 0, \phi)$ of the TiO_6-octahedra around the tetragonal c-axis, which are alternating $(0, 0, \pm\phi)$ along all three cubic directions. Due to these staggered rotations, the translations $(2n_1 + 1)(a, 0, 0)$, $(2n_2 + 1)(0, a, 0)$ and $(2n_3 + 1)(0, 0, a)$ are lost. The symmetry reduction from $Pm\bar{3}m \rightarrow I4/mmm$ leads to three orientation DSs, i.e. $1_1 = (\phi, 0, 0)$, $2_1 = (0, \phi, 0)$, $3_1 = (0, 0, \phi)$ and three translational DSs, denoted as $1_2 = (-\phi, 0, 0)$, $2_2 = (0, -\phi, 0)$, $3_2 = (0, 0, -\phi)$.

Following the definition of equation (2.8) the spontaneous strains in the different orientation DSs in Voigt notation are

$$\varepsilon_{(s)}(1_1) = (\varepsilon_{1s}, \varepsilon_{1s}, \varepsilon_{3s}, 0, 0, 0)$$
$$\varepsilon_{(s)}(2_1) = (\varepsilon_{3s}, \varepsilon_{1s}, \varepsilon_{1s}, 0, 0, 0) \tag{2.41}$$
$$\varepsilon_{(s)}(3_1) = (\varepsilon_{1s}, \varepsilon_{3s}, \varepsilon_{1s}, 0, 0, 0),$$

where $\varepsilon_{1s} = \frac{a - a_0}{a_0}$ and $\varepsilon_{3s} = \frac{c - a_0}{a_0}$. The spontaneous strain for a given orientation domain state is independent of its translational state, i.e. $\varepsilon_{(s)}(i_2) = \varepsilon_{(s)}(i_1)$, $i = 1, 2, 3$. The domain walls (DW) between the orientation DsS, which satisfy the conditions of elastic compatibility [97], form planes with DW normals $\mathbf{n} = [101], [10\bar{1}]$ for the domain pair (DP) $(1_1/2_1)$, $\mathbf{n} = [011], [0\bar{1}1]$ for $(1_1/3_1)$ and $\mathbf{n} = [110], [\bar{1}10]$ for $(2_1/3_1)$.

The influence of domain wall motion on the elastic or dielectric properties (for ferroelectrics) of materials is a rather complex mesoscopic problem which involves many length and time scales. It is a matter of active experimental and theoretical research and is far from being completely understood. Nevertheless, as we shall show here, some of the data can be described well in terms of a relatively simple model.

Figure 2.14 displays the temperature dependence of Young's modulus of SrTiO$_3$ measured [98] by DMA in three-point bending geometry at a frequency of 10 Hz.

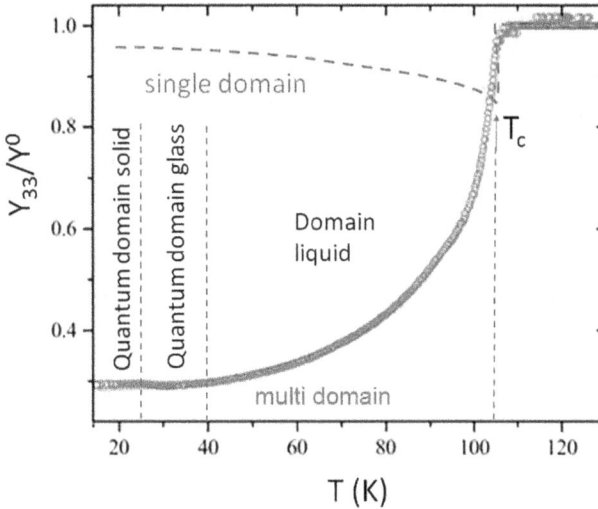

Figure 2.14. Temperature dependence of the Young's modulus Y_{33} of SrTiO$_3$ measured [98] at very low frequency (10 Hz) using a dynamic mechanical analyser in three-point bending geometry. The data can be excellently fitted (black line) with $Y_{33} = (1 + S^{DW} + \Delta S^{LK})^{-1}$, where S^{LK} is the usual Landau type contribution for an improper ferroelastic phase transition and $S^{DW} \propto \phi(T)^2$ is the DW contribution to the elastic compliance. The single domain behaviour (blue dashed line) is shown for comparison. The classification of the different temperature regions is taken from [101]. We will discuss its meaning later. (Reproduced with permission from [98]. Copyright 2000 the American Physical Society.)

The high frequency (40–200 MHz) elastic moduli of SrTiO$_3$ were measured by Rehwald [99] using the ultrasonic pulse-echo method. In this frequency range, the domain walls do not contribute to the elastic behaviour and the data reflect the behaviour as expected for an improper ferroelastic PT, i.e. a small negative cusp at T_c. The huge (superelastic) softening at low frequencies is due to the stress induced motion of ferroelastic domain walls. In [100] we have calculated the contribution of DWs to the softening of the elastic modulus for some perovskites, including SrTiO$_3$. In the following we will present the main ideas of the model. For simplicity we consider only two different ferroelastic DSs, e.g. 1_1 and 2_1 with corresponding domain widths x^+ and x^-. Application of an external stress σ increases the width of one domain at the expense of the other. The resulting macroscopic strain due to the domain wall shift $\Delta x(\sigma)$ is given as

$$\varepsilon^{DW} = 2\frac{\varepsilon_s}{d}\Delta x(\sigma), \tag{2.42}$$

where $\Delta x = x^+ - x^-$ and d is the average distance between domain walls. Since we are interested in the elastic response at small external stress, we can assume that d is independent of σ. Quite generally, the domain thickness d depends on temperature, but for the present perovskites d is almost independent of T. The DW contribution to the elastic compliance can then be calculated as

$$S^{DW} = 2\frac{\varepsilon_s}{d}\frac{\partial}{\partial\sigma}\Delta x(\sigma). \tag{2.43}$$

Usually $\Delta x(\sigma)$ is calculated from the simple equation of motion [102]

$$m\frac{d^2\Delta x}{dt^2} + k\Delta x = 2\varepsilon_s\sigma(t), \tag{2.44}$$

where m = effective mass and k = effective spring constant, yielding

$$\Delta x(t) = \frac{2\varepsilon_s}{k - m\omega^2}\sigma(t). \tag{2.45}$$

The DW contribution to the elastic compliance then takes the form

$$S^{DW} = \frac{4\varepsilon_s^2}{(k - m\omega^2)d}. \tag{2.46}$$

Although equation (2.46) yields a resonance [102] of the DW contribution at $\nu_0(m, k)$, most materials show a relaxational behaviour of the DW susceptibility. We will come back to this later. Here we would like to stress the following observation. Since for improper ferroelastic PTs the spontaneous strain is always proportional to the square of the order parameter ϕ, i.e. $\varepsilon_s \propto \phi^2$, the DW contribution to the compliance at very low frequency yields

$$S(T)^{DW} \propto \frac{\phi^4(T)}{d}. \tag{2.47}$$

However, for all improper ferroelastic materials we studied [69, 98, 100, 103], e.g. $SrTiO_3$, $LaAlO_3$, $KMnF_3$, etc, we found that $S^{DW} \propto \phi^2(T)$ which is in sharp contrast to equation (2.47). To account for this difference, it should be noted that equation (2.47) was derived with the assumption that the DWs have zero thickness. To explain the observed discrepancy between theory and experiment, we take into account the finite thickness w of DWs. Including the repulsion between extended domain walls and taking into account the long-range elastic interactions between ferroelastic needle shaped domains, the free energy density reads [100]

$$
\begin{aligned}
F = F_{LG} + q\varepsilon_s^2(x^+ + x^-) + \frac{1}{x^+ + x^-} \\
[E_w + b\exp(-x^+/w) + b\exp(-x^-/w)] + 2\varepsilon_s \frac{x^+}{x^+ + x^-}\sigma,
\end{aligned}
\tag{2.48}
$$

where F_{LG} is the Landau–Ginzburg expansion of the free energy density, describing the corresponding PT. b determines the strength of the repulsion between DWs and E_w is the DW energy. The free energy density (2.48) can be used to calculate the DW contribution to the elastic susceptibility as

$$
S^{DW} = \frac{\varepsilon_s^2 w^2}{bd\exp(-d/w)}.
\tag{2.49}
$$

In [100] it is shown that $d\exp(-d/w)$ is nearly independent of temperature for many perovskites. Since for second order PTs [5] $w(T)^2 \propto \phi^{-2}$, we obtain $S^{DW} \propto \phi(T)^2$ in perfect agreement [69, 100] with experimental results. Most clearly this agreement can be visualized for the case of $LaAlO_3$, which corresponds to the few examples where the domain wall width was measured [95] in a broad temperature range. In figure 2.15 we have redrawn the data of figure 6 of [103] showing the resulting dependence of the DW susceptibility $S^{DW} \propto \phi(T)^2 \propto (T_c - T)$, as expected from

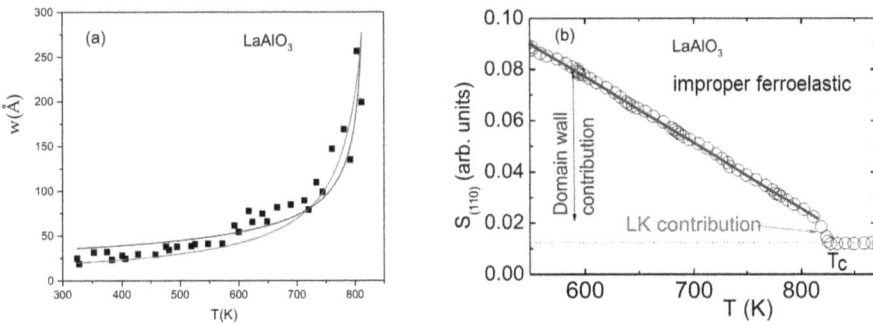

Figure 2.15. (a) Temperature dependence of the wall widths w (squares). The solid red line represents a least-squares fit to $w(T) \propto (T_c - T)^{-1}$ with $T_c = 850$ K. The solid blue line is a fit to $w(T) \propto (T_c - T)^{-\frac{1}{2}}$ ($T_c = 820$ K), which is the behaviour expected from Landau–Ginzburg theory [5]. (b) Temperature dependence of the compliance $S_{(110)}$ of $LaAlO_3$ with applied stress in the (110)-direction. The main contribution to $S_{(110)}$ results from the motion of domain walls and is excellently fitted with $S^{DW} \propto \phi(T)^2 \propto (T_c - T)$. ((a) Reproduced with permission from [95]. Copyright 1999 AIP Publishing. (b) Reproduced with permission from [103]. Copyright 2022 Elsevier.)

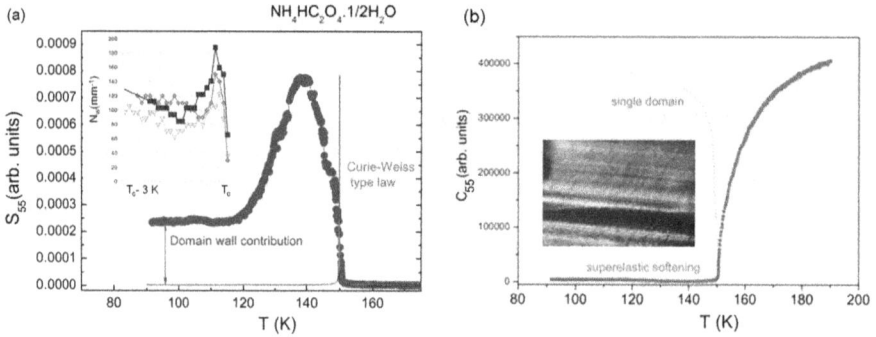

Figure 2.16. (a) Temperature dependence of the shear elastic compliance S_{55} of $NH_4HC_2O_4 \cdot \frac{1}{2}H_2O$ (AHO) measured with DMA in three-point bending geometry (blue points). The red line shows the Curie–Weiss type law of a nomodomain crystal. The inset shows the temperature dependence of the number of domain walls N_w as determined from optical polarizing microscope measurements. (b) The temperature dependence of C_{55} of AHO is very similar to the temperature dependence of Y' in figure 2.10 for lowest applied stress of ca. 20 MPa, showing Curie–Weiss behaviour above and a very low plateau below T_c. ((a) Reproduced with permission from [69]. Copyright 2012 AIP Publishing.)

equation (2.49). It should be noted that this linear temperature dependence of S^{DW} is only obtained if $w(T) \propto (T_c - T)^{-\frac{1}{2}}$. Although the data of [95] do not allow the exponent of $w(T) \propto (T_c - T)^{-\alpha}$ to be determined precisely, the present analysis is in favour of $\alpha = \frac{1}{2}$, which is also supported by Landau–Ginzburg theory [5].

The DW contribution to the elastic compliance for a proper or a pseudo-proper ferroelastic phase transition is quite different from the improper ferroelastic case. For example, for a pseudo-proper ferroelastic phase transition $\varepsilon_s \propto \phi(T)$. With $w(T) \propto \phi(T)^{-1}$ we obtain $S^{DW} \propto d^{-1}$, which is proportional to the number N_w of DWs. Figure 2.16 shows that this relation is quite nicely fulfilled [69] for the second order $Pmnb \rightarrow P2_1/n$ phase transition in $NH_4HC_2O_4 \cdot \frac{1}{2}H_2O$ (AHO). Also in NdP_5O_{14}, which exhibits a proper ferroelastic phase transition from $Pncm \rightarrow P2_1/c$, the relation $S_{55} \propto N_w$ is excellently fulfilled (see figure 2 of [69]).

Although these results are quite promising for contributing to a better understanding of the influence of domain walls on macroscopic susceptibilities, a full understanding of this topic is still not achieved. Some open questions concern, for example, the dynamical behaviour of DWs at different temperatures and different applied forces, a topic that will be given some attention in the next section.

2.6.1.1 Domain glass

Quite a number of ferroelastic materials show disappearance of the DW contribution to the susceptibility (see e.g. figure 6 of [103]) upon lowering temperature well below T_c, depending on the time scale of the measurements and the excitation magnitude. This re-hardening in Y' is usually accompanied by a peak in the imaginary part of the complex susceptibility (dielectric or elastic, depending on the type of PT), which appears around a temperature T_f, where $\omega\tau(T_f) \approx 1$. Here, $\tau(T)$ is a (mean-)relaxation time, characterizing the relaxation behaviour of the

DWs. It could be, for example, included in the equation of motion (2.44) above, by adding a term $\gamma\frac{dx}{dt}$. Usually $\tau(T)$ displays a thermally activated behaviuor in some temperature range, i.e. $\tau(T) = \tau_0 e^{E/k_B T}$. For a given frequency ω, in the temperature range where $\omega\tau(T) > 1$, the domain walls can no longer follow the applied field (stress, or electric field) and they are said to be frozen. There is a question about the origin of this domain freezing process. For example, in LaAlO$_3$ [105–107] the dynamic elastic response at low forces and low frequencies (0.1–30 Hz) can be described well by a Debye-like relaxation process with a distribution of activation energies and corresponding relaxation times (the domain-sliding regime). The mean relaxation time follows an Arrhenius behaviour $\tau = \tau_0 \exp(E/k_B T)$, with $\tau_0 = 5 \times 10^{-8}$ s and $E = 0.66$ eV.

On the other hand, there are also examples where the mean relaxation time instead seems to follow a Vogel–Fulcher type behaviour, i.e. $\tau = \tau_0 \exp[E/(T - T_{VF})]$. The DWs below the phase transition in PbZrO$_3$ (PZO) display such a behaviour [78]. PbZrO$_3$ exhibits an improper ferroelastic phase transition from a paraelectric phase with cubic symmetry $Pm\bar{3}m$ ($Z = 1$) to an antiferroelectric orthorhombic phase $Pbam$ ($Z = 8$) at $T_c \approx 503$–510 K, depending on crystal quality. The low frequency ($f = 0.05$–100 Hz) elastic moduli $Y_{[110]}$ measured with applied stress at an angle of 45° with respect to the DW normal display (figure 2.17) a clear softening below T_c due to the sliding of the DWs. As shown above (figure 2.11), the high frequency ($f = 1.5$ MHz) elastic moduli show completely different temperature dependencies.

Recently, low frequency DMA measurements [108] of KSCN and RbSCN yielded a very interesting domain wall dynamics with a mean relaxation time deviating strongly from a simple Arrhenius behaviour. The temperature dependence of the domain wall relaxation time in both systems can be quite well fitted with a Vogel–Fulcher law, which indicates domain glass behaviour.

Figure 2.18 compares the relaxation times of different systems [107], showing that for some materials τ differs from pure Arrhenius behaviour. However, such deviations are not very pronounced. In fact, the data of PZO can be nearly equally well fitted with Arrhenius as well as with Vogel–Fulcher behaviour [78]. The Arrhenius fit yields $\tau_0 = 5 \times 10^{-11}$ s, $E = 0.58 \pm 0.02$ eV and the Vogel–Fulcher fit leads to $\tau_0 = 1.2 \times 10^{-7}$ s, $E = 0.23 \pm 0.01$ eV and $T_{VF} = 120$ K. To clarify the question (Vogel–Fulcher versus Arrhenius, etc), measurements of the dynamic elastic susceptibility over a wider range of frequency are required, which is a rather challenging task.

Not long ago, Ekhard Salje coined the term 'domain glass' [109]. Indeed, large scale computer simulations [110] of a simple model showed that domain boundary movements under shear deformation leads to a Vogel–Fulcher behaviour at high temperatures ($T > 1.2T_{VF}$). Below the Vogel–Fulcher temperature, no thermal activation was found and the time evolution of the domain pattern became athermal. Below T_{VF} the movement of domain boundaries is dominated by the nucleation and growth of needle domains. Their movement occurs in fast jerks. The probability to observe jerks of energy E follows a power-law spectrum ($P(E) \sim E^{-\alpha}$) with energy exponents close to $\alpha \approx 2$. Our recent measurements of

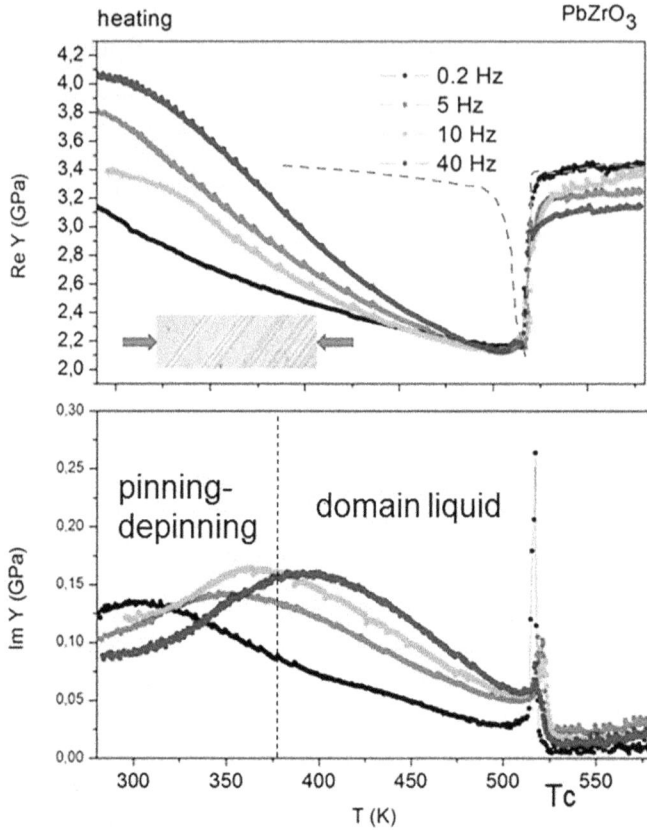

Figure 2.17. The temperature dependencies of real and imaginary parts of Young's modulus of PbZrO₃ measured in the [110]-direction at various frequencies is dominated in the domain liquid regime by domain wall motion induced softening. The single domain behaviour is sketched by the blue dashed line. The inset shows a polarizing microscopy picture of the domains and DWs to sketch the orientation of the applied force with respect to the DW direction. The mapping of the different regions, i.e. domain liquid and pinning–depinning regions, is according to results from slow compression experiments [111].

DW motion in PbZrO₃ and LaAlO₃ under extremely slow compressive stress [111] confirmed such a behaviour (figure 2.19). Well below T_c, at $T \leqslant 370$ K (for PZO) the DW movement occurs in jerks (the pinning–depinning regime), following a power-law energy distribution with $\alpha \approx 1.6$. At $T > 370$ K the energy distribution changes to exponential, i.e. $P(E) \propto e^{-E/T}$, in agreement with results from computer simulations [110].

All this reflects typical aspects of glasses. However, in contrast to strain glasses [112], a domain glass does not need extrinsic ingredients such as point defects, or other random fields, etc, to form. The domain boundary patterns can develop glass-like states while the underlying matrix remains fully crystalline without any defect induced disorder. Interestingly enough, such a situation occurs in SrTiO₃ single crystals. As figure 2.14 shows, the movement of DWs in SrTiO₃ in response to an

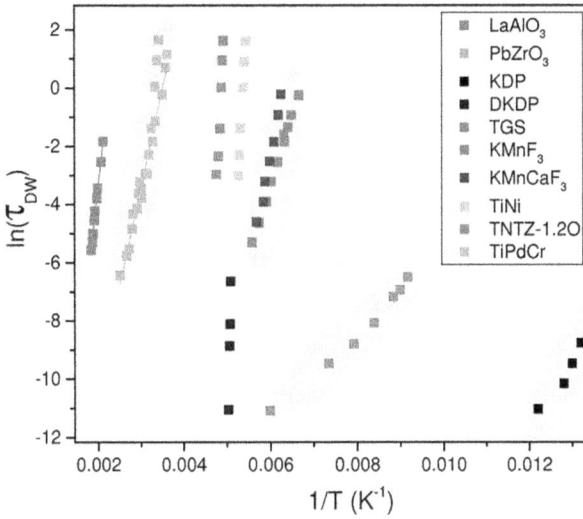

Figure 2.18. Comparison of the temperature dependencies of DW relaxation times for different materials, plotted in semilogarithmic scale. (Reproduced from [107]. CC BY 4.0.)

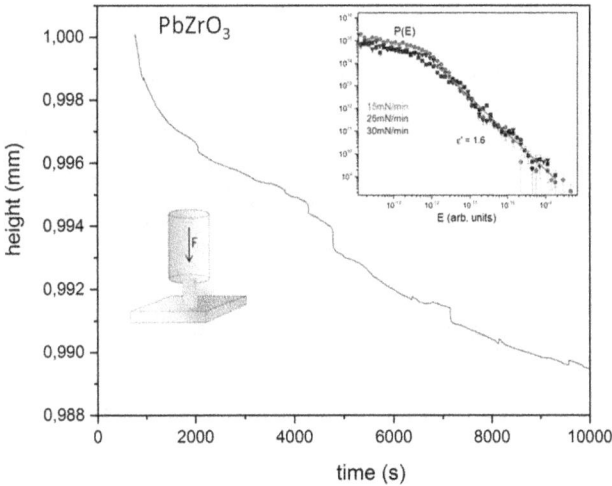

Figure 2.19. Slow compression experiments (ca. 0.5 mN s^{-1}) of PbZrO$_3$ at $T = 370$ K showing tiny jumps (ca. 10–100 nm) in sample height (jerks). At $T \leqslant 370$ K they are power-law distributed (see the inset) with $P(E) \propto E^{-\alpha}$ and $\alpha \approx 1.6$. (Adapted from [111]. CC BY 4.0.)

applied dynamic stress causes a huge superelastic softening. Young's modulus can be excellently fitted over a wide temperature range (between 105 K and 40 K) with $Y_{33} = (1 + S^{DW} + \Delta S^{LK})^{-1}$, where S^{LK} is the usual Landau type contribution for an improper ferroelastic phase transitionand $S^{DW} \propto \phi^2(T)$. Kustov *et al.* [101] called this region a 'domain liquid', characterized by linear viscous flow of unpinned twin boundaries (TBs). It should be noted that in the resonance experiments (at 90 kHz) of [101] on $SrTiO_3$ the domain liquid region ended at a higher temperature (ca. 80 K) compared to the low frequency data. Between ca. 40 K and 70 K jerky damping occurs in the resonance experiment above a critical strain amplitude of about $\varepsilon_{cr} \approx 2 \times 10^{-7}$ due to thermally activated depinning of twin boundaries, and thus this region is called the 'pinning region'. The depinning strain in $SrTiO_3$ is extremely low, it is about 300 times lower than in $LaAlO_3$. In previous DMA experiments [98, 113] the strain amplitude was much too high to induce pinning–depinning jerks, and as a result the region of viscous flow (figure 2.14) of twin boundaries extends down to about 40 K. This is the temperature where in both experiments an additional softening of the Young's modulus occurs, but interestingly enough this softening increases non-linearly with applied stress [113]. In a resonance experiment [101] performed at a frequency of 90 kHz, the damping rises sharply below 40 K and the non-linear twin boundary response switches abruptly from jerky to continuous. This non-linear internal friction response is typical for glassy dynamics (barrier heights diverge with decreasing strain amplitude), which was also observed below ca. 40 K in recent RUS measurements [114]. Kustov *et al.* [101] called the region between 25 K and 40 K the 'quantum domain glass' region. Below approximately 25 K the experiments [98, 101, 113] yield increasing stiffness and completely frozen relaxation of a subsystem of domain boundaries, which led the authors referring to this new state as a 'quantum domain solid'. It should, however, be noted that the vast majority of twin boundaries are still mobile at 25 K and contribute to the superelastic softening down to the lowest measured temperatures of about 10 K. Based on ultrasonic measurements [115] in the frequency regime of some tens of MHz, the authors found a typical Arrhenius relaxation time dependence of twin boundary movement, i.e. $\tau = \tau_0 \exp(E_a/k_B T)$ with $\tau_0 = 1.5 \times 10^{-11}$ s and $E_a = 40$ meV. Assuming a conventional Debye equation for the domain wall response, $\Delta S^{DW}/(1 + \omega^2\tau^2)$, one would expect domain freezing ($\omega\tau > 1$) in the range of about 20 K for measurement frequencies 1– 10 Hz. Measurements of the elastic behaviour of $SrTiO_3$ below 20 K as a function of temperature, frequency and applied dynamic stress could thus contribute to a better understanding of these long-standing problems.

To date, there is no clear picture of the low temperature behaviour of $SrTiO_3$. Nevertheless, there is broad agreement that the observed anomalies in elastic and dielectric susceptibilites are related to the presence of domain walls. Indeed, Scott *et al* [116] proposed that the domain mobility becomes enhanced below ca. 40 K due to changes in the structure of the twin walls related to the polarity of the TBs. Here we show that—instead of the twin boundaries—the onset of a ferroelectric phase transition in hard antiphase boundaries (APBs) at about 40 K may be related to the appearance of glassy behaviour in STO at this temperature. In fact, the twin boundaries actually become polar [117] at the much lower temperature of about 25 K. The internal structures and properties of domain walls in $SrTiO_3$ have been studied by various methods, i.e. by Landau–Ginzburg–Devonshire free energy expansions [117–119], *ab initio* calculations [120, 121] and using machine-learned force fields [122]. As already mentioned above, the symmetry reduction in $SrTiO_3$ leads to six different DSs, three orientational and three translational ones.

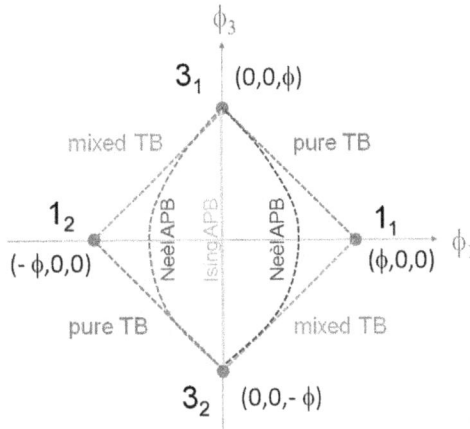

Figure 2.20. Four different domain states of SrTiO$_3$ out of six are shown here in order parameter space. Order parameter profiles follow an extremal path between the domain states.

The primary order parameter of the phase transition is given by the rotation angle $\vec{\phi}$ of the oxygen octahedra, which in a given domain state takes only one component, say $3_1 = (0, 0, \phi)$, $3_2 = (0, 0, -\phi)$, etc (figure 2.20).

Tagantsev *et al.* [118] have shown that so-called 'hard' antiphase boundaries (APBs) in SrTiO$_3$ are of the Néel-type immediately below $T_c = 105$ K, while 'soft' APBs are of the Ising type at all temperatures below T_c. For example, for the domain wall $(3_1/3_2)$ with domain wall normal $\mathbf{n} = [100]$ (hard APB), the order parameter develops an additional component $\phi_2(x)$, i.e. it follows the path $(0, \phi_2(x), \phi_3(x))$ with $\phi_2(\pm\infty) = 0$ and $\phi_3(\pm\infty) = \pm\phi$ (figure 2.20). In this work it was also shown [118] that hard APBs become ferroelectric below $T_c^{\text{APB}} \approx 40$ K with $P_3(x) \neq 0$ inside the APB. This is in contrast to soft APBs ($\mathbf{n} = [001]$) that do not develop a ferroelectric polarization. Later, Morozovska *et al.* [117] found that hard ferroelastic TBs also become ferroelectric, but at a lower temperature of about $T_c^{\text{FEL}} \approx 25$ K. Similar to soft APBs, soft ferroelastic twin boundaries also do not show a switchable domain wall polarization at any temperature.

Figure 2.21 shows the temperature dependence of the in-plane polarization P_3 at the centre of a hard APB $(3_1/3_2)$, $\mathbf{n} = [100]$ in SrTiO$_3$ calculated from the Landau–Ginzburg–Devonshire (LGD) free energy expansion of [119].

Without rotopolar coupling terms between the polarization and the gradients of the order parameter the phase transition occurs at $T_c^{\text{APB}} \approx 45$ K in good agreement with the findings of [118] and [117]. Rotopolar terms, being linear in P_3, smear the phase transition, similar to what happens for an Ising-type phase transition in the presence of a magnetic field. Interestingly, the region of glass-like dynamics in SrTiO$_3$ starts at the same temperature (ca. 40 K) where a ferroelectric phase transition takes place [101] inside the hard APBs. To find some clue as to whether this is just coincidence, let us take a closer look at the structure and properties of hard APBs.

Figure 2.22 shows two different situations of order parameter and polarization profiles of two hard APBs $(3_1/3_2/3_1)$ calculated by minimizing the LGD free energy

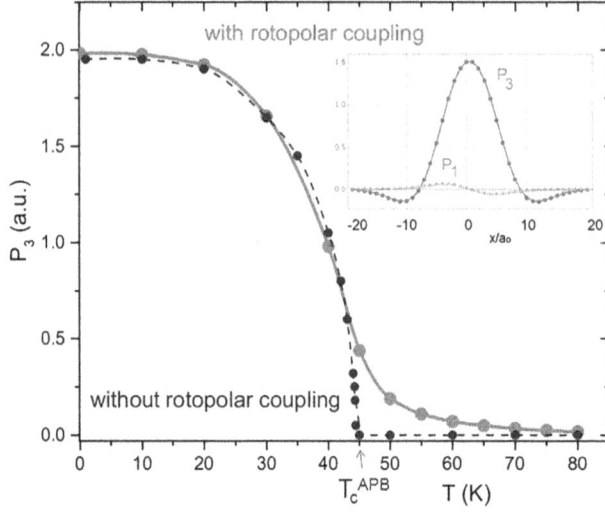

Figure 2.21. Temperature dependence of the maximum of the in-plane polarization $P_3(x = 0)$ of a hard APB in SrTiO$_3$ with and without rotopolar coupling terms using the free energy expansion of [119]. The inset shows typical profiles of the in-plane polarization $P_3(x)$ (symmetric) and the out-of plane polarization $P_1(x)$ (asymmetric).

of [119], i.e. with parallel ($3_1 \downarrow 3_2 \downarrow 3_1$) and anti-parallel ($3_1 \downarrow 3_2 \uparrow 3_1$)$P_3$ components at consecutive APBs. During the minimization process, the system flips between these two degenerate configurations. The polarization switching is related to the rotopolar coupling terms [121, 123] $\propto P_3 \phi_1 \frac{\partial \phi_3}{\partial x}$, which implies that $P_3(x) \propto \phi_1(x) \frac{\partial \phi_3(x)}{\partial x}$. As a result, any change of sign of ϕ_1 (change of chirality) between two consecutive APBs leads to parallel polarizations P_3 of consecutive APBs (figure 2.22), whereas P_3 become anti-parallel for sign-preserving values of consecutive ϕ_1, i.e. if the chirality does not change. Clearly, this imposes a huge number of degenerate states of different chirality ($\pm \phi_1$) and polarization ($\pm P_3$) values (figure 2.23) if the crystal contains enough hard APBs and the flipping between these states could probably explain the glassy behaviour of SrTiO$_3$ below ca. 40 K. It should be noted that *a priori* the APBs are not susceptible to stress variations, but certainly they interact with the twin boundaries (figure 2.23), which were proven to move under external stress. Since the twin boundaries are polar (not ferroelectric) due to rotopolar and/or flexo-electric coupling even at temperatures above 25 K [117], they can also be moved by application of an external electric field [116]. How the APBs interact with the twin boundaries, and how exactly this is related to the glassy behaviour is far from being understood. It is also quite remarkable that the quantum domain solid regime [101] starts at a temperature (ca. 20 K) at which the hard twin boundaries become ferroelectric. To obtain deeper insights into the structure and ferroelectric properties of easy and hard twin boundaries in SrTiO$_3$, LGD and *ab initio* based machine learning calculations similar to those of [119, 122] are currently underway.

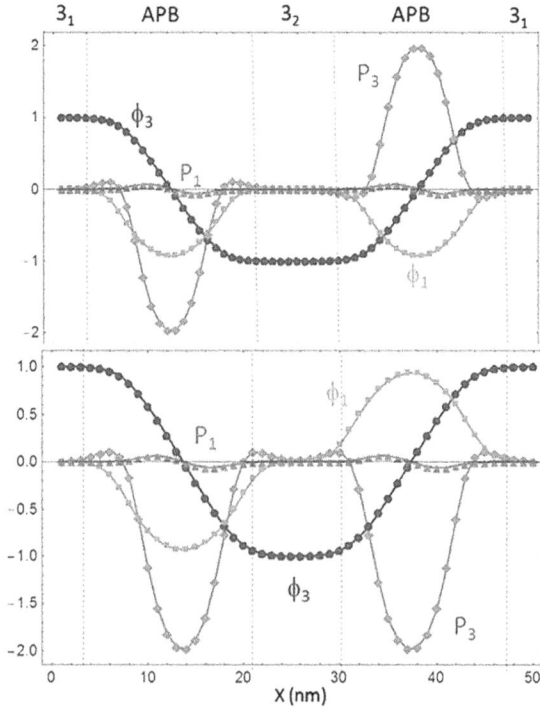

Figure 2.22. Order parameter profiles $\phi_1(x)$ (orange) and $\phi_3(x)$ (blue) and polarization profiles of out-of plane $P_1(x)$ (red) and in-plane $P_3(x)$ (green) polarization components of two hard APBs, calculated from the LGD free energy expansion of [119].

2.6.2 Precursor structures

Structural PTs are usually accompanied by dynamic precursor structures. Prior to ferroelastic PTs, such fluctuations often form fine-scaled twinning (in the nm range), so-called 'tweed' structures [9, 124]. Upon cooling below T_c, these short-range ordered metastable structures turn into a pattern of twins, i.e. domains with walls between them. The tweed structure at $T > T_c$ is most appropriately described as structural modulations $Q(\mathbf{k})$ with wave vectors $\mathbf{k}$ along the elastically soft directions. As was shown in previous works [23], the elastically soft directions coincide with the twin planes. For example, as shown in section 2.4.1, the soft mode in LaNbO$_4$ corresponds to a transverse acoustic mode, with the $\mathbf{q}$-vector propagating along the direction which is inclined by about $23°$ in the [110]-plane, which indeed coincides with the orientation of the ferroelastic domain walls.

Quite generally, in ferroelastic materials, where strain is the order parameter, the accessible phase space of the acoustic soft mode can have various dimensionalities [125]. Type I behaviour refers to transitions in which the wave vector $\mathbf{k}$ and the displacement vector $\mathbf{u}$ of the soft acoustic mode are restricted to specific directions

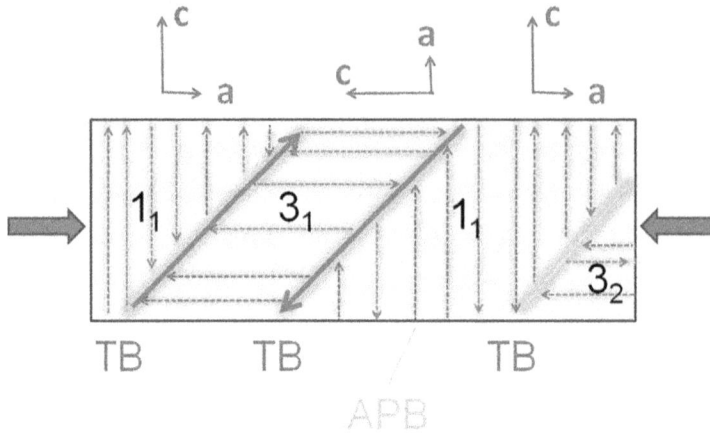

Figure 2.23. Schematic picture of a SrTiO$_3$ crystal with the orientations of hard APBs (orange) and twin boundaries (blue) in the respective domains. Below $T_c^{APB} \approx 40$ K the APBs become ferroelectric (polarization denoted by red arrows) and below $T_c^{TB} \approx 25$ K the hard twin boundaries (polarization denoted by blue arrows) too. Easy twin boundaries, here shown between ($1_1/3_2$), do not exhibit ferroelectricity in SrTiO$_3$ down to 0 K.

(one-dimensional soft mode). In type II ferroelastic PTs **k** is restricted to specific directions, but **u** can be anywhere within the plane perpendicular to **k** (or vice versa). In this case the soft acoustic mode is two-dimensional. In a pioneering work, Folk, Iro and Schwabl [126] have shown, using renormalization group theory, that for type I ferroelastics in three dimensions ($d = 3$, most proper FELs are of this type) the critical exponents are described by mean-field (Landau) theory. For a two-dimensional soft **k**-space (type II systems), logarithmic corrections appear. Indeed, one of the aspects that make ferroelastic PTs interesting is the fact that Landau theory works so well in these systems, in most cases not only qualitatively, but even quantitatively. A beautiful example is the pseudo-proper ferroelastic transition in LaP$_5$O$_{14}$ [56], where all elastic moduli (figure 8 of [56]) follow the predictions of Landau theory without any deviations from mean-field behaviour. In particular, no precursor rounding effects are observed in the paraelastic phase.

For improper ferroelastic PTs or for co-elastic transitions the situation is quite different. As discussed above, here the primary order parameter is not a strain component, but is coupled to some strain components in a non-linear way. Examples are shown in section 2.4.3. In Landau theory, improper ferroelastics show no elastic softening in the high-symmetry paraelastic phase, making them a perfect choice for the investigation of softening mechanisms beyond mean-field theory. The first experimental observation of unexpected precursor elastic softening (see figure 2.12) in the hexagonal paraphase of co-elastic ($\alpha \to \beta$) quartz was made [127, 128] more than 50 years ago. Meanwhile, other examples have been found. For example, in terbium molybdate (TMO = twin boundary$_2$(MoO$_4$)$_2$) acoustic precursor effects were thoroughly studied [129] near the improper ferroelastic–ferroelectric phase transition using Brillouin scattering and Landau–Ginzburg theory. Based on the typical Landau free energy expansion in the order parameter q, and a coupling term $\frac{\lambda}{2}\varepsilon q^2$ typical for an improper FEL PT, they derived expressions for the real and imaginary parts of the elastic susceptibility, which had previously been obtained by Levanyuk [130], and described the observed downwards bending of the elastic moduli very well. Later, Fossum [131] derived the fluctuation contribution to the complex elastic stiffness tensor more generally in presence of the coupling terms $\lambda_{ijl}\varepsilon_1 q_j q_l$, using the classical fluctuation–dissipation theorem:

$$C_{mn}(\mathbf{q},\,\omega) = \frac{1}{k_{\mathrm{B}}T} \int d^d\mathbf{r} \int_0^\infty \exp(-i\mathbf{q}\mathbf{r})\exp(i\omega t){<}\delta\dot{\sigma}_m(\mathbf{r},\,t)\delta\sigma_n(0,\,0){>}, \quad (2.50)$$

where $\mathbf{q}$ and ω are the wave vector and frequency of the acoustic wave, and the fluctuations of internal stresses are given as

$$\delta\sigma_m(\mathbf{r},\,t) = \lambda_{mjl}\Big(q_j(\mathbf{r},\,t)q_l(\mathbf{r},\,t) - \big\langle q_j(\mathbf{r},\,t)q_l(\mathbf{r},\,t)\big\rangle\Big). \quad (2.51)$$

Decomposing $q_j = q_{0j} + \delta q_j$, where q_{0j} is the equilibrium value of the OP, one obtains after Fourier transformation and assuming a relaxation behaviour of the order parameter fluctuations ($\delta q(\mathbf{k},\,t) = \delta q(\mathbf{k},\,0)\exp(-t/\tau_k)$), the final form [131] of the contributions to the complex dynamic elastic stiffness tensor as

$$C_{mn}(0,\,\omega) = C_{mn}^0 + \Delta C_{mn}^{\mathrm{LK}} - \frac{2k_{\mathrm{B}}T}{(2\pi)^d}\lambda_{mij}\lambda_{nkl}\int_0^{k_m} \frac{d^d\mathbf{k}}{1 + i\frac{\omega\tau(\mathbf{k})}{2}}\chi_{ij}(\mathbf{k},\,0)\chi_{kl}(-\mathbf{k},\,0), \quad (2.52)$$

where $\Delta C_{mn}^{\mathrm{LK}}$ is the classical LK contribution to the elastic stiffness, which produces a negative step such as an anomaly (for a 2–4 Landau potential) or a cusp (for a 2–4–6 potential) in the elastic modulus. The last term accounts for the contribution of the order parameter fluctuations $\Delta C_{mn}^{\mathrm{crit}}$. The final result depends crucially on the dispersion $\omega(\mathbf{k})$ of the soft mode, which is related to the susceptibility $\chi_{ij}(\mathbf{k},\,0)$. Using the well-known mean-field relations [131] for isotropic fluctuations,

$$\chi_{ij}(\mathbf{k},\,0) = \frac{\chi_{ij}(0,\,0)}{1 + k^2\xi^2}$$

$$\tau(\mathbf{k}) = \frac{\tau(0)}{1 + k^2\xi^2}, \quad (2.53)$$

where $\chi_{ii}(0,\,0) = (\frac{\partial^2 F}{\partial q_i \partial q_i})^{-1} = (A(T - T_c) + \cdots)^{-1}$ and the correlation length of the order parameter fluctuations $\xi^2 = g\chi_{ii}(0,\,0)$ (g is the gradient term in the Landau–Ginzburg free energy). After integration of equation (2.52) over $d = 1,\,2,\,3$ one arrives at the well-known relation [131], i.e.

$$\Delta C_{mn}^{\mathrm{crit}} \propto (T - T_c)^{-\alpha} \quad \text{with} \quad \alpha = 2 - \frac{d}{2}. \quad (2.54)$$

One should, however, keep in mind that this result is obtained in the limit $T \to T_c$, and for $\omega\tau(k) < 1$. Moreover, a very simple isotropic order parameter susceptibility was used here. In real systems, the detailed temperature dependence of ΔC^{crit} is difficult to obtain, because it requires an exact knowledge of mode mixing properties, including potential local modes which are not captured by conventional spectroscopy. Nevertheless, it turns out [23, 135] that, in quite a few cases, the elastic precursor softening can be conveniently described by a power law, $\Delta C_{ij}^{\text{crit}} = A_{ij}(T - T_c)^{-\kappa}$, where A_{ij} and κ are material dependent parameters. It was already mentioned above that the precursor softening near the improper ferroelastic–ferroelectric phase transition in terbium molybdate (TMO = twin boundary$_2$(MoO$_4$)$_2$) is well fitted over a temperature range of about 250 K (figure 4 of [129]) using equation (2.52). To fit the data, the authors calculated the fluctuation integral in equation (2.52) after inserting the measured soft mode dispersion.

Using a similar approach [133], the observed precursor softening in NH$_4$LiSO$_4$ near the uniaxial ferroelectric phase transition at $T_c = 460$ K was described. 3D isotropic fluctuation behaviour ($\alpha = 0.5$) was found over more than 40 K above T_c, followed by a crossover to dipolar fluctuations at ca. 0.5 K above T_c. KMnF$_3$ is another example where isotropic fluctuations (with $\alpha \approx 0.5$) have been found [132] to contribute to the elastic softening near the cubic to tetragonal PT, while domain wall movements [136] in Ca-doped samples are best described in terms of Vogel–Fulcher (VF) relaxations with a VF energy of 0.23 eV.

Quite interestingly, a number of data on elastic precursor softening that were previously described by a power law, were recently equally well fitted by a VF type temperature dependence, i.e.

$$\Delta C_{ik} = B_{ik} \exp\left(\frac{E_a/k_{\text{B}}}{T - T_{\text{VF}}}\right), \tag{2.55}$$

where B_{ik} is a material parameter, E_a is the activation energy, and T_{VF} is the VF temperature.

For example, in BaTiO$_3$ single crystals ($T_c \approx 400$ K) an elastic precursor softening (figure 2.24) measured with RUS [134] was fitted by a VF equation $f^2 \propto \Delta C \propto \exp(E_a/k_{\text{B}}(T - T_{\text{VF}}))$ with $E_a \approx 2800$ K and $T_{\text{VF}} = 90$ K over a temperature range of about 250 K.

In contrast to equation (2.54), which results from order parameter fluctuations, equation (2.55) is not the result of a formal theory. Such behaviour is typical for glasses and for local clusters in order–disorder phase transitions. Indeed, polar nanoclusters of static [137] and dynamic nature [138] have been identified in the paraelectric phase of BaTiO$_3$.

Very recently a comprehensive study was made [135], measuring the temperature evolution of the elastic constants in Ba$_{1-x}$Sr$_x$TiO$_3$ for six compositions x, by electrostatically exciting flexural modes of the bars at about 1–2 kHz. Interestingly, the observed elastic precursor softening could be equally well fitted by a power law $\Delta C \propto (T - T_c)^{-\kappa}$ with a characteristic exponent κ decreasing from 1.5 in SrTiO$_3$ to 0.2 in BaTiO$_3$, or by a VF law (2.55) with extremely low VF energies E_a of only a few kelvins, which increase from SrTiO$_3$ to BaTiO$_3$ indicating a change from a displacive to a weakly order–disorder character of the elastic precursor.

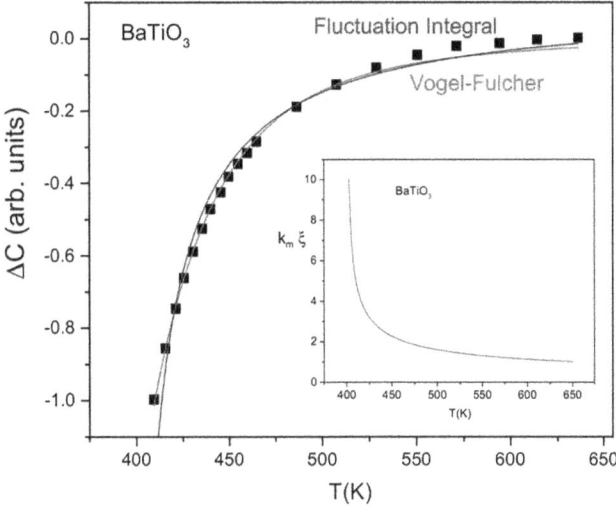

Figure 2.24. Precursor softening in a BaTiO$_3$ single crystal measured with RUS (data extracted from [134]). The red line shows a fit with a VF relation $\propto\exp[E/(T - T_{VF})]$ yielding $E \approx 2800$ K and $T_{VF} = 90$ K. The blue curve is a fit using equation (2.56) (see the text).

The amplitude of the precursor elastic softening increases continuously from SrTiO$_3$ to BaTiO$_3$.

Here, we would like to stress that when interpreting precursor effects, one should bear in mind that the power laws originating from order parameter fluctuations are strictly valid only in a temperature region sufficiently close to T_c. Therefore we think that one should resort to different phenomenological fit procedures, such as Vogel–Fulcher relations, etc, only after attempting to understand the data based on the fluctuation contribution as obtained from the full solution of the integral of equation (2.52). For example, 3D integration of equation (2.52) assuming isotropic fluctuations, equation (2.53), yields in the limit $\omega\tau < 1$

$$\Delta C^{\text{crit}} \propto b^2\xi\left[\frac{k_m\xi}{1 + k_m^2\xi^2} - \arctan(k_m\xi)\right]. \tag{2.56}$$

With $\xi \propto (T - T_c)^{-\frac{1}{2}}$ it fits (figure 2.24) the precursor softening of BaTiO$_3$ very well over more than 250 K. The power-law behaviour is only reached for $k_m\xi > >1$, which (see the inset of figure 2.24) occurs only very close to T_c.

Very interesting precursor effects largely outside the asymptotic limit were also found when approching the phase transition of potassium thiocyanate (KSCN). KSCN crystals undergo an order–disorder improper ferroelastic phase transition [87, 88] at $T_c = 415$ K from $I4/mcm$ to $Pbcm$ (figure 2.25) at the critical wave vector $\mathbf{k}_c = (00\frac{2\pi}{c})$. This wave vector describes a wave whose wavelength is equal to the lattice parameter c, but the centring translation ($\frac{1}{2}\frac{1}{2}\frac{1}{2}$) of the tetragonal $I4/mcm$ is lost. The SCN-molecules, which are head–tail disordered in the tetragonal phase become ordered below T_c as depicted in figure 2.25. Simultaneously with the

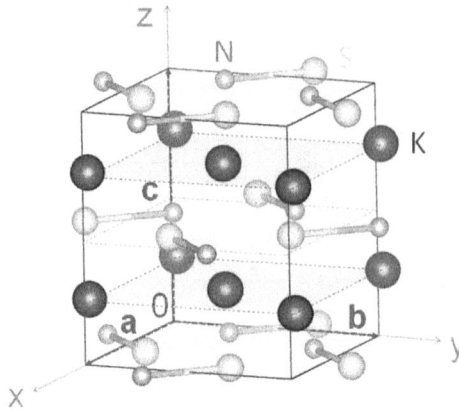

Figure 2.25. Arrangement of atoms in the orthorhombic *Pbcm* structure of KSCN. The vectors **a**, **b**, **c** of the primitive unit cell are marked by red dashed arrows. C atoms are omitted for clarity in the figures. (Image produced using VESTA [40].)

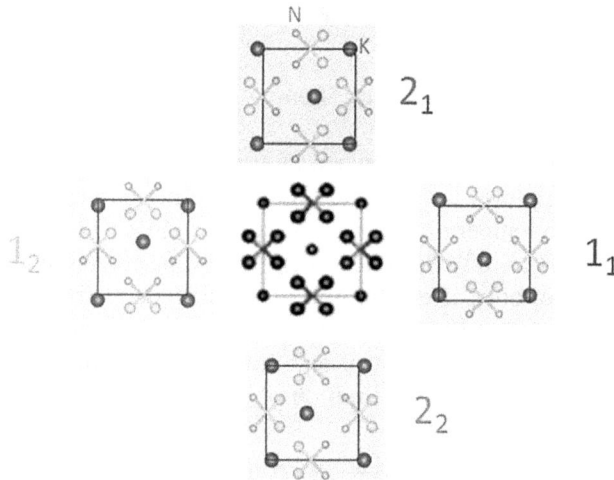

Figure 2.26. Two-dimensional scheme of the tetragonal high-symmetry phase of KSCN in the centre and of the four different DSs of the orthorhombic low symmetry phase. K = purple, S = yellow, N = grey. C atoms not shown here.

SCN-molecules, the K atoms move between four off-centre sites in the tetragonal phase. Below T_c they take one of these four off-centre positions (figure 2.26) in a given DS.

Due to the symmetry reduction, four DSs are possible. We will denote them as 1_1, 1_2, 2_1, 2_2 (figure 2.26). The DSs 1_1 and 2_1 are called orientational DSs and are related by 90° rotations, whereas 1_1, 1_2 are translational DSs that are related by fractional lattice shifts of $(\frac{1}{2}\frac{1}{2}\frac{1}{2})$. The corresponding domain walls form ferroelastic TBs for $(1_1/2_1)$ and antiphase domain boundaries (APBs) for $(1_1/1_2)$ and $(2_1/2_2)$.

Domain boundaries of the type $(1_1/2_2)$ are mixed ones. A detailed treatment of domains and domain boundaries of KSCN can be found in [139].

MD simulations [140, 141] of a two-dimensional model of KSCN yielded pronounced precursor effects on both sides of the transition. Heating a monodomain sample—say 1_1—from room temperature up to T_c, one observes a flipping of the domain state 1_1 (figure 2.27) into the three other DSs 1_2, 2_1 and 2_2. The density of these fluctuations increases continuously with increasing temperature, obtaining an equal number of all four orientation states above T_c. These results are in very good agreement with diffuse neutron scattering data [142], which yielded an increase of the diffuse intensity ($\propto$ number of local clusters) approaching the phase transition from both sides. However, an interesting precursor effect was found [143] below T_c. It turned out that while the diffuse intensity increases when approaching T_c from both sides, the width remains constant in the whole long-range-ordered phase below T_c. This behaviour is rather unusual, since according to Landau–Ginzburg theory the diffuse intensity $I_D(\mathbf{q} = 0)$ and the correlation length ξ are connected as $\xi \propto \sqrt{I_D(\mathbf{q} = 0)}$. Such behaviour is indeed nicely fulfilled when cooling the crystal from above T_c. Here the width of the diffuse intensity decreases, indicating an increase of the correlation length $\xi \propto (T - T_c)^{-\frac{1}{2}}$ as expected from Landau–Ginzburg theory.

Molecular-dynamics simulations [141] indicate that the size of the precursor clusters is stabilized by inhomogenous strains originating from order parameter–strain interactions of the type $\lambda q^2 \varepsilon$. By extending the usual Landau–Ginzburg free energy expansion [144] to include strain gradients, it was shown that below T_c the precursor clusters can indeed be suppressed by the inhomogeneous strain fields that arise from the order parameter fluctuations. However, we think that a final understanding of this problem has not yet been reached.

An interesting question also concerns the time scale of the order parameter fluctuations in KSCN. NMR measurements [145–148] yielded a dynamic symmetry breaking of the tetragonal high-temperature phase on the time scale of about 10^{-7} s. Thus, the relaxation time for the SCN fluctuations (figure 2.27) is so slow ($\tau > 10^{-7}$ s) that low symmetry NMR angular rotation patterns are observed in the high-temperature high-symmetry phase. However, despite all these efforts, the temperature dependence of the order parameter relaxation time τ could not be

| 0.8 T_c | 0.85 T_c | 0.96 T_c | 1.1 T_c |

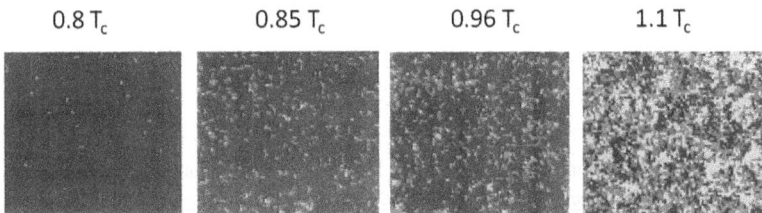

Figure 2.27. MD simulations [140, 141] of a two-dimensional model of KSCN, showing the fluctuations prior to the order–disorder phase transition at $T_c = 415$ K. The colours correspond to the domain states shown in figure 2.26. (Adapted with permission from [140]. Copyright 1994 the American Physical Society.)

extracted from NMR measurements and we thus performed dynamic elastic measurements. Figure 2.28 shows resonant ultrasound spectroscopy (RUS) data of KSCN in a temperature range 300–430 K and at frequencies of 100–600 kHz.

Figure 2.29 displays the temperature dependencies of f^2 and Q^{-1} of a characteristic resonance selected from figure 2.28. Similar as for ultrasonic measurements [149] of KSCN, also the resonance frequencies (at few 100 kHz) show temperature variations with $f^2(T) \propto q^2(T)$, i.e. due to a coupling of the type $bq^2\varepsilon^2$. The fact that $f^2(T)$ does not display any negative anomaly at T_c, as would be expected from the $\lambda q^2 \varepsilon$ term, shows that $\omega\tau(T) > 1$ in the temperature range around T_c, implying $\tau > 10^{-6}$ s.

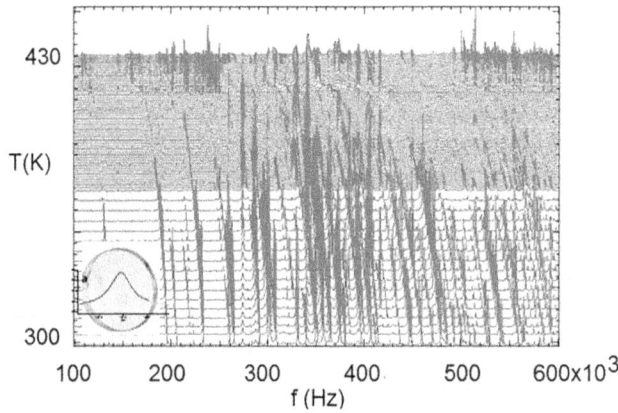

Figure 2.28. A stack of RUS spectra collected during heating of a KSCN single crystal. Each spectrum has been offset up the y-axis in proportion to the temperature at which it was collected between 300 and ~430 K. The inset shows the typical asymmetric Lorentzian line shape of the peaks, which is used to extract Q^{-1} data.

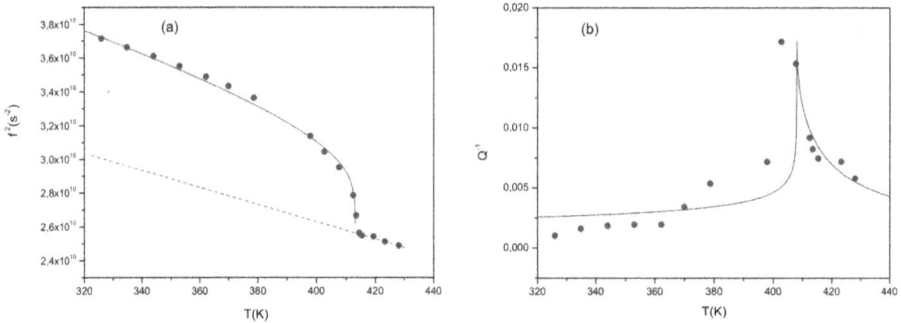

Figure 2.29. Evolution with temperature of (a) f^2 and (b) Q^{-1} values for one selected typical resonance in RUS spectra (2.28). As can be seen in figure 2.28 all resonances show similar temperature dependencies. Lines represent fits using equations (2.58) and (2.60). The temperature dependence of Q^{-1} results from dynamic order parameter fluctuations (see the text). Interestingly enough, in the limit $\omega\tau > 1$ the fluctuations contribute to the real part only as const $\times$ T, whereas a peak occurs in the imaginary part of the elastic response. This is quite different from the LK mechanism, which is effective only below T_c.

In contrast to the high frequency elastic data, the Young's modulus measured with DMA (see figure 2 of [150]) in the frequency range 0.1–50 Hz exhibits a negative dip anomaly typical for an improper ferroelastic PT. Thus, $\omega\tau < 1$, implying $\tau < 10^{-3}$ s. From the dynamic elastic measurements we can thus conclude that the order parameter relaxation time of KSCN varies with temperature within the range 10^{-6} s $< \tau(T) < 10^{-3}$ s. With this in mind, we can calculate the fluctuation contribution to the complex elastic constant using equation (2.52) in the limit $\omega\tau(k) > 1$ as

$$\Delta C^{\mathrm{crit}} = -\lambda^2 k_m^3 \frac{k_{\mathrm{B}}T}{3\pi^2} \frac{\Gamma^2}{\omega^2} \tag{2.57}$$

which is equal above and below T_c. The overall temperature variation of the elastic constant can thus be described as

$$f^2(T) \propto C(T, \omega) = C^0 + \frac{\Delta C^{\mathrm{LK}}}{\omega^2\tau^2} + 2bq^2(T) + \Delta C^{\mathrm{crit}}(T). \tag{2.58}$$

Equation (2.57) contains a linear temperature dependence and since $\omega\tau > 1$ the LK-contribution approaches zero. As a result, the overall temperature variation of the elastic constants is $\propto bq^2(T)$ below T_c and displays a linear T-dependence above T_c in perfect agreement (figure 2.29(a)) with the experimental curve. To fit the data (figure 2.29(a)) we calculated [91] the temperature dependence of the order parameter $q(T)$ from the relation

$$q(T) = \tanh\left[\frac{T_0}{T}(q + Aq^2)\right] \tag{2.59}$$

with $A = 0.44$.

For the imaginary part of the elastic constant we obtain

$$Q^{-1}(T, \omega) \propto C''(T, \omega) = \frac{\Delta C^{\mathrm{LK}}}{\omega\tau} + \lambda^2 \frac{k_{\mathrm{B}}T}{2\pi^3} \frac{\Gamma}{\omega} \int_0^{k_m} dk\chi(k)k^2. \tag{2.60}$$

In the limit $\omega\tau > 1$ also the LK-damping vanishes, implying that Q^{-1} is solely affected by order parameter fluctuations, which finally leads to

$$Q^{-1}(T, \omega) \propto C''(T, \omega) = \lambda^2 \frac{k_{\mathrm{B}}T}{2\pi^3 g} \frac{\Gamma}{\omega}[k_m - \xi(T)^{-1} \arctan(k_m\xi(T))], \tag{2.61}$$

where $\xi = \sqrt{g\chi(T, 0)}$, with the OP-susceptibility calculated from the compressible pseudospin model [91] as

$$\chi^{-1}(T, 0) = \frac{4k_{\mathrm{B}}T_0}{T}(1 + 3Aq^2(T))[T - (1 - q^2(T))T_0]. \tag{2.62}$$

Inserting equation (2.62) into equation (2.61) yields a clear cusp in Q^{-1} at the phase transition. Although the data in figure 2.29(b) are not perfectly fitted, these results show that dynamical symmetry-breaking clusters can be detected in the imaginary part of the elastic susceptibility even if their dynamics are slow. We expect that the dynamic elastic behaviour of $PbZrO_3$ single crystals shown, for example, in figure 2.11(a) can be described in a similar way.

These examples shows that when treating precursor effects, one must pay attention to the temperature range being considered. For example, the asymptotic limit $k_m\xi > 1$ of equation (2.61) leads to $Q^{-1} = \text{const.} \times T$, which in contrast to the full equation does not describe the observed (figure 2.29(b)) increase above and below T_c.

2.7 Summary

In the present chapter, we give an overview of a number of materials exhibiting ferroelastic phase transitions. Using several examples we demonstrate that strains play a fundamental role in almost all types of phase transitions, either as the driving order parameter (acoustic soft mode, e.g. in $LaNbO_4$ or TeO_2) or by coupling with another driving mechanism, which may be structural (soft mode at $q = 0$, e.g. in LaP_5O_{14} or $BiVO_4$; or soft mode at $q \neq 0$ as in $LaAlO_3$ or $SrTiO_3$), ferroelectric (e.g. $BaTiO_3$, $PbTiO_3$, NH_4LiSO_4) or antiferroelectric (e.g. $PbZrO_3$), antiferrodistortive (e.g. KSCN, $SrTiO_3$, $EuTiO_3$ [151]), magnetic (ferro/antiferromagnetic, as e.g. in CoF_2, $YMnO_3$, Fe_xO [152]), or electronic (e.g. $LaCoO_3$[153], $BaFe_2As_2$, etc). Understanding strain coupling mechanisms is important, since they do not only influence the bulk properties of a material, but are effective [8, 154–156] on quite different length and time scales in the bulk, as well as in domain walls. Dynamic elastic susceptibilities are very sensitive to spatial and temporal microstructural changes. They can be determined by a number of different methods (torsion pendulum, DMA, RUS, ultrasonics, etc) which, however, work in relatively narrow frequency windows. However, by putting data from different methods together one can cover 14 decades in frequency. Dynamic elastic measurements are also to be performed under different external conditions, e.g. by changing temperature (many examples are shown here), pressure (see e.g. figures 2.4, 2.6 and 2.7), stress (figures 2.10 and 2.19), and electric [157] or magnetic fields [80].

By making use of the different stress and strain amplitudes applied at the different methods [79] one can, for example, study the motion of domain walls in different regimes (static $\rightarrow$ creep $\rightarrow$ sliding). For example the relatively high stress and low frequency conditions of a DMA experiment cause the forward and back movement of the tips of needle twins, while the relatively low stress and high frequency conditions experienced by a resonating sample in an RUS experiment probably favour a local bowing mechanism. In some materials the motion of domain walls freezes out with a mean relaxation time that deviates from simple Arrhenius behaviour [109]. We think that further measurements in a broad frequency range are required to see if the domain wall dynamics follows a Vogel–Fulcher law, or if it can be described by a distribution of relaxation times with gradual broadening of the spectrum with decreasing temperature as anticipated by Tagantsev [158] for relaxor ferroelectrics.

Another topic of similar complexity concerns the precursor effects observed in elastic measurements prior to a ferroelastic phase transition. Also here, the elastic constants are frequently fitted in the precursor regime of the paraphase by a Vogel–Fulcher law [135], although there is no formal theory behind that would support such assumption. Recent MD simulations [159] of a proper ferroelastic phase transition also yielded precursor effects that can be described by a Vogel–Fulcher law. Certainly, there is much to do here before one can draw definite conclusions about the origin and physics of such precursor effects. For example resonant piezoelectric spectroscopy (RPS), an adaptation of the RUS method, is very useful for the investigation of local polar precursor structures [138], polar twin walls [116], etc, with high sensitivity.

Despite the number of open questions, it should be stressed that many of the observed pattern of dynamic elastic measurements can be conveniently described by Landau theory or Landau–Ginzburg–Devonshire theory, revealing not only the nature and magnitude of strain coupling with driving order parameters that can be due to any transition mechanism but also variations of the absolute relaxation times of the strain–order parameter coupling. Knowing these parameters is not only important for bulk samples but also feeds into considerations for thin film technologies where strain effects are expected to be even more effective.

Acknowledgments

This research was funded by the Austrian Science Fund (FWF) 10.55776/PIN2246224 and 10.55776/PAT2916124.

References

[1] Tolédano J C and Tolédano P 1987 *The Landau Theory of Phase Transitions: Application to Structural, Incommensurate, Magnetic and Liquid Crystal Systems* World Scientific Lecture Notes in Physics vol 3 *(Singapore: World Scientific)*

[2] Cowley R A 1980 Structural phase transitions, I. Landau theory *Adv. Phys.* **29** 1–110

[3] Carpenter M A, Salje E K H and Graeme-Barber 1998 A spontaneous strain as a determinant of thermodynamic properties for phase transitions in minerals *Eur. J. Mineral.* **10** 621–91

[4] Wadhawan V K 1982 Ferroelasticity and related properties of crystals *Phase Transit. Multinatl. J.* **3** 3–103

[5] Salje E K H 1993 *Phase Transitions in Ferroeleastic and Co-elastic Crystals* (Cambridge: Cambridge University Press)

[6] Li F, Jin L, Xu Z and Zhang S 2014 Electrostrictive effect in ferroelectrics: an alternative approach to improve piezoelectricity *Appl. Phys. Rev.* **1** 011103

[7] Meier Q N *et al* 2017 Global formation of topological defects in the multiferroic hexagonal manganites *Phys. Rev. X* **7** 041014

[8] Viehland D D and Salje E K H 2014 Domain boundary dominated systems: adaptive structures and functional twin boundaries *Adv. Phys.* **63** 267

[9] Salje E K H 2012 Ferroelastic materials *Annu. Rev. Mater. Res.* **42** 265

[10] Salje E K H and Zhang H 2009 Domain boundary engineering *Phase Transit.* **82** 452

[11] Meier D and Selbach S M 2022 Ferroelectric domain walls for nanotechnology *Nat. Rev. Mater.* **7** 157

[12] Salje E K H 2020 Ferroelastic domain walls as templates for multiferroic devices *J. Appl. Phys.* **128** 164104

[13] Zhao H J and Íñiguez J 2019 Creating multiferroic and conductive domain walls in common ferroelastic compounds *npj Comput. Mater.* **5** 92

[14] Landau L D and Lifshitz E M 1969 *Statistical Physics* (Oxford: Pergamon)

[15] Hohenberg P C and Krekhov A P 2015 An introduction to the Ginzburg Landau theory of phase transitions and nonequilibrium patterns *Phys. Rep.* **572** 1

[16] Lu G, Ding X, Sun J and Salje E K H 2022 Wall-wall and kink-kink interactions in ferroelastic materials *Phys. Rev.* B **106** 144105

[17] Marais S, Salje E, Heine V and Bratkovsky A 1994 Strain-related microstructures in materials: a computer simulation study of a simple model *Phase Transit.* **48** 15

[18] Togo A, Oba F and Tanaka I 2008 First-principles calculations of the ferroelastic transition between rutile-type and $CaCl_2$ -type SiO_2 at high pressures *Phys. Rev.* B **78** 134106

[19] Peng J, Zhang S, Refson K and Dove M T 2022 The ferroelastic phase transition in hydrogen cyanide studied by density functional theory *J. Phys.: Condens. Matter* **34** 095402

[20] He R, Wu H, Zhang L, Wang X, Fu F, Liu S and Zhong Z 2022 Structural phase transitions in $SrTiO_3$ from deep potential molecular dynamics *Phys. Rev.* B **105** 064104

[21] Parlinski K, Hashi Y, Tsunekawa S and Kawazoe Y 1997 Computer simulation of ferroelastic phase transition in $LaNbO_4$ *J. Mater. Res.* **12** 2428

[22] Tolédano P and Li G 2021 Antiferroelectricity in pure ferroelastics *Phys. Rev.* B **103** L220101

[23] Carpenter M A and Salje E K H 1998 Elastic anomalies in minerals due to structural phase transitions *Eur. J. Mineral.* **10** 693–812

[24] Aizu K 1969 Possible species of 'ferroelastic' crystals and of simultaneously ferroelectric and ferroelastic crystals *J. Phys. Soc. Japan* **27** 387–96

[25] Aizu K 1970 Possible species of ferromagnetic, ferroelectric, and ferroelastic crystals *Phys. Rev.* B **2** 754–72

[26] Aizu K 1969 Considerations of crystals which are 'antiferroelastic' as well as paraelectric, ferroelectric or antiferroelectric *J. Phys. Soc. Japan* **27** 1171–8

[27] Aizu K 1970 Considerations of partially ferroelastic and partially antiferroelastic crystals and partially ferroelectric and partially antiferroelectric crystals *J. Phys. Soc. Japan* **28** 717–22

[28] Tolédano P and Guennou M 2016 Theory of antiferroelectric phase transitions *Phys. Rev.* B **94** 014107

[29] Litvin D B 1984 Tensorial classification of non-magnetic ferroic crystals *Acta Cryst.* A **40** 255

[30] Aizu K 1972 Electrical, mechanical and electromechanical orders of state shifts in nonmagnetic ferroic crystals *J. Phys. Soc. Japan* **32** 1287

[31] Tolédano J C and Tolédano P 1980 Order parameter symmetries and free-energy expansions for purely ferroelastic transitions *Phys. Rev.* B **21** 1139–72

[32] Tolédano J C and Tolédano P 1976 Order parameter symmetries for improper ferroelectric nonferroelastic transitions *Phys. Rev.* B **14** 3097–109

[33] Carpenter M A 2000 Strain and elasticity at structural phase transitions in minerals *Rev. Mineral. Geochem.* **39** 35–64

[34] Slonczewski J C and Thomas H 1970 Interaction of elastic strain with the structural transition of strontium titanate *Phys. Rev.* B **1** 3599

[35] Rehwald W 1973 The study of structural phase transitions by means of ultrasonic experiments *Adv. Phys.* **22** 721

[36] Lüthi B and Rehwald W 1981 Ultrasonic studies near structural phase transitions *Top. Curr. Phys.* **23** 131

[37] Janovec V, Dvorak V and Petzelt J 1975 Symmetry classification and properties of equitranslation structural phase transitions *Czech J. Phys.* B **25** 1362–96

[38] Wallace D C 1972 *Thermodynamics of Crystals* (New York: Wiley)

[39] Tröster A, Schranz W and Miletich R 2002 How to couple Landau theory to an equation of state *Phys. Rev. Lett.* **88** 055503

[40] Momma K and Izumi F 2011 VESTA 3 for three-dimensional visualization of crystal, volumetric and morphology data *J. Appl. Crystallogr.* **44** 1272–6

[41] Aizu K 1970 Determination of the state parameters and formulation of spontaneous strain for ferroelastics *J. Phys. Soc. Japan* **28** 706–16

[42] Carpenter M A 2007 Elastic anomalies accompanying phase transitions in $(Ca,Sr)TiO_3$ perovskites: Part I. Landau theory and a calibration for $SrTiO_3$ *Am. Mineral.* **92** 309

[43] Janovec V and Přívratská J 2006 *International Tables for Crystallography* vol D ed A Authier (New York: Wiley) ch 3,4 pp 449–505

[44] Haun M J, Furman E, Jang S J, McKinstry H A and Cross L E 1987 Thermodynamic theory of $PbTiO_3$ *J. Appl. Phys.* **62** 3331

[45] Wojdel J C and Íñiguez J 2014 Ferroelectric transitions at ferroelectric domain walls found from first principles *Phys. Rev. Lett.* **112** 247603

[46] Rychetsky I, Schranz W and Tröster A 2023 Landau–Ginzburg–Devonshire theory of the chiral phase transition in 180° domain walls of $PbTiO_3$ *Phys. Rev.* B **108** 104107

[47] Tröster A, Ehsan S, Belbase K, Blaha P, Kreisel J and Schranz W 2017 Finite-strain Landau theory applied to the high-pressure phase transition of lead titanate *Phys. Rev.* B **95** 064111

[48] Tröster A, Schranz W, Karsai F and Blaha P 2014 Fully consistent finite-strain Landau theory for high-pressure phase transitions *Phys. Rev.* X **4** 031010

[49] Guennou M, Bouvier P, Kreisel J and Machon D 2010 Pressure-temperature phase diagram of $SrTiO_3$ up to 53 GPa *Phys. Rev.* B **81** 054115

[50] Toledano P, Fejer M M and Auld B A 1983 Nonlinear elasticity in proper ferroelastics *Phys. Rev.* B **27** 5717

[51] Folk R, Iro H and Schwabl F 1979 Critical dynamics of elastic phase transitions *Phys. Rev.* B **20** 1229

[52] Hara K, Sakai A, Tsunekawa S, Sawada A, Ishibashi Y and Yagi T 1985 A soft acoustic mode in the ferroelastic phase transition of $LaNbO_4$ *J. Phys. Soc. Japan* **54** 1168–72

[53] Peercy P S and Fritz I J 1974 Pressure-induced phase transition in paratellurite (TeO_2) *Phys. Rev. Lett.* **32** 466

[54] Peercy P S, Fritz I J and Samara G A 1975 Temperature and pressure dependences of the properties and phase transition in paratellurite (TeO_2): ultrasonic, dielectric and Raman and Brillouin scattering results *J. Phys. Chem. Solids* **36** 1105

[55] Worlton T G and Beyerlein R A 1975 Structure and order parameters in the pressure-induced continuous transition in TeO_2 *Phys. Rev.* B **12** 1899

[56] Errandonea G 1980 Elastic and mechanical studies of the transition in $LaP5O_{14}$: a continuous ferroelastic transition with a dassical Landau-type behavior *Phys. Rev.* B **21** 5221

[57] Radzhabov A K and Charnaya E V 2001 Acoustic study of the ferroelastic phase transition in LiCsSO$_4$ crystal *Phys. Solid State* **43** 732

[58] Pinczuk A, Webber B and Dacol F H 1979 Mechanisms of the ferroelastic transition in BiV0$_4$ *Solid St. Comm.* **29** 515

[59] Zhang Y, Fu S, Wang B and Lin J-F 2021 Elasticity of a pseudoproper ferroelastic transition from stishovite to post-stishovite at high pressure *Phys. Rev. Lett.* **126** 025701

[60] Andrault D, Fiquet G, Guyot F and Hanfland M 1998 Pressure-induced Landau-type transition in stishovite *Science* **282** 720

[61] Hemley R J, Shu J, Carpenter M A, Hu J, Mao H K and Kingma K J 2000 Strain/order parameter coupling in the ferroelastic transition in dense SiO$_2$ *Solid State Commun.* **114** 527

[62] Carpenter M A, Hemley R J and Mao H K 2000 High-pressure elasticity of stishovite and the $P4_2/mnm \rightleftharpoons Pnnm$ phase transition *J. Geophys. Res.* **105** 10807

[63] Martinelli A, Bernardini F and Massidda S 2016 The phase diagrams of iron-based superconductors: theory and experiments *C. R. Physique* **17** 5

[64] Rotter M, Tegel M, Johrendt D, Schellenberg I, Hermes W and Pöttgen R 2008 Spin-density-wave anomaly at 140 K in the ternary iron arsenide BaFe$_2$ As$_2$ *Phys. Rev.* B **78** 020503

[65] Goto T, Kurihara R, Araki K, Mitsumoto K, Akatsu M, Nemoto Y, Tatematsu S and Sato M 2011 Quadrupole effects of layered iron pnictide superconductor Ba(Fe 0.9 Co 0.1)$_2$ As$_2$ *J. Phys. Soc. Japan* **80** 073702

[66] Nandi S *et al* 2010 Anomalous suppression of the orthorhombic lattice distortion in superconducting Ba(Fe$_{1-x}$ Co$_x$)$_2$ As$_2$ single crystals *Phys. Rev. Lett.* **104** 057006

[67] Cano A, Civelli M, Eremin I and Paul I 2010 Interplay of magnetic and structural transitions in iron-based pnictide superconductors *Phys. Rev.* B **82** 020408

[68] Carpenter M A, Evans D M, Schiemer J A, Wolf T, Adelmann P, Böhmer A E, Meingast C, Dutton S E, Mukherjee P and Howard C J 2019 Ferroelasticity, anelasticity and magnetoelastic relaxation in Co-doped iron pnictide: Ba(Fe$_{0.957}$ Co$_{0.043}$)$_2$ As$_2$ *J. Phys.: Condens. Matter* **31** 155401

[69] Schranz W, Kabelka H, Sarras A and Burock M 2012 Giant domain wall response of highly twinned ferroelastic materials *Appl. Phys. Lett.* **101** 141913

[70] Fernandes R M, Chubukov A V and Schmalian J 2014 What drives nematic order in iron-based superconductors? *Nat. Phys.* **10** 97

[71] Böhmer A E *et al* 2022 Nematicity and nematic fluctuations in iron-based superconductors *Nat. Phys.* **18** 1412

[72] Böhmer A E and Meingast C 2016 Electronic nematic susceptibility of iron-based super-conductors *C. R. Phys.* **17** 90

[73] Fernandes R M, VanBebber L H, Bhattacharya S, Chandra P, Keppens V, Mandrus D, McGuire M A, Sales B C, Sefat A S and Schmalian J 2010 Effects of nematic fluctuations on the elastic properties of iron arsenide superconductors *Phys. Rev. Lett.* **105** 157003

[74] Böhmer A E, Burger P, Hardy F, Wolf T, Schweiss P, Fromknecht R, Reinecker M, Schranz W and Meingast C 2014 Nematic susceptibility of hole-doped and electron-doped BaFe$_2$ As$_2$ iron-based superconductors from shear modulus measurements *Phys. Rev. Lett.* **112** 047001

[75] Carpenter M A, Sinogeikin S V and Bass J D 2010 Elastic relaxations associated with the $Pm\bar{3}m$ –$R\bar{3}c$ transition in LaAlO$_3$: II. Mechanisms of static and dynamical softening *J. Phys.: Condens. Matter* **22** 035404

[76] Yang D, Chatterji T, Schiemer J A and Carpenter M A 2016 Strain coupling, micro-structure dynamics, and acoustic mode softening in germanium telluride *Phys. Rev. B* **93** 144109

[77] Carpenter M A, Salje E K H, Costa M B, Majchrowski A and Roleder K 2022 Static and dynamic strain relaxation associated with the paraelectric–antiferroelectric phase transition in $PbZrO_3$ *J. Alloys Compounds* **898** 162804

[78] Puchberger S, Soprunyuk V, Majchrowski A, Roleder K and Schranz W 2016 Domain wall motion and precursor dynamics in $PbZrO_3$ *Phys. Rev. B* **94** 214101

[79] Carpenter M A 2015 Static and dynamic strain coupling behaviour of ferroic and multiferroic perovskites from resonant ultrasound spectroscopy *J. Phys.: Condens. Matter* **27** 263201

[80] Evans D M, Schiemer J A, Schmidt M, Wilhelm H and Carpenter M A 2017 Defect dynamics and strain coupling to magnetization in the cubic helimagnet $Cu_2 OSeO_3$ *Phys. Rev. B* **95** 094426

[81] Driver S L, Herrero-Albillos J, Bonilla C M, Bartolomé F, García L M, Howard C J and Carpenter M A 2014 Multiferroic (ferroelastic/ferromagnetic/ferrimagnetic) aspects of phase transitions in RCo_2 Laves phases *J. Phys.: Condens. Matter* **26** 056001

[82] Carpenter M A, Salje E K H, Graeme-Barbeer A, Wruck B, Dove M and Knight K 1998 Calibration of excess thermodynamic properties and elastic constant variations associated with the $\alpha \leftrightarrow \beta$ phase transition in quartz *Am. Mineral.* **83** 2

[83] Thomson R I, Chatterji T and Carpenter M A 2014 CoF_2: a model system for magnetoelastic coupling and elastic softening mechanisms associated with paramagnetic $\leftrightarrow$ antiferromagnetic phase transitions *J. Phys.: Condens. Matter* **26** 146001

[84] Thomson R I, Chatterji T, Howard C J, Palstra T T M and Carpenter M A 2014 Elastic anomalies associated with structural and magnetic phase transitions in single crystal hexagonal $YMnO_3$ *J. Phys.: Condens. Matter* **26** 045901

[85] Liu Z, Lu T, Ye J, Wang G, Dong X, Withers R and Liu Y 2018 Antiferroelectrics for energy storage applications: a review *Adv. Mater. Technol.* **3** 1800111

[86] Zhai J, Yao Y, Li X, Hung T F, Xu Z K, Chen H, Colla E V and Wu T B 2002 Dielectric properties of oriented $PbZrO_3$ thin films grown by sol-gel process *J. Appl. Phys.* **92** 3990

[87] Yamada Y and Watanabe T 1963 The phase transition of crystalline potassium thiocyanate, KSCN. II. X-ray study *Bull. Chem. Soc. Jpn.* **36** 1032

[88] Yamamoto S, Sakuno M and Shinnaka Y 1987 Structure analysis of the phase transition in KSCN *J. Phys. Soc. Jpn.* **56** 4393

[89] Kabelka H, Fuith A, Warhanek H, Kamba S, Petzelt J, Lebedev S P and Volkov A A 1992 The dielectric behaviour of KSCN *Ferroelectrics* **135** 303

[90] Schranz W, Tröster A and Rychetsky I nd Microscopic model of the dielectric permittivity in KSCN (unpublished)

[91] Schranz W, Warhanek H, Blinc R and Zeks B 1989 Pseudospin model for KSCN *Phys. Rev. B* **40** 7141

[92] Tolédano P and Khalyavin D D 2019 Symmetry-determined antiferroelectricity in $PbZrO_3$, $NaNbO_3$, and $PbHf O_3$ *Phys. Rev. B* **99** 024105

[93] Tagantsev A K, Cross L E and Fousek J 2010 *Domains in Ferroic Crystals and Thin Films* (Berlin: Springer)

[94] Aird A and Salje E K H 1998 Sheet superconductivity in twin walls: experimental evidence of WO_{3-x} *J. Phys. Condens. Matter* **10** L377

[95] Chrosch J and Salje E K H 1999 Temperature dependence of the domain wall width in LaAlO$_3$ *J. Appl. Phys.* **85** 722

[96] Nataf G F, Guennou M, Gregg J M, Meier D, Hlinka J, Salje E K H and Kreisel J 2020 Domain-wall engineering and topological defects in ferroelectric and ferroelastic materials *Nat. Rev. Phys.* **2** 634–48

[97] Sapriel J 1975 Domain-wall orientations in ferroelastics *Phys. Rev.* B **12** 5128

[98] Kityk A V, Schranz W, Sondergeld P, Havlik D, Salje E K H and Scott J F 2000 Low-frequency superelasticity and nonlinear elastic behavior of SrTiO$_3$ crystals *Phys. Rev.* B **61** 946

[99] Rehwald W 1970 Low temperature elastic moduli of strontium titanate *Solid State Commun.* **8** 1483

[100] Schranz W 2011 Superelastic softening in perovskites *Phys. Rev.* BB **83** 094120

[101] Kustov S, Liubimova I and Salje E K H 2020 Domain dynamics in quantum-paraelectric SrTiO$_3$ *Phys. Rev. Lett.* **124** 016801

[102] Sonin A S and Strukov B A 1977 *Einführung in die Ferroelektrizität* (Heidelberg: Springer)

[103] Harrison R J and Redfern S A T 2002 The influence of transformation twins on the seismic-frequency elastic and anelastic properties of perovskite: dynamical mechanical analysis of single crystal LaAlO$_3$ *Phys. Earth Planet. Inter.* **134** 253

[104] Wang Y, Sun W, Chen X, Shen H and Lu B 1987 Internal friction associated with the domain walls and the second-order ferroelastic transition in LNPP *Phys. Stat. Sol.* a **102** 279–85

[105] Harrison R J, Redfern S A T and Salje E K H 2004 Dynamical excitation and anelastic relaxation of ferroelastic domain walls in LaAlO$_3$ *Phys. Rev.* B **69** 144101

[106] Puchberger S, Soprunyuk V, Schranz W and Carpenter M A 2018 Segmental front line dynamics of randomly pinned ferroelastic domain walls *Phys. Rev. Materials* **2** 013603

[107] Puchberger S 2018 Properties and dynamics of ferroelastic domain walls in perovskites *PhD Thesis* University of Vienna

[108] Soprunyuk V, König P, Tröster A, Schranz W, Carpenter M A and Salje E K H 2025 Domain glass dynamics of potassium thiocyanate KSCN *J. Appl. Phys.* **137** 145101

[109] Salje E K H, Ding X and Aktas O 2014 Domain glass *Phys. Status Solidi* B **251** 2061–6

[110] Salje E K H, Ding X, Zhao Z, Lookman T and Saxena A 2011 Thermally activated avalanches: jamming and the progression of needle domains *Phys. Rev.* B **83** 104109

[111] Puchberger S, Soprunyuk V, Schranz W, Tröster A, Roleder K, Majchrowski A, Carpenter M A and Salje E K H 2017 The noise of many needles *APL Mater.* **5** 046012

[112] Ji Y, Wang D, Wang Y, Zhou Y, Xue D, Otsuka K, Wang Y and Ren X 2017 Ferroic glasses *npj Comput. Mater.* **3** 43

[113] Kityk A V, Schranz W, Sondergeld P, Havlik D, Salje E K H and Scott J F 2000 Nonlinear elastic behaviour of SrTiO$_3$ crystals in the quantum paraelectric regime *EPL* **50** 41

[114] Pesquera D, Carpenter M A and Salje E K H 2018 Glasslike dynamics of polar domain walls in cryogenic SrTiO$_3$ *Phys. Rev. Lett.* **121** 235701

[115] Balashova E V, Lemanov V V, Kunze R, Martin G and Weihnacht M 1996 Ultrasonic study of the tetragonal and Müller phase *Ferroelectrics* **183** 75

[116] Scott J F, Salje E K H and Carpenter M A 2012 Domain wall damping and elastic softening in SrTiO$_3$: evidence for polar twin walls *Phys. Rev. Lett.* **109** 187601

[117] Morozovska A N, Eliseev M D, Glinchuk L-Q C and Gopalan V 2012 Interfacial polarization and pyroelectricity in antiferrodistortive structures induced by a flexoelectric effect and rotostriction *Phys. Rev.* B **85** 094107

[118] Tagantsev A K, Courtens E and Arzel L 2001 Prediction of a low-temperature ferroelectric instability in antiphase domain boundaries of strontium titanate *Phys. Rev.* **B 64** 224107

[119] Tröster A, Pils J, Bruckner F, Rychetsky I, Verdi C and Schranz W 2023 Hard antiphase domain boundaries in strontium titanate: a comparison of Landau–Ginzburg–Devonshire and *ab initio* results *Phys. Rev.* **B 108** 144108

[120] Kvasov A, Tagantsev A K and Setter N 2016 Structure and pressure-induced ferroelectric phase transition in antiphase domain boundaries of strontium titanate from first principles *Phys. Rev.* **B 94** 054102

[121] Schiaffino A and Stengel M 2017 Macroscopic polarization from antiferrodistortive cycloids in ferroelastic $SrTiO_3$ *Phys. Rev. Lett.* **119** 137601

[122] Tröster A, Verdi C, Dellago C, Rychetsky I, Kresse G and Schranz W 2022 Hard antiphase domain boundaries in strontium titanate unravelled using machine-learned force fields *Phys. Rev. Materials* **6** 094408

[123] Schranz W, Schuster C, Tröster A and Rychetsky I 2020 Polarization of domain boundaries in $SrTiO_3$ studied by layer group and order-parameter symmetry *Phys. Rev.* **B 102** 184101

[124] Schmahl W W *et al* 1989 Twin formation and structural modulations in orthorhombic and tetragonal $YBa_2(Cu_{1-x}Co_x)_3O_{7-\delta}$ *Philos. Mag. Lett.* **60** 241–8

[125] Cowley R A 1976 Acoustic phonon instabilities and structural phase transitions *Phys. Rev.* **B 13** 4877

[126] Folk R, Iro H and Schwabl F 1976 Critical statics of elastic phase transitions. Critical statics of elastic phase transitions *Z. Phys.* **B 25** 69

[127] Miller P B and Axe J D 1967 Internal strain and Raman-active vibrations in solids *Phys. Rev.* **163** 924

[128] Axe J D and Shirane G 1970 Study of the α–β quartz phase transformation by inelastic neutron scattering *Phys. Rev.* **B 1** 342

[129] Yao W, Cummins H Z and Bruce R H 1981 Acoustic anomalies in terbium molybdate near the improper ferroelastic–ferroelectric phase transition *Phys. Rev.* **B 24** 424

[130] Levanyuk A P 1966 Contribution to a phenomenological theory of sound absorption near second order phase transition points *Sov. Phys. JETP* **22** 901

[131] Fossum J O 1985 A phenomenological analysis of ultrasound near phase transitions *J. Phys. C: Solid State Phys.* **18** 5531

[132] Salje E K H and Zhang H 2009 Domain boundary pinning and elastic softening in $KMnF_3$ and $KMn_{1-x}Ca_xF_3$ *J. Phys.: Condens. Matter* **21** 035901

[133] Kubinec P, Schranz W and Kabelka H 1992 Acoustic anomalies in NH_4LiSO_4 near the uniaxial ferroelectric phase transition at $T_c \approx 460$ K *J. Phys: Condens. Matter* **4** 7009

[134] Salje E K H, Carpenter M A, Nataf G F, Picht G, Webber K, Weerasinghe J, Lisenkov S and Bellaiche L 2013 Elastic excitations in $BaTiO_3$ single crystals and ceramics: mobile domain boundaries and polar nanoregions observed by resonant ultrasonic spectroscopy *Phys. Rev.* **B 87** 014106

[135] Cordero F, Trequattrini F, da Silva P S, Venet M, Aktas O and Salje E K H 2023 Elastic precursor effects during $Ba_{1-x}Sr_xTiO_3$ ferroelastic phase transitions *Phys. Rev. Res.* **5** 013121

[136] Schranz W, Sondergeld P, Kityk A V and Salje E K H 2009 Dynamic elastic response of $KMn_{1-x}Ca_xF_3$: elastic softening and domain freezing *Phys. Rev.* **B 80** 094110

[137] Bencan A, Oveisi E, Hashemizadeh S, Veerapandiyan V K, Hoshina T, Rojac T, Deluca M, Drazic G and Damjanovic D 2021 Atomic scale symmetry and polar nanoclusters in the paraelectric phase of ferroelectric materials *Nat. Commun.* **12** 3509

[138] Aktas O, Carpenter M A and Salje E K H 2013 Polar precursor ordering in $BaTiO_3$ detected by resonant piezoelectric spectroscopy *Appl. Phys. Lett.* **103** 142902

[139] Schranz W, Rychetsky I and Hlinka J 2019 Polarity of domain boundaries in nonpolar materials derived from order parameter and layer group symmetry *Phys. Rev.* B **100** 184105

[140] Parliński K 1994 Computer simulation of domain formation in the order–disorder phase transition of the KSCN model *Phys. Rev.* B **50** 59

[141] Lodziana Z and Parliński K 1996 Computer simulation of diffuse scattering from KSCN crystal *Phase Transit.* **58** 273

[142] Blaschko O, Schwarz W, Schranz W and Fuith A 1991 Order–disorder phase transition in potassium thiocyanate *Phys. Rev.* B **44** 9159

[143] Blaschko O, Schranz W, Fally M, Krexner G and Lodziana Z 1998 Strain stabilized precursor clusters in potassium thiocyanate *Phys. Rev.* B **58** 8362

[144] Tröster A, Schranz W, Krexner G, Kityk A V and Lodziana Z 2000 Suppression of the order parameter correlation length by inhomogeneous strains *Phys. Rev. Lett.* **85** 2765

[145] Blinc R *et al* 1991 ^{39}K NMR study of the antiferroelectric phase transition inpotassium thiocyanate *Phys. Rev.* B **43** 569

[146] Blinc R, Dolinsek J, Apih T, Schranz W, Fuith A and Warhanek H 1995 2D 39K NMR study of the phase transition in KSCN *Solid State Commun.* **93** 609

[147] Blinc R, Apih T, Friedelj A, Dolinsek J, Seliger J, Fuith A, Schranz W, Warhanek H and Ailion D C 1997 ^{14}N NMR investigation of KSCN *Europhys. Lett.* **39** 627

[148] Blinc R, Apih T, Seliger J, Fuith A, Schranz W and Warhanek H 1999 Dynamic symmetry breaking in KSCN: a ^{39}K and ^{14}N NMR study *Phase Transit.* **67** 617

[149] Schranz W, Fuith A, Tröster A and Kroupa J 2005 Dynamic acoustic anomalies in KSCN near the improper ferroelastic phase transition *Ferroelectrics* **314** 189

[150] Schranz W and Havlik D 1994 Heat-diffusion central peak in the elastic susceptibility of KSCN *Phys. Rev. Lett.* **73** 2575

[151] Goian V *et al* 2012 Antiferrodistortive phase transition in $EuTiO_3$ *Phys. Rev.* B **86** 054112

[152] Zhang Z, Church N, Lappe S-C, Reinecker M, Fuith A, Jackson I, Saines P J, Harrison R J, Schranz W and Carpenter M A 2012 Elastic and anelastic anomalies associated with the antiferromagnetic ordering transition in wüstite, Fe_xO *J. Phys.: Condens. Matter* **24** 215404

[153] Zhang Z, Koppensteiner J, Schranz W, Prabhakaran D and Carpenter M A 2011 Strain coupling mechanisms and elastic relaxations associated with spin state transitions in $LaCoO_3$ *J. Phys.: Condens. Matter* **23** 145401

[154] Takae K and Kawasaki T 2022 Emergent elastic fields induced by topological phase transitions: impact of molecular chirality and steric anisotropy *Proc. Natl. Acad. Sci.* **119** 2118492119

[155] Yudin P, Duchon J, Pacherova O, Klementova M, Kocourek T, Dejneka A and Tyunina M 2021 Ferroelectric phase transitions induced by a strain gradient *Phys. Rev. Res.* **3** 033213

[156] Lajzerowicz J 1981 Domain wall near a first order phase transition: role of elastic forces *Ferroelectrics* **35** 219

[157] Darling T W, Allured B, Tencate J A and Carpenter M A 2010 Electric field effects in RUS measurements *Ultrasonics* **50** 145

[158] Tagantsev A K 1994 Vogel–Fulcher relationship for the dielectric permittivity of relaxor ferroelectrics *Phys. Rev. Lett.* **72** 1100

[159] Lu G, Cordero F, Hideo K, Ding X, Xu Z, Chu R, Howard C J, Carpenter M A and Salje E K H 2024 Elastic precursor softening in proper ferroelastic materials: a molecular dynamics study *Phys. Rev. Res.* **6** 013232

IOP Publishing

Ferroelastic Materials

Guillaume F Nataf, Blai Casals and Ekhard K H Salje

Chapter 3

Symmetry aspects of ferroelastic domains and domain walls

Jirka Hlinka and Petr Ondrejkovič

3.1 Introduction

Ferroelastic materials, or ferroelastics, are commonly considered as materials undergoing ferroelastic phase transitions. More precisely, there is a characteristic crystal lattice distortion in these materials which can be understood as one of the symmetry-breaking order parameters. For many purposes, it is convenient to also extend the definition to the materials in which the existence of a real ferroelastic phase transition has not been documented experimentally, but where the structure of the hypothetical parent paraelastic phase can be inferred theoretically.

The previous chapter has already provided a comprehensive overview of the state-of-the-art of the physics of ferroelastic materials. Various examples of ferroelastic and related materials have been discussed from the point of view of the phenomenological Landau theory of the associated ferroelastic phase transition. The Landau theory is particularly useful for ferroelastic materials because practically all investigated systems are either improper or pseudo-proper ferroelastics, so that the primary order parameter is not the spontaneous strain itself. Since the Landau theory describes the symmetry breaking in terms of irreducible representations of the parent symmetry phase, it inherently also provides a description of the ferroic domain states. When the Landau theory is complemented by energy contributions related to the spatial gradients of the order parameters, the resulting Ginzburg–Landau models can, in principle, also reveal insights into the properties of domain walls. These aspects were also already mentioned in the previous chapter and will be recalled in the following chapters.

The Landau theory is an excellent mathematical framework that allows one to capture simultaneously all of the already introduced qualitative and quantitative properties related to the proximity of the system to the phase transition point in a systematic way. However, one of its drawbacks is that formulation of the specific

doi:10.1088/978-0-7503-6089-0ch3

model generally requires a rather detailed characterization of the phase transition as well as a certain level of educated physics intuition in appreciation of the importance of relevant order parameters and related energy contributions to the Landau free-energy functional.

In this sense, it is generally useful to examine the possible domains and domain wall types in the system of interest in a more abstract way, considering only the symmetry aspects of the parent and ferroelastic phase pair. This method can be used as a complementary approach, allowing one to distinguish what is a pure consequence of the symmetry and what is reflecting a specific phase transition mechanism. It can also guide the physics intuition needed to formulate a suitable Landau theory model, or to serve as a tool for classifying different materials into classes, identifying a whole set of symmetry-imposed domain properties.

Such a group-theoretical approach has been systematically developed in the works of Janovec and co-workers [1, 2]. The theory, with its results and multiple examples, is described in full detail in [2]. The purpose of this short chapter is to provide readers with its main assumptions, definitions, and a selection of general results useful for the investigations of ferroelastic materials. We hope that this will generate the motivation and background for utilizing this beautiful theoretical approach.

3.2 Domain structure from the symmetry perspective and an alternative definition of ferroelastic materials

The domain structure is a natural manifestation of symmetry-breaking phase transitions in crystals. By this statement we do not mean that the domain structure was necessarily formed at the symmetry-breaking phase transition. The statement applies equally to the domain structures formed by mechanical or other treatments or domain structures formed in materials that never passed through a phase transition point. What we primarily have in mind by this statement is the very basic assumption that domains refer to the formally equivalent structural variants (domain states) existing in the form of compact regions (domains) covering a given crystal (or a crystal grain) volume, and that the equivalence of these variants can be described by a set of some 'lost' symmetry operations. These 'lost' symmetry operations together with the symmetry operations of the domain states themselves generate a higher symmetry supergroup G, which in turn defines the complete set of the identified structural variants. While there is some potential arbitrariness in this definition, in all practical cases this group naturally corresponds to the symmetry of the already known real or virtual, but clearly identified, higher symmetry parent phase. In principle, G could be guessed or derived from the structural relations between individual domains of the low-symmetry phase and the symmetry group F_A of one particular domain state (structural variant) A.

Therefore, whenever we talk about domains, we in fact implicitly assume that the individual mutual relations between the structures of domains attached to a common crystal or crystal grain reference correspond to the individual elements of this high-symmetry group G. In other words, talking about a domain always implies the existence of two symmetry groups, F_A and G, such that the former is a subgroup of the latter $F_A < G$. In a way, the rest of this chapter is a discussion of what can be derived from the knowledge of these two symmetry groups.

Having in mind the proposed implicit symmetry content of the term 'domain structure', the following alternative definition of ferroelastic materials can be given: 'If some domains in the same crystal or crystal grain of the material are systematically distinguishable by the spontaneous strain tensor, then the material is ferroelastic.' This definition implies that under a suitably oriented homogeneous stress, some domains (that are distinguishable by the spontaneous strain tensor) will have a different energy. In other words, due to the canonical linear stress–strain coupling, this definition reasonably matches with the requirements of experiment-based operational definitions imposing that ferroelastic domains 'can be switched' by a mechanical stimulus or that the material should 'show ferroelastic hysteresis loops'. However, our aim is not to reproduce the experimental definition fully. We rather wish to expose the advantages of the symmetry-based definition. First of all, the distinction by the spontaneous strain can be replaced in the definition by any other macroscopic domain property transforming as a second order symmetric polar tensor. For example, a ferroelastic material is a crystalline material with domains that can be distinguished within the given common crystal grain by the tensor of their optical refractive index. Interestingly, it can also be shown that a ferroelastic material is a crystalline material with domains distinguishable by the components of a neutral macroscopic bidirector function of the domain structure [6]. Since the neutral bidirector only defines a crystallographic axis, this result basically implies that a ferroelastic material is a crystalline material possessing domains distinguishable by the orientation of some specific material axis, such as the unique axis of the

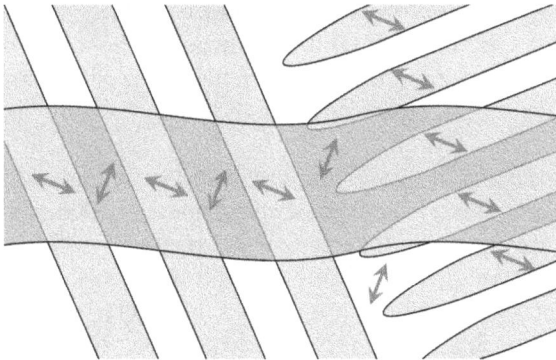

Figure 3.1. Typical architecture of the ferroelastic domain pattern. Ferroelastic crystals are defined by the possibility to form ferroelastic domains. At the same time, the domain structure containing domains that can be distinguished by the orientation of some unique axis in the structure of the symmetry-broken phase with respect to the parent structure proves that the material is a ferroelastic crystal. The existence of the unique axis in the symmetry-broken structure implies the possibility to define a macroscopic property with a neutral bidirector property, indicated in the image by the double-sided arrow. For example, in the case of triclinic domains, the possible choice of the unique bidirector could be the direction of the maximum spontaneous strain or the smallest refractive index. Frequently, ferroelastic materials also contain non-ferroelastic walls (ferroic or translational walls). Such walls divide the given ferroelastic domain into subdomains, distinguished by parameters other than the ferroelastic order parameter.

optical indicatrix or the axis of the largest thermal dilatation (see figure 3.1). The unique axis of the structure that identifies ferroelastic domain pairs sometimes coincides with the unique symmetry axis of the structure of the symmetry-broken phase but in other cases the orientation of the unique axis is not defined by symmetry alone.

3.3 Ferroelastic species

Since the ferroelastic phase transition involves symmetry-breaking spontaneous strain, it can also be very simply defined as a symmetry-breaking phase transition connected with a change of the crystal family. At the same time, it represents a subset of the larger set of macroscopic symmetry-breaking phase transitions, involving a change of the crystal class (that is, the corresponding crystallographic point group). This broader class of phase transitions, also called ferroic phase transitions in the sense of the definitions adopted in [2], represents a very convenient general framework for discussing macroscopic domain phenomena and highlights the specificity of ferroelastic domain phenomena.

For the purpose of classification of ferroelastic domain structures and the focus of this chapter, we shall limit the classification by considering only the 32 non-magnetic crystallographic point groups. Ferroic phase transitions involving changes of the non-magnetic crystallographic point groups can be classified into the 212 classes called Aizu species, representing qualitatively distinct symmetry reductions [3, 4]. The list of all these species is given in figures 3.2–3.5. Each species is defined by a group–subgroup pair $G > F$, where F stands for the point group of the representative domain (the same as F_A defined above, but ignoring the irrelevant index referring to a particular domain state) of the ferroic phase and G is the point group of the parent phase. To keep the notation minimal, we describe each species using only the international symbols of the two groups, whenever the correspondence between elements of F and G is obvious or irrelevant. In other cases, the mutual correspondence of elements is specified by additional subscripts added to the symbol of F (see figure 3.6 for the adopted labeling convention). Let us also stress that in figures 3.2–3.5, we provide the list of species in the same strict systematic sequential order, as defined in [3].

The macroscopic symmetry-breaking species can be divided and named according to the crystal family of the high-symmetry phase. Triclinic, monoclinic and orthorhombic species are given in figure 3.2 and distinguished by the background color. Tetragonal species are given in figure 3.3, hexagonal species in figure 3.4 and cubic species in figure 3.5. These tables can be inspected, for example, to verify whether a phase transition with a crystallographic point group symmetry breaking is ferroelastic or not, what the total number of macroscopic domain states is (structural variants distinguishable by their macroscopic properties), or whether none, all or some of the macroscopic states have distinct polarization. At the same time, these tables can be used for inverse considerations, such as deducing the plausible point groups when some partial information about the domain structure is known. In addition, for a given species, a more extended and detailed list of spontaneous tensor properties can be found in the supplement to [3]. The power of this symmetry-based

No.	G	F	n	↑↕		No.	G	F	n	↑↕
1	$\bar{1}$	1	2	●○		11	$mm2$	2	2	○●
2	2	1	2	●●		12	$mm2$	1	4	●●
3	m	1	2	●●		13	mmm	$mm2$	2	●○
4	$2/m$	m	2	●○		14	mmm	222	2	○○
5	$2/m$	2	2	●○		15	mmm	$2/m$	2	○●
6	$2/m$	$\bar{1}$	2	○●		16	mmm	m	4	●◐
7	$2/m$	1	4	●◐		17	mmm	2	4	◐◐
8	222	2	2	●●		18	mmm	$\bar{1}$	4	○●
9	222	1	4	●●		19	mmm	1	8	●◐
10	$mm2$	m	2	●●						

Figure 3.2. List of triclinic, monoclinic and orthorhombic species of broken macroscopic crystallographic symmetry. The first column shows the sequential order of species adopted from [3, 6, 7], organized according to the increasing symmetry of the high-symmetry group G and decreasing symmetry of the group with the broken symmetry F. The second and third columns show the high-symmetry point group G and the low-symmetry point group F. The next column provides the number n of distinct macroscopic domain states. The last-but-one column indicates whether a pair of distinct macroscopic domain states differs always (filled circle), sometimes (half-filled circle) or never (open circle) in one of the components of a macroscopic quantity that transforms as a polar vector, such as the spontaneous polarization. The last column indicates the same information with respect to the domain distinction based on the components of the spontaneous strain tensors (or any other macroscopic quantity that defines a unique macroscopic neutral bidirector). The filled circle, half-filled circle and open circle in the last column identify the fully ferroelastic, partially ferroelastic and non-ferroelastic species, respectively.

No.	G	F	n	↑↕		No.	G	F	n	↑↕
20	4	$2_{|}$	2	○●		43	$\bar{4}2m$	222	2	○●
21	4	1	4	●●		44	$\bar{4}2m$	m_{-}	4	●●
22	$\bar{4}$	$2_{|}$	2	●●		45	$\bar{4}2m$	$2_{|}$	4	◐◐
23	$\bar{4}$	1	4	●●		46	$\bar{4}2m$	2_{-}	4	●●
24	$4/m$	$\bar{4}$	2	○○		47	$\bar{4}2m$	1	8	●●
25	$4/m$	4	2	●○		48	$4/mmm$	$\bar{4}2m$	2	○○
26	$4/m$	$2_{|}/m$	2	○●		49	$4/mmm$	$4mm$	2	●○
27	$4/m$	$m_{|}$	4	●◐		50	$4/mmm$	422	2	○○
28	$4/m$	$2_{|}$	4	◐◐		51	$4/mmm$	$4/m$	2	○○
29	$4/m$	$\bar{1}$	4	○●		52	$4/mmm$	$\bar{4}$	4	○○
30	$4/m$	1	8	●◐		53	$4/mmm$	4	4	◐○
31	422	4	2	●○		54	$4/mmm$	mmm	2	○●
32	422	222	2	○●		55	$4/mmm$	$mm2_{|}$	4	◐◐
33	422	$2_{|}$	4	◐◐		56	$4/mmm$	$mm2_{-}$	4	●◐
34	422	2_{-}	4	●◐		57	$4/mmm$	222	4	○◐
35	422	1	8	●●		58	$4/mmm$	$2_{|}/m$	4	○●
36	$4mm$	4	2	○○		59	$4/mmm$	$2_{-}/m$	4	○●
37	$4mm$	$mm2_{|}$	2	○●		60	$4/mmm$	$m_{|}$	8	●◐
38	$4mm$	m_{-}	4	●●		61	$4/mmm$	m_{-}	8	●◐
39	$4mm$	$2_{|}$	4	○◐		62	$4/mmm$	$2_{|}$	8	◐◐
40	$4mm$	1	8	●●		63	$4/mmm$	2_{-}	8	◐◐
41	$\bar{4}2m$	$\bar{4}$	2	○○		64	$4/mmm$	$\bar{1}$	8	○●
42	$\bar{4}2m$	$mm2_{|}$	2	●●		65	$4/mmm$	1	16	●◐

Figure 3.3. List of tetragonal species of broken macroscopic crystallographic symmetry. The organization of the table and the meaning of the symbols are the same as in figure 3.2. Note that the low-symmetry group labels include additional subscripts specifying whether the given symmetry axis or normal of the mirror symmetry plane is parallel (|) or perpendicular (_) to the principal symmetry axis of the high-symmetry phase, which is the four-fold axis (see [3]).

No.	G	F	n	↑↕		No.	G	F	n	↑↕
66	3	1	3	●●		106	622	1	12	●●
67	$\bar{3}$	3	2	●○		107	$6mm$	6	2	○○
68	$\bar{3}$	$\bar{1}$	3	○●		108	$6mm$	$3m$	2	○○
69	$\bar{3}$	1	6	●◐		109	$6mm$	3	4	○○
70	32	3	2	●○		110	$6mm$	$mm2_|$	3	○●
71	32	2_-	3	●●		111	$6mm$	m_-	6	●●
72	32	1	6	●●		112	$6mm$	$2_|$	6	○●
73	$3m$	3	2	○○		113	$6mm$	1	12	●●
74	$3m$	m_-	3	●●		114	$\bar{6}m2$	$\bar{6}$	2	○○
75	$3m$	1	6	●●		115	$\bar{6}m2$	$3m$	2	●○
76	$\bar{3}m$	$3m$	2	●○		116	$\bar{6}m2$	32	2	○○
77	$\bar{3}m$	32	2	○○		117	$\bar{6}m2$	3	4	◐○
78	$\bar{3}m$	$\bar{3}$	2	○○		118	$\bar{6}m2$	$mm2_-$	3	●●
79	$\bar{3}m$	3	4	◐○		119	$\bar{6}m2$	$m_|$	6	●●
80	$\bar{3}m$	$2_-/m$	3	○●		120	$\bar{6}m2$	m_-	6	●●
81	$\bar{3}m$	m_-	6	●◐		121	$\bar{6}m2$	2_-	6	◐●
82	$\bar{3}m$	2_-	6	●◐		122	$\bar{6}m2$	1	12	●●
83	$\bar{3}m$	$\bar{1}$	6	○●		123	$6/mmm$	$\bar{6}m2$	2	○○
84	$\bar{3}m$	1	12	●◐		124	$6/mmm$	$6mm$	2	●○
85	6	3	2	○○		125	$6/mmm$	622	2	○○
86	6	$2_|$	3	○●		126	$6/mmm$	$6/m$	2	○○
87	6	1	6	●●		127	$6/mmm$	$\bar{6}$	4	○○
88	$\bar{6}$	3	2	●○		128	$6/mmm$	6	4	◐○
89	$\bar{6}$	$m_|$	3	●●		129	$6/mmm$	$\bar{3}m$	2	○○
90	$\bar{6}$	1	6	●●		130	$6/mmm$	$3m$	4	◐○
91	$6/m$	$\bar{6}$	2	○○		131	$6/mmm$	32	4	○○
92	$6/m$	6	2	●○		132	$6/mmm$	$\bar{3}$	4	○○
93	$6/m$	$\bar{3}$	2	○○		133	$6/mmm$	3	8	◐○
94	$6/m$	3	4	◐○		134	$6/mmm$	mmm	3	○●
95	$6/m$	$2_|/m$	3	○●		135	$6/mmm$	$mm2_|$	6	◐◐
96	$6/m$	$m_|$	6	●◐		136	$6/mmm$	$mm2_-$	6	●◐
97	$6/m$	$2_|$	6	◐◐		137	$6/mmm$	222	6	○○
98	$6/m$	$\bar{1}$	6	○●		138	$6/mmm$	$2_|/m$	6	○●
99	$6/m$	1	12	●◐		139	$6/mmm$	$2_-/m$	6	○●
100	622	6	2	●○		140	$6/mmm$	$m_|$	12	●◐
101	622	32	2	○○		141	$6/mmm$	m_-	12	●◐
102	622	3	4	◐○		142	$6/mmm$	$2_|$	12	◐◐
103	622	222	3	○●		143	$6/mmm$	2_-	12	◐◐
104	622	$2_|$	6	◐●		144	$6/mmm$	$\bar{1}$	12	○●
105	622	2_-	6	●●		145	$6/mmm$	1	24	●◐

Figure 3.4. List of hexagonal species of broken macroscopic crystallographic symmetry. The organization of the table and the meaning of the symbols are the same as in figure 3.2. The low-symmetry groups have additional subscripts showing whether the given symmetry axis or normal of the mirror symmetry plane is parallel (|) or perpendicular (_) to the principal symmetry axis of the high-symmetry phase, which is the three-fold or six-fold axis (see [3]).

approach relies on the fact that all these properties do not depend explicitly on the primary order parameter or phase transition mechanism, so that the group–subgroup pair $F < G$ relationships are sufficient to define these aspects. Finally, it is important to emphasize that the concept of Aizu species provides a very elegant and general way of classifying and organizing all macroscopic symmetry-breaking

No.	G	F	n	↑↓	No.	G	F	n	↑↓
146	23	3	4	●●	180	$\bar{4}3m$	1	24	●●
147	23	2_+22	3	○●	181	$m\bar{3}m$	$\bar{4}3m$	2	○○
148	23	2_+	6	●●	182	$m\bar{3}m$	432	2	○○
149	23	1	12	●●	183	$m\bar{3}m$	$m\bar{3}$	2	○○
150	$m\bar{3}$	23	2	○○	184	$m\bar{3}m$	23	4	○○
151	$m\bar{3}$	$\bar{3}$	4	○●	185	$m\bar{3}m$	$\bar{3}m$	4	○●
152	$m\bar{3}$	3	8	●◐	186	$m\bar{3}m$	$3m$	8	●◐
153	$m\bar{3}$	m_+mm	3	○●	187	$m\bar{3}m$	32	8	○◐
154	$m\bar{3}$	m_+m2_+	6	●◐	188	$m\bar{3}m$	$\bar{3}$	8	○○
155	$m\bar{3}$	2_+22	6	○◐	189	$m\bar{3}m$	3	16	◐◐
156	$m\bar{3}$	$2_+/m$	6	○●	190	$m\bar{3}m$	$4/mmm$	3	○●
157	$m\bar{3}$	m_+	12	●◐	191	$m\bar{3}m$	$\bar{4}2_+m$	6	○◐
158	$m\bar{3}$	2_+	12	◐◐	192	$m\bar{3}m$	$\bar{4}2\backslash m$	6	○◐
159	$m\bar{3}$	$\bar{1}$	12	○●	193	$m\bar{3}m$	$4mm$	6	●●
160	$m\bar{3}$	1	24	●◐	194	$m\bar{3}m$	422	6	○◐
161	432	23	2	○○	195	$m\bar{3}m$	$4/m$	6	○◐
162	432	32	4	○●	196	$m\bar{3}m$	$\bar{4}$	12	○○
163	432	3	8	●◐	197	$m\bar{3}m$	4	12	◐◐
164	432	422	3	○●	198	$m\bar{3}m$	m_+mm	6	○●
165	432	4	6	●◐	199	$m\bar{3}m$	$m\backslash mm$	6	○●
166	432	2_+22	6	○●	200	$m\bar{3}m$	m_+m2_+	12	◐◐
167	432	$2\backslash22$	6	○●	201	$m\bar{3}m$	$m\backslash m2_+$	12	◐◐
168	432	2_+	12	◐●	202	$m\bar{3}m$	$mm2\backslash$	12	●◐
169	432	$2\backslash$	12	●●	203	$m\bar{3}m$	2_+22	12	○◐
170	432	1	24	●●	204	$m\bar{3}m$	$2\backslash22$	12	○◐
171	$\bar{4}3m$	23	2	○○	205	$m\bar{3}m$	$2_+/m$	12	○●
172	$\bar{4}3m$	$3m$	4	●●	206	$m\bar{3}m$	$2\backslash/m$	12	○●
173	$\bar{4}3m$	3	8	◐◐	207	$m\bar{3}m$	m_+	24	●◐
174	$\bar{4}3m$	$\bar{4}2_+m$	3	○●	208	$m\bar{3}m$	$m\backslash$	24	●◐
175	$\bar{4}3m$	$\bar{4}$	6	○○	209	$m\bar{3}m$	2_+	24	◐◐
176	$\bar{4}3m$	$m\backslash m2_+$	6	●●	210	$m\bar{3}m$	$2\backslash$	24	◐◐
177	$\bar{4}3m$	2_+22	6	○●	211	$m\bar{3}m$	$\bar{1}$	24	○●
178	$\bar{4}3m$	$m\backslash$	12	●●	212	$m\bar{3}m$	1	48	●◐
179	$\bar{4}3m$	2_+	12	◐◐					

Figure 3.5. List of cubic species of broken macroscopic crystallographic symmetry. The organization of the table and the meaning of the symbols are the same as in the preceding three figures. Note that low-symmetry groups have additional indices showing whether the given symmetry axis or normal of the mirror symmetry plane is parallel (+) to a four-fold axis of the parent cubic phase or bisecting (\) the right angle between the two tetragonal axes of the parent cubic phase (see [3]).

phase transitions in crystals. Selected examples of ferroelastic phase transitions and their attribution to the species are listed in table 3.1.

3.4 Domain pairs and compatibility planes

A typical experimental signature of ferroelastic materials is a lamellar domain structure, in which domains take the form of thin lamellae, which belong to two alternating macroscopic domain states with distinct spontaneous strain tensors, and are separated by planar and parallel domain boundaries. The fundamental reason behind this observation is the geometrical consequence of the mechanical

Table 3.1. List of selected ferroelastic materials with a symmetry-breaking phase transition from a parent phase G to a subgroup F with assigned species numbers. Subscripts 'sg' and 'pg' stands for the space and point groups, respectively. The first compound in the table is the very first known ferroelectric crystal [8], also known as Rochelle salt.

Chem. formula	G_{sg}	F_{sg}	G_{pg}	F_{pg} (oriented)	Species No.	Transition point	Ref.
$KNaC_4H_4O_6 \cdot 4H_2O$	$P2_12_12_1$	$P2_1$	222	2	8	255 K	[9, 10]
$LiNH_4C_4H_4O_6 \cdot H_2O$	$P2_12_12$	$P2_1$	222	2	8	98 K	[11]
$LiNH_4SO_4$	$Pbn2_1$	$P2_1$	$mm2$	2	11	283 K	[12]
$La_{1-x}Nd_xP_5O_{14}$	$Pmcm$	$P2_1/c$	mmm	$2/m$	15	414 K	[13]
LaP_5O_{14}	$Pncm$	$P2_1/c$	mmm	$2/m$	15	400 K	[14]
$LiCsSO_4$	$Pmcn$	$P2_1/n$	mmm	$2/m$	15	203 K	[15]
NdP_5O_{14}	$Pmna$	$P2_1/b$	mmm	$2/m$	15	420 K	[16]
$LaNbO_4$	$I4_1/a$	$I2/a$	$4/m$	$2_1/m$	26	800 K	[17]
$BiVO_4$	$I4_1/a$	$I2/a$	$4/m$	$2_1/m$	26	528 K	[18]
TeO_2	$P422$	$P2222$	422	222	32	0.9 GPa	[19, 20]
$Tb_2(MoO_4)_2$	$P\bar{4}2_1m$	$Pba2$	$\bar{4}2m$	$mm2_1$	42	433 K	[21]
$BaFe_2As_2$	$I4/mmm$	$Fmmm$	$4/mmm$	mmm	54	140 K	[22]
SiO_2	$P4_2/mmm$	$Pnnm$	$4/mmm$	mmm	54	50–55 GPa	[23]
$LaAlO_3$	$Pm\bar{3}m$	$R\bar{3}c$	$m\bar{3}m$	$\bar{3}m$	185	795 K	[24]
$KMnF_3$	$Pm\bar{3}m$	$I4/mcm$	$m\bar{3}m$	$4/mmm$	190	187 K	[25]
$SrTiO_3$	$Pm\bar{3}m$	$I4/mcm$	$m\bar{3}m$	$4/mmm$	190	105 K	[26]
$PbTiO_3$	$Pm\bar{3}m$	$P4mm$	$m\bar{3}m$	$4mm$	193	765 K	[27]
$PbZrO_3$	$Pm\bar{3}m$	$Pbam$	$m\bar{3}m$	$m_\backslash mm$	199	505 K	[28]

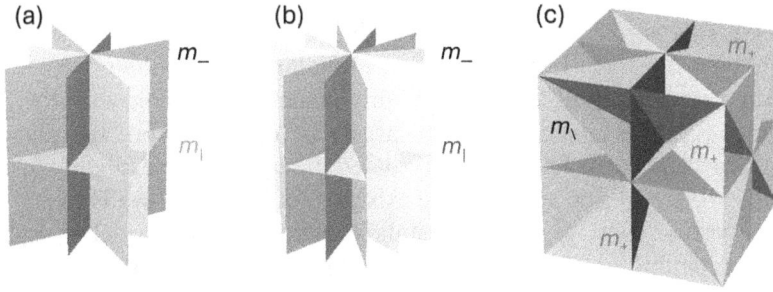

Figure 3.6. Distinction and labeling of the non-equivalent types of mirror symmetry planes in certain (a) tetragonal, (b) hexagonal and (c) cubic species of broken macroscopic crystallographic symmetry. The mirror symmetry planes perpendicular to the highest-order symmetry axis are drawn in color and labeled as $m_|$ in tetragonal and hexagonal species, and m_+ in the cubic species. The other type of mirror symmetry planes shown in gray are labeled as m_- in tetragonal and hexagonal species, and $m_\backslash$ in the cubic species. When the high-symmetry group contains both families of the mirror symmetry planes, it is important to keep track of the type of mirror planes that remain in its subgroup. However, in the symbols of the species, we use the additional superscript only in the few cases where it is not obvious. Similar notation is applied for the two-fold axes (as in [3]).

compatibility of a given pair of ferroelastic domain states, as clarified in the seminal papers of [29, 30].

The main ideas can be outlined as follows [2, 29, 30]. Assuming that two macroscopic domain states A and B are separated in space by a narrow but coherent interface, it is natural to require that the interface is equally deformed from both its sides. This leads to a condition stating that the strain tensor difference $\Delta\varepsilon = \varepsilon_A - \varepsilon_B$ is zero within the plane tangential to the interface. Consequently, the contraction of this difference tensor by any vector $\mathbf{x}$ within this plane should be zero. Formally, we can express this condition as

$$\Delta\varepsilon_{ij} x_i x_j = 0. \tag{3.1}$$

Considering that the equivalence of the domain states implies that the trace of the $\Delta\varepsilon$ is zero, there are just three main possibilities. First, if $\Delta\varepsilon$ itself is a zero tensor, domain states have the same spontaneous strain and this mechanical condition imposes no restriction. The orientation of the compatibility plane is thus completely arbitrary (it is the so-called W_∞-wall case). Second, if both $\Delta\varepsilon$ and its determinant are nonzero, there are no compatible planes at all. In practice, such walls may still exist as coherent interfaces, for example when the $\Delta\varepsilon$ tensor happens to be marginally small, or when the deviation from the spontaneous strain in the vicinity of the wall can be sustained by lattice elasticity, or when the lattice is strained by other coherent or incoherent interfaces, or in general, whenever another additional strain relaxation mechanism is present.

Finally, the most commonly explored case occurs when the $\Delta\varepsilon$ tensor is nonzero but its determinant is zero. In this case, there are two mutually perpendicular compatibility planes satisfying the compatibility condition. The orientation of each compatibility plane from this pair is either fully determined by the symmetry (the W_f-wall case) or it partly depends on the magnitude of the strain tensor elements (the

S-wall case). This third case corresponds to the mechanically compatible ferroelastic states that typically form hierarchical or simple lamellar twinning structures. The existence and orientation of the compatibility planes depend solely on the relations of spontaneous strain tensors that can largely be deduced from the macroscopic symmetry considerations. The symmetry-based predictions are explicitly given in [2] and [30] for all ferroelastic species. The existence of W_f and S-walls along with the related properties such as their crystallographic orientation, clapping angles, electrical compatibility, etc, can be found with the help of the set of detailed tables provided in [2]. The approach of [2] classifies the domain pairs according to the factor group of F_A of the ferroic domain state A and the rotational part of the switching operation g (from the factor group of G), which defines the point group symmetry operations of partner domain state $F_B = gF_A$.

A more in-depth discussion of various symmetry aspects of domain pairs and the related domain structures obviously requires us to consider the full space groups F_A and F_B of domains A and B. An important category of domain pairs is that of the transposable pairs, characterized by the existence of a switching operation g in G, which interchanges both domain states mutually. This operation, sometimes labeled as g^*, is called a transposing operation. In general, domain pairs are systematically characterized by three basic, generally distinct, symmetry groups: the group of the ordered domain pair F_{AB}, which is the simple intersection of F_A and F_B; the group of the unordered pair J_{AB}, which includes also the operations that interchange the two domain states; and the twinning group K_{AB}, which is a minimal subgroup of G containing F_A and g.

The existence of transposing operations allows J_{AB} and K_{AB} to be interpreted as a dichromatic groups. It can be shown that mechanically incompatible domain state pairs are non-transposable but the opposite implication does not hold. In general, it is convenient to classify domain pairs into species of the $F_A < K_{AB}$ macroscopic symmetry breaking. Among others, a ferroelastic domain pair can be defined solely by the symmetry: if the twinning group of the domain pair and the ferroic group of a domain belong to different crystal families, the domain pair is a ferroelastic one.

It is worth mentioning that the regular lamellar twins are sometimes so well defined that they can be considered as superdomains. While the macroscopic symmetry becomes a much less rigorous concept when applied to these mesoscopic textures, the application of the outlined symmetry-based approach, and the mechanical compatibility in particular, often allows classification and insights in the superdomain boundaries and related subjects [37–40].

3.5 Symmetry of ferroelastic domain walls

Domain walls can be intuitively understood as interfaces between the neighboring domains. In the case of domains related to structural phase transitions, the thickness of domain walls is often on the order of just a few atomic planes. When examined at scales not much larger than their thickness, not only ferroelastic W_f-walls or S-walls, but practically all domain walls related to structural phase transitions can be considered as planar objects with a well-defined 2D translation symmetry.

Therefore, the convenient groups for the description of the domain wall symmetry are so-called layer groups [2, 33].

Before defining the symmetry of a domain wall, it is instructive to introduce the notation for the domain walls. In a very general manner, we can specify the wall by the 'oriented' domain walls symbols $[A|\mathbf{n}, \alpha > B]$. Here A and B stand for specific identifiers of the domain states. Since the number N of all domain states is given by the $|G|/|F|$ ratio of the orders of the space groups G and F, the identifier should distinguish between the N possible domain states. Let us stress that N may or may not be equal to the number n of all macroscopically distinguishable domain states, determined by the ratio of the orders of the corresponding factor groups. If $n = N$, the number of all domains states can be directly read out from the tables in figures 3.2–3.5 and all domains states can be distinguished by macroscopic tensor properties only [3].

In general, not every pair of distinct macroscopic domain pairs in a ferroelastic material differs in spontaneous strain. Only in the case of a so-called ferroelastic domain pair can A and B be distinguished by spontaneous strain. Nonetheless, spontaneous strain usually does not determine the properties of the ferroelastic domain pair completely. To fully define the domain pair, additional information may thus be required. In other words, additional macroscopic parameters such as the spontaneous polarization of the domain, or microscopic parameters such as the phase of a Brillouin zone boundary frozen phonon mode may be needed.

Equally important is the oriented domain wall normal $\mathbf{n}$ that points in real space towards the domain that is ascribed to the state of the right-hand side of the domain wall symbol $[A|\mathbf{n}, \alpha > B]$, as it is mnemonically indicated by the embedded 'ket' bracket. This definition highlights the fact that it is not only the domain states and the orientation of the domain wall interface that define the domain wall and its properties, but also the sequential order of the domain states. In other words, $[A|\mathbf{n}, \alpha > B]$ and $[B|\mathbf{n}, \alpha > A]$ are distinct objects and can differ even in scalar properties such as energy, thickness, or charge. Finally, the domain wall properties may depend on an additional parameter α, specifying its exact location in the structure [2, 33], or identifying the possible competing degenerate or non-equivalent internal domain wall states, for example in the case of Bloch ferroelectric domain walls [32, 35, 36].

The symmetry of a uniquely defined planar domain wall $[A|\mathbf{n}, \alpha > B]$ is given by a $T_{AB}^{\alpha}|hkl>$ layer group, composed of all parent phase space group operations, that result in an indistinguishable domain wall geometry. The highest possible symmetry $T_{AB}^{0}(hkl)$ and its factor group can be determined from symmetry arguments only, considering the orientation of the domain wall plane and its exact centering in the crystal structure. The procedure is described in detail in [2].

Even more complete information is contained in the four-color antisymmetry group $\bar{J}_{AB}|(hkl)$ of the domain wall pair $[A|\mathbf{n}, \alpha > B]$ and $[B|\mathbf{n}, \alpha > A]$, which can be obtained as a sectional layer group of a (hkl) plane in J_{AB}. It is interesting that $T_{AB}^{0}(hkl)$ is not obtained as a sectional group and that it differs from both $\bar{J}_{AB}|(hkl)$ and $\bar{F}_{AB}|(hkl)$, which is the sectional layer group of F_{AB}.

The macroscopic (point group) symmetry of $T_{AB}^{0}(hkl)$ was determined for all mechanically compatible domain walls explicitly in [2]. Among other findings, it has been realized that macroscopic symmetry classes (point groups) of all mechanically compatible ferroelastic domain walls are equal to $mm2$ or its subgroups. This has the strong implication that all mechanically compatible ferroelastic domain walls are polar and have a wall-specific second harmonic generation activity [2].

Although the domain wall symmetry has been addressed in some recent works [33, 34], the impact of the results is limited because detailed structural data on perfect domain walls are still lacking due to the challenging requirement imposed on both the currently available experimental and computational tools.

3.6 Summary

We have shown that ferroelastic materials can be conveniently defined in terms of a particular type of macroscopic symmetry breaking, involving the change of the crystal family, and explained how and why this is inherently manifested in the domain structure. We have given several definitions that can be used to identify the ferroelastic materials, and we have also introduced the concept of macroscopic symmetry-breaking species, that allows us to classify different types of ferroelastic materials systematically or to compare them with many other types of ferroic materials. We believe the species perspective is useful and will be used in the near future more frequently. Interactive tools for exploring 212 ordinary species and 1602 magnetic species of macroscopic symmetry breaking are conveniently available online [41]. We have also summarized the main results of the symmetry approach to the domain pairs and general results concerning the mechanical compatibility of ferroelastic domain states, that constitutes the origin of the rich plethora of ferroelastic domain structures and related phenomena. Finally, we have discussed the symmetry theory of the ideal planar domain wall, which still awaits more extensive exploitation in detailed studies of ferroelastic domain walls.

Acknowledgments

This work was supported by the European Union's Horizon 2020 research and innovation programme under grant agreement no. 964931 (TSAR) and by the Ferroic Multifunctionalities project, supported by the Ministry of Education, Youth, and Sports of the Czech Republic, Project No. CZ.02.01.01/00/22_008/ 0004591, co-funded by the European Union.

References

[1] Janovec V 1981 Symmetry and structure of domain walls *Ferroelectrics* **35** 105

[2] Janovec V and Přívratská J 2003 *International Tables for Crystallography, Volume D: Physical properties of crystals* vol D; A Authier (Boston, MA: Kluwer) p 449

[3] Hlinka J, Privratska J, Ondrejkovic P and Janovec V 2016 Symmetry guide to ferroaxial transitions *Phys. Rev. Lett.* **116** 177602

[4] Aizu K 1970 Possible species of ferromagnetic, ferroelectric, and ferroelastic crystals *Phys. Rev. B* **2** 754

[5] Hlinka J 2014 Eight types of symmetrically distinct vectorlike physical quantities *Phys. Rev. Lett.* **113** 165502

[6] Erb K C and Hlinka J 2020 Vector, bidirector, and Bloch skyrmion phases induced by structural crystallographic symmetry breaking *Phys. Rev.* B **102** 024110

[7] Erb K C and Hlinka J 2018 Symmetry guide to chiroaxial transitions *Phase Transit.* **91** 953

[8] Valasek J 1921 Piezo-electric and allied phenomena in Rochelle salt *Phys. Rev.* **17** 475

[9] Mitsui T 1958 Theory of the ferroelectric effect in Rochelle salt *Phys. Rev.* **111** 1259

[10] Oja T and Casabella P A 1969 Nuclear-magnetic-resonance studies of ferroelectricity in normal and irradiated Rochelle salt *Phys. Rev.* **177** 830

[11] Sawada A, Udagawa M and Nakamura T 1977 Proper ferroelastic transition in piezoelectric lithium ammonium tartrate *Phys. Rev. Lett.* **39** 829

[12] Hildmann B O, Hahn T h, Cross L E and Newnham R E 1975 Lithium ammonium sulphate, a polar ferroelastic which is not simultaneously ferroelectric *Appl. Phys. Lett.* **27** 103

[13] Wang Y, Sun W, Chen X, Shen H and Lu B 1987 Internal friction associated with the domain walls and the second-order ferroelastic transition in LNPP *Phys. Stat. Sol.* a **102** 279

[14] Errandonea G 1980 Elastic and mechanical studies of the transition in LaP_5O_{14}: a continuous ferroelastic transition with a classical Landau-type behavior *Phys. Rev.* B **21** 5221

[15] Asahi T and Hasebe K 1988 X-ray study of $LiCsSO_4$ in connection with its ferroelastic phase transition *J. Phys. Soc. Jpn.* **57** 4184

[16] Budin J P, Milatos-Roufos A, Chinh N D and Le Roux G 1975 Ferroelastic behavior of $Nd_xLa_{1-x}P_5O_{14}$ crystals *J. Appl. Phys.* **46** 2867

[17] Hara K, Sakai A, Tsunekawa S, Sawada A, Ishibashi Y and Yagi T 1985 A soft acoustic mode in the ferroelastic phase transition of $LaNbO_4$ *J. Phys. Soc. Jpn.* **54** 1168

[18] David W I F, Glazer A M and Hewat A W 1979 The structure and ferroelastic phase transition of $BiVO_4$ *Phase Transit.* **1** 155

[19] Peercy P S and Fritz I J 1974 Pressure-induced phase transition in paratellurite (TeO_2) *Phys. Rev. Lett.* **32** 466

[20] Peercy P S, Fritz I J and Samara G A 1975 Temperature and pressure dependences of the properties and phase transition in paratellurite (TeO_2): ultrasonic, dielectric and Raman and Brillouin scattering results *J. Phys. Chem. Solids* **36** 1105

[21] Brixner L H, Bierstedt P E, Sleight A W and Licis M S 1971 Precision parameters of some $Ln_2(MoO_4)_3$-type rare earth molybdates *Mat. Res. Bull.* **6** 545

[22] Rotter M, Tegel M, Johrendt D, Schellenberg I, Hermes W and Pöttgen R 2008 Spin-density-wave anomaly at 140 K in the ternary iron arsenide $BaFe_2As_2$ *Phys. Rev.* B **78** 020503(R)

[23] Andrault D, Fiquet G, Guyot F and Hanfland M 1998 Pressure-induced Landau-type transition in stishovite *Science* **282** 720

[24] Müller K A, Berlinger W and Waldner F 1968 Characteristic structural phase transition in perovskite-type compounds *Phys. Rev. Lett.* **21** 814

[25] Salazar A, Massot M, Oleaga A, Pawlak A and Schranz W 2007 Critical behavior of the thermal properties of $K Mn F_3$ *Phys. Rev.* B **75** 224428

[26] Hayward S A and Salje E K H 1999 Cubic-tetragonal phase transition in $SrTiO_3$ revisited: Landau theory and transition mechanism *Phase Transit.* **68** 501

[27] Haun M J, Furman E, Jang S J, McKinstry H A and Cross L E 1987 Thermodynamic theory of $PbTiO_3$ *J. Appl. Phys.* **62** 3331

[28] Zhang H, Thong H-C, Bastogne L, Gui C, He X and Ghosez P 2024 Finite-temperature properties of the antiferroelectric perovskite $PbZrO_3$ from a deep-learning interatomic potential *Phys. Rev.* B **110** 054109

[29] Fousek J and Janovec V 1969 The orientation of domain walls in twinned ferroelectric crystals *J. Appl. Phys.* **40** 135

[30] Sapriel J 1975 Domain-wall orientations in ferroelastics *Phys. Rev.* B **12** 5128

[31] Diéguez O *et al* 2013 Domain walls in a perovskite oxide with two primary structural order parameters: first-principles study of BiFeO$_3$ *Phys. Rev.* B **87** 024102

[32] Marton P, Rychetsky I and Hlinka J 2010 Domain walls of ferroelectric BaTiO$_3$ within the Ginzburg–Landau–Devonshire phenomenological model *Phys. Rev.* B **81** 144125

[33] Schranz W, Rychetsky I and Hlinka J 2019 Polarity of domain boundaries in nonpolar materials derived from order parameter and layer group symmetry *Phys. Rev.* B **100** 184105

[34] Goncalves M A P, Graf M, Pasciak M and Hlinka J 2025 Non-reciprocal neutral ferroelectric domain walls in BiFeO$_3$ arXiv: 2501.00534

[35] Stepkova V, Marton P and Hlinka J 2012 Stress-induced phase transition in ferroelectric domain walls of BaTiO$_3$ *J. Phys. Condens. Matter* **24** 212201

[36] Rychetsky I, Schranz W and Troster A 2021 Symmetry and polarity of antiphase boundaries in PbZrO$_3$ *Phys. Rev.* B **104** 224107

[37] Neuber E *et al* 2018 Architecture of nanoscale ferroelectric domains in GaMo$_4$S$_8$ *J. Phys. Condens. Mater.* **30** 445402

[38] Bednyakov P S and Hlinka J 2023 Charged domain walls in BaTiO$_3$ crystals emerging from superdomain boundaries *Adv. Electron. Mater.* **9** 2300005

[39] Bednyakov P S, Rafalovskyi I and Hlinka J 2025 Fragmented charged domain wall below the tetragonal-orthorhombic phase transition in BaTiO$_3$ *Appl. Phys. Lett.* **126** 012909

[40] Tovaglieri L, Torruella M H P, Hsu C-Y, Korosec L, Alexander D T L, Paruch P, Triscone J-M and Lichtensteiger C 2025 Investigating domain structures and superdomains in ferroelectric PbTiO$_3$ based heterostructures on DyScO$_3$ *APL Mater.* **13** 021118

[41] Hlinka J and Ondrejkovic P 2025 Website for crystallographic point group symmetry-breaking species https://species.fzu.cz

IOP Publishing

Ferroelastic Materials

Guillaume F Nataf, Blai Casals and Ekhard K H Salje

Chapter 4

Response of ferroelastic domains to an applied field

Kevin Nadaud, Mojca Otoničar, Susan Trolier-McKinstry and Jurij Koruza

4.1 Introduction

The motion of ferroelastic domain walls has a profound influence on the material's properties, such as the elastic stiffness, the relative permittivity, and the piezoelectric coefficients. It is important to note, however, that ferroelastic domain walls do not necessarily move smoothly through real solids where 0D, 1D, 2D and 3D defects are omnipresent. Rather, they move through a complex energy profile with pinning centers that impede wall motion and induce hysteresis in the presence of oscillating fields. These movements are probed through a diverse range of electrical measurements: polarization and strain hysteresis loops, Rayleigh analysis, third-harmonic responses, and first order reversal curves. Structural characterizations, e.g. using x-ray diffraction and electron microscopy, complement these approaches.

4.2 Driving forces for ferroelastic wall motion

Ferroic materials are those in which transformations between spontaneous, crystallographically defined variants can be achieved by applied forces [1, 2]. The free energy difference, ΔG between ferroic twins can be written as

$$\Delta G = \Delta \varepsilon_{ij} \sigma_{ij} + \Delta P_i E_i + \Delta M_i H_i, \tag{4.1}$$

where ε_{ij} is strain, P_i is polarization, and M_i is magnetization. The subscripts denote Einstein notation.

As shown in table 4.1, the primary ferroics include ferroelectrics, ferromagnets (and ferrimagnets), and ferroelastics. In these materials, the orientation states differ in the first derivatives of the free energy with respect to electric field (E_i), magnetic field (H_i), and elastic field, e.g. stress (σ_{ij}). Therefore, ferroelastic domains which differ in strain can be controlled with an applied stress, ferroelectric domains differing in polarization can be controlled with an applied electric field, and

doi:10.1088/978-0-7503-6089-0ch4
4-1

Table 4.1. Primary, secondary, and some tertiary ferroics [3, 4].

Primary ferroics	Orientation states differ in...	Switching force
Ferromagnet	Spontaneous magnetization	Magnetic field
Ferroelectric	Spontaneous polarization	Electric field
Ferroelastic	Spontaneous strain	Mechanical stress
Secondary ferroics	**Orientation states differ in...**	**Switching force**
Ferrobimagnetic	Magnetic susceptibility	Magnetic field
Ferrobielectric	Dielectric susceptibility	Electric field
Ferrobielastic	Elastic compliance	Mechanical stress
Ferromagnetoelastic	Piezomagnetic coefficients	Magnetic field and mechanical stress
Ferroelastoelastic	Piezoelectric coefficients	Electric field and mechanical stress
Ferromagnetoelectric	Magnetoelectric coefficients	Magnetic field and electric field
Tertiary ferroics	**Orientation states differ in...**	**Switching force**
Ferrotrielastic	Non-linear elastic compliances	Mechanical stress
Ferroelectrobielastic	Stress dependence of piezoelectric coefficient	Electric field and mechanical stress
etc		

ferromagnetic domains differing magnetism can be selected using an applied magnetic field. In much the same way, the driving forces for domain wall motion in secondary ferroics are given by the second derivatives of the free energy difference, while tertiary ferroics differ in the third derivatives. Examples are included in table 4.1.

Fundamentally, ferroelastic domain states differ in spontaneous strain. Ferroelasticity can either occur on its own, or in conjunction with other ferroic order parameters, typically in response to phase transformation from a higher symmetry state. On going through the phase transition temperature, one or more symmetry elements are lost; these symmetry elements define the allowed domain states in a given material's system [5]. The boundaries between different domain states are domain walls. In many cases, the motion of the domain walls has a profound influence on the material's properties. For example, the motion of ferroelastic walls can influence the elastic stiffness, the relative permittivity, and the piezoelectric coefficients. It is important to note, however, that ferroelastic domain walls do not necessarily move smoothly through real (and hence defective) solids. Rather, they move through a complex energy profile with pinning centers that impede wall motion and induce hysteresis in the presence of oscillating fields. Thus, ferroelastic materials are interesting both in terms of fundamental materials science and as a design tool critical for optimizing device performance.

4.3 Crystallography of some important ferroelastic oxides

As noted above, ferroelastic domains differ in the state of spontaneous strain, but the differences are small enough that selection of the preferred domain state can be done with applied forces that do not destroy the material (induce cracking or dielectric breakdown, for example). Often, this condition is achieved in cases where the material undergoes a displacive phase transition from a higher symmetry prototype phase to a lower symmetry phase. As this occurs, the material will often nucleate transformation twins that differ in the spontaneous strain. Applied stresses can favor one of these ferroelastic states over the other, inducing motion of the ferroelastic wall.

4.3.1 Perovskites

An example of this is illustrated in figure 4.1 for the case of the perovskite $PbTiO_3$. Above the Curie temperature, the material is cubic, and the Ti atom is at the center of the unit cell. On reducing the temperature, the material undergoes a displacive phase transformation, in which no bonds are broken, but where the Ti displaces towards one of the adjacent oxygen atoms, producing six possible polarization states. As this occurs, the unit cell elongates parallel to the polarization direction and

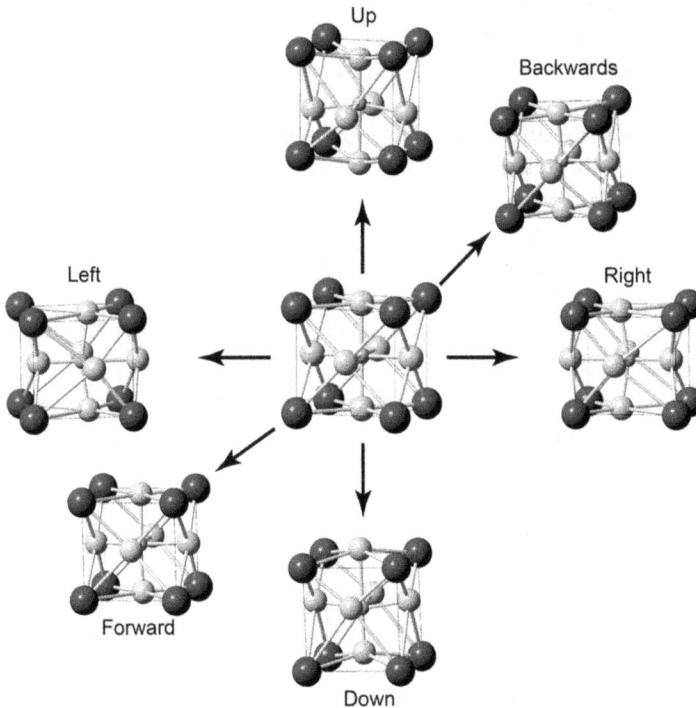

Figure 4.1. Cubic prototype of $PbTiO_3$ (center), with the six ferroelectric domain states shown. Pb is shown in purple, Ti in light blue and O in gray. These six ferroelectric domains correspond to three ferroelastic domains, since there is no shape difference between the pairs up–down, forward–backwards, and left–right.

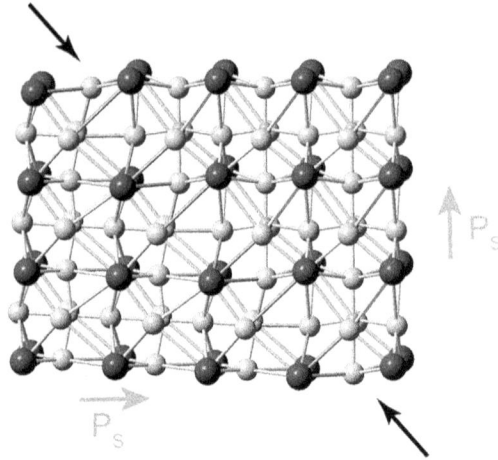

Figure 4.2. Schematic of a hypothetical infinitely thin domain wall between ferroelastic domains in PbTiO$_3$ marked by the black arrows. The direction of the spontaneous polarization, P_S, and hence the spontaneous strain in each of the domains is marked. Note that in practice, the width of ferroelastic domain walls can be several unit cells.

contracts laterally. This is the origin of the spontaneous strain that also makes PbTiO$_3$ ferroelastic. That is, when a compressive stress is applied along the long axis of the material, the domain state is less favorable than one with the short axis aligned with the compressive stress, inducing motion of the ferroelastic walls (see figure 4.2). The three available ferroelastic domains in this case arise because of three-fold rotation axis along the cube diagonal, <111>, of the cubic prototype which was lost on transforming to the tetragonal phase. Compatibility between available ferroelastic domains will govern the allowed orientation of domain walls. There are a very large number of ferroelastic materials with the cubic perovskite phase as a prototype, including materials with tetragonal, orthorhombic, rhombohedral, and monoclinic distortions [6].

The order parameter of the phase transition determines whether the transition is proper or improper. It should be noted that in PbTiO$_3$ the order parameter for the phase transition is the spontaneous polarization. Thus, PbTiO$_3$ is a proper ferroelectric and an improper ferroelastic. In the same way, multiaxial ferroelectrics are typically also ferroelastic.

Ferroelastic domains in perovskite materials can also result from tilt transitions in the structure (i.e. self-consistent rotations of the octahedra). An example of this is shown in figure 4.3 for GdScO$_3$. At elevated temperatures, GdScO$_3$ has the prototype perovskite structure, but at lower temperatures, the Gd^{3+} ion is too small to occupy the 12 coordinated A site. To accommodate a reduction in coordination number, the ScO$_6$ octahedra rotate without breaking any of the Sc–O bonds, such that the Sc–O–Sc bond angle is not 180°. The rotations can occur in different directions, producing the potential ferroelastic states.

Ferroelasticity in the absence of ferroelectricity has also been reported in many of the hybrid inorganic/organic perovskites [7], as discussed in chapter 9.

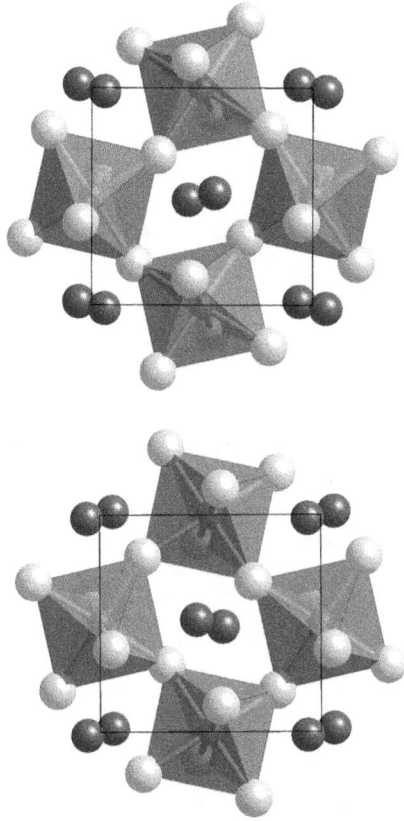

Figure 4.3. Two transformation twins in GdScO$_3$. The Gd ions are shown as the dark spheres, the ScO$_6$ octahedra are shown; O is gray. The outline of the unit cell is shown as thin black lines.

4.3.2 Crystallography of other selected ferroelastics

Many other crystal structures support ferroelasticity. As an example, we consider two of the scheelite-related polymorphs of BiVO$_4$, as shown in figure 4.4. At ~255°C [8], there is a displacive phase transition which is correlated with the motion of the Bi atoms in the unit cell. The resulting twinned domain structures are mobile under mechanical stress. Again, the atom displacements between the two domain states is modest, such that the applied strain can select between ferroelastic twins without inducing bond-breaking. A combination of domain wall motion with crack deflection on mechanical failure increases the fracture toughness of BiVO$_4$ relative to the prototype paraelastic state [9]. Isomorphs of the BiVO$_4$ ferroelastics include $(Na_{0.5x}Bi_{1-0.5x})(Mo_xV_{1-x})O_4$ [10] among others.

Ferroelasticity is also widely observed in minerals. Examples of this are the MAlSi$_3$O$_8$ and MAl$_2$Si$_2$O$_8$ feldspars, where M is a medium sized ion such as Na$^+$ or Ca^{2+}. At elevated temperatures, the amplitude of the thermal motion of the M ion is large enough to keep the aluminosilicate framework propped open. However, as the temperature decreases, the amplitude of the M ion decreases, and the

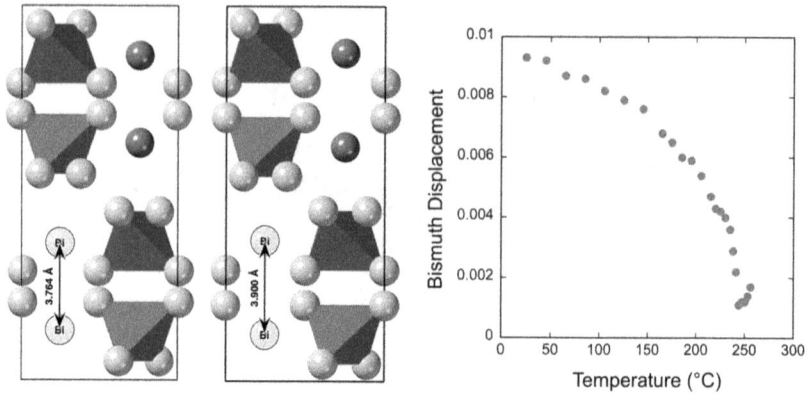

Figure 4.4. Scheelite-based crystal structures of ferroelastic $BiVO_4$. The VO_4 tetrahedron are shown; oxygen is light gray and the Bi^{3+} ions are shown as purple spheres. The figure on the left shows the room temperature monoclinic crystal structure, the middle figure shows the high temperature tetragonal structure. Structural data are from Sleight *et al.* [11]. The Bi–Bi separation distances are marked. The plot on the right shows the displacement of the Bi ion from a high symmetry position as a function of temperature, as reported by David and Wood [12]. This displacement is linked to the ferroelasticity.

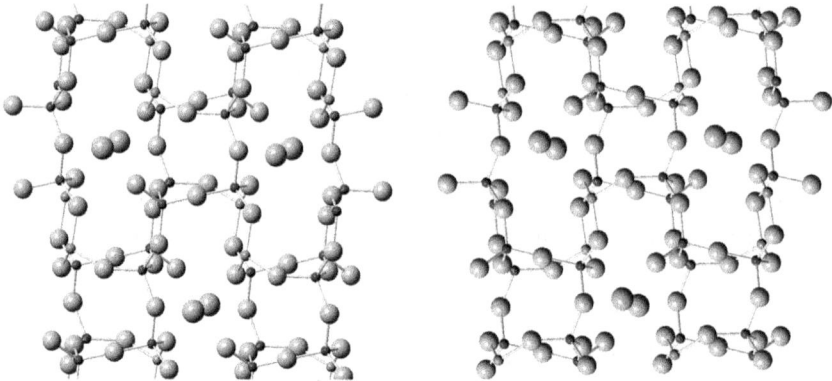

Figure 4.5. Ferroelastic twins in albite feldspar $NaAlSi_3O_8$. Si is shown as the small dark blue ions and Al as the small lighter blue ions. O is shown in gray and the Na^+ ions as the larger yellow ions residing in the large interstices in the aluminosilicate framework.

aluminosilicate framework slumps to decrease the effective coordination of the M ion. This, in turn decreases the symmetry [13]. As shown in figure 4.5, there are twins which differ in spontaneous strain, and which enable ferroelasticity.

Gadolynium molybdate is an improper ferroelectric driven by a proper ferroelastic transition. Figure 4.6 shows the twins of the structure, where structural data were taken from [14]. It is clear that the molybdenum and the gadolinium polyhedra are corner-connected, and that the ferroelastic states differ in the direction of the rotation. The atom displacements are modest, and no bond-breaking is required to switch between the twins.

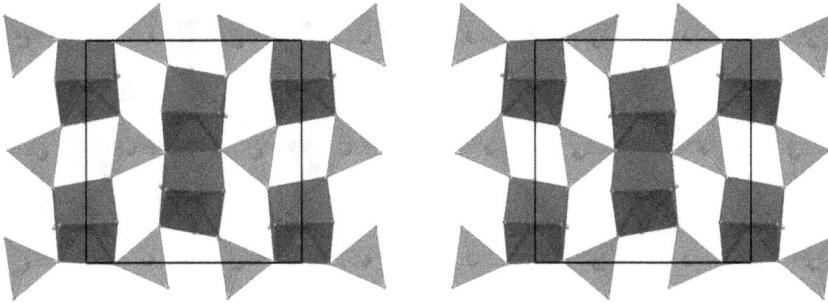

Figure 4.6. Ferroelastic twins of $Gd_2(MO_4)_3$. The molybdenum tetrahedra are shown corner-connecting the larger Gd polyhedra.

Details on the structural origin of ferroelasticity are available elsewhere for other materials, such as $Pb_3(PO_4)_2$, $Pb_3(VO_4)_2$, $Pb_3(AsO_4)_2$, and solid solutions of these materials on either the Pb or the tetrahedral sites. Likewise, many other minerals and organic compounds are also ferroelastic.

4.4 Mechanisms for domain wall movement

4.4.1 Hysteresis loop and domain reversal

Domain walls can be moved by applying an electric or a stress field. Depending on the characteristics of the signal (magnitude, frequency, shape), the movement can be different. When applying an electric field, the polarization of the unit cells can be switched in one direction, which thus reduces the number of domain walls in the material. The common way to study the reversal of the polarization as a function of the applied electric field is the $P(E)$ hysteresis cycle (figure 4.7(a)). The experimental apparatus to obtain the $P(E)$ loop is presented in section 4.5.2. It is also possible to measure the deformation versus pressure $S(\sigma)$ loop [15], but this is less common.

At the initial state, the material presents a multidomain configuration since it is the most favorable energetically [16]. When applying a small electric field, the polarization increases linearly with it (AB segment). In this range, the field is not high enough to reverse the polarization of domains with opposite polarization. When the field is sufficiently high, the domains start to switch in the direction that is as close as possible to the direction of the electric field. The polarization increases rapidly and is strongly non-linear (segment BC). Once the domains are aligned (point C), the polarization increases only linearly (segment CD).

When the applied electric field decreases, some domains can switch back to their initial orientation, but at zero field, the polarization is not null (P_r^+). To obtain a polarization equal to zero, the field needs to be reversed (point F). If the electric field is decreased again, the polarization aligns in the opposite direction (point G). The value of the field at which the net polarization is zero, the coercive field E_C, is obtained by the intercept with the field axis and does not correspond to an absolute threshold value [17]—the application of a continuous field below the coercive field, but for a long duration, may switch the polarization.

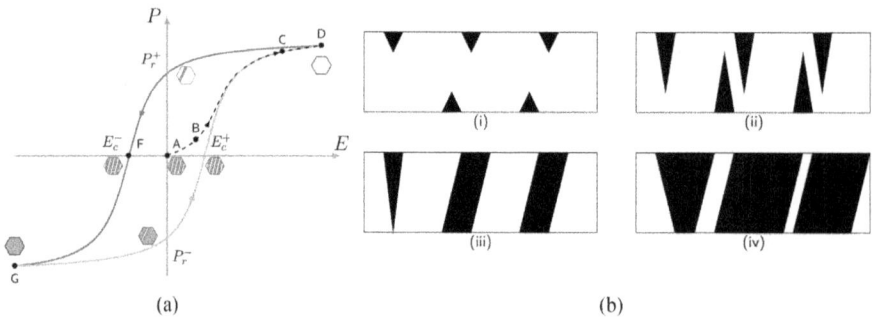

Figure 4.7. (a) Example of polarization versus electric field loop. (b) Possible sequence of polarization switching in ferroelectrics: (i) nucleation of domains (ii) growth of domains, (iii) motion of domain walls (iv) coalescence of domains. ((a) Adapted with permission from [17]. Copyright 2006 Elsevier. (b) Adapted with permission from [18]. Copyright 1996 Taylor and Francis.)

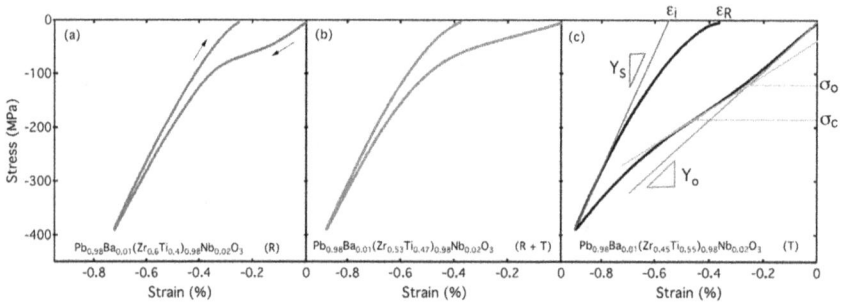

Figure 4.8. Room temperature stress–strain behavior of unpoled Ba- and Nb-doped polycrystalline Pb $(Zr_{1-x}Ti_x)O_3$ across the morphotropic phase boundary (MPB): (a) rhombohedral $Pb(Zr_{0.6}Ti_{0.4})O_3$, (b) Pb $(Zr_{0.53}Ti_{0.47})O_3$ at the MPB, and (c) tetragonal $Pb(Zr_{0.45}Ti_{0.55})O_3$. The characterizing parameters are labeled in (c); the arrows indicate the loading direction. The phase of each material is labeled. (Reproduced with permission from [19]. Copyright 2017 IOP Publishing.)

A possible sequence of domain switching is described in figure 4.7(b). Similar evolution of the domain structure occurs when measuring the strain as a function of stress [19, 20]. During the initial loading, the material exhibits a linear elastic behavior, with an initial Young's modulus Y_0. When increasing the loading, ferroelastic reorientation occurs through irreversible domain wall movement. The non-linear switching effect continues until a saturation is reached, where all the ferroelastic domains switch (similar to the polarization loop). In the saturation region, the deformation is linear since no additional ferroelastic switching occurs (figure 4.8).

The exact movement of the domain walls is very complex, and no universal mechanism is valid for all ferroelectrics [16, 18]. Nevertheless, two prevalent models are used to describe the switching kinetics.

    1. One of the most common switching mechanisms used to describe ferro-electrics is based on the results obtained by Kolmogorov and Avrami, often

called the Kolmogorov–Avrami–Ishibashi (KAI) model [21]. This model assumes that domains are initiated by independent nucleation centers. At the beginning of the growth, domains are not overlapping. Thus, the reversed polarization is the sum of the polarization of all the switched domains. When domains start overlapping, the total polarization variation is lower than the sum of each individual domain, which is taken into account for in the model. This mechanism is successfully applied to Al,ScN [22], epitaxial PZT/SrRuO$_2$ thin film [23] and polycrystalline PZT/Pt thin films [21], but not to hafnia-based materials [22, 24]. This model assumes that the main limitation for the switching time is domain expansion.

2. A second model can also be used to describe the polarization reversal. It is called the nucleation-limited switching (NLS) model [21]. It assumes that the switching is limited by nucleation process, contrary to the KAI model which assumes that domain growth is the limiting element. This model accounts for the broadening of the switching time when the applied electric field is decreased.

To determine which model accurately describes the switching mechanism, pulses with various amplitude and duration are used (see section 4.5.2).

The $P(E)$ loop is usually measured at a given frequency of the driving field, which may affect the obtained values of polarization and coercive field [25, 26]. This increase of the coercive field has been observed both in ceramics and thin films. It reveals that the kinetics of the switching involves domain walls that are moving in a viscous medium [27]. When the measurement frequency increases, the coercive field increases, following a power law dependency $E_C \propto f^\beta$. The value of β is used to determine the effective dimensionality of the switching process [26, 27].

When performing the measurement of the coercive field while varying the frequency by several orders of magnitude (three or four at least), two regimes with different values of β are visible, and a crossover frequency can be observed [23, 28]. These two regimes may come from the competition between two different processes such as domain wall motion and nucleation, or thermally activated and viscous domain wall motion [28]. The presence of 90° domain walls may also affect the frequency dependence of the response in ferroelectric materials since their mobility is lower than that of 180° domain walls [29].

The domain dynamics is also affected by temperature, as (i) a faster growth of domains is obtained at high temperature [30] and (ii) the coercive field decreases with increasing temperature.

4.4.2 The role of defects

The mobility of ferroic domain walls depends primarily on the material's chemical composition and crystal structure, which define the Peierls potential and thus the activation barrier. Additionally, domain walls encounter various obstacles when moving through the material, which influence their mobility and consequently the material's macroscopic functional properties. These obstacles are either naturally occurring or artificially imposed defects with different dimensionalities, present within the crystal structure or the microstructure [31]. The cause for their interaction with ferroic domain walls is the difference between defects' electric charge, size, or magnetic properties. The defects thus directly modify the energy landscape by forming local energy minima and influencing the barrier heights for domain wall movement. On the other hand, certain defects can also act as points for domain nucleation and thus facilitate switching. Since domain walls can contribute up to 80% of the material's total functional response, understanding the role of defects is of utmost importance and their adjustment represents a key design tool for materials scientists.

Defects are generally classified based on their dimensionality (figure 4.9). Point defects (zero-dimensional) are the most frequently occurring and widely studied type of imperfection. Electronic imperfections include free electrons, holes, and polarons, which are formed by charge injection, radiation, or as a compensating consequence of ionic defects. On the other hand, ionic imperfections include foreign ions, vacancies, or interstitials, and originate from precursor impurities, processing, or are intentionally introduced by chemical doping. In a wider sense, point defect complexes also belong to this category. It should be recalled that equilibrium concentrations of intrinsic vacancies exist in all materials above 0 K, since these defects are stabilized by the configurational entropy and are thus thermodynamically stable.

Despite their name, the influence of these defects spreads beyond a 'point', as they are associated with local electric and strain fields. The electric fields originate from the charge mismatch of the aliovalent impurity ions or the defect complexes between the impurity ion and an oppositely charged vacancy [32]. On the other hand, the strain field is caused by the ion size discrepancy and the deformation due to the vacant lattice site [33]. Due to these fields, defects commonly act as pinning centers for ferroelectric [34, 35], ferroelastic [36], and ferromagnetic [37] domain walls. Direct evidence has been provided by local observations of domain wall pinning and bowing (figure 4.10(a)) [35], as well as measurements of the domain wall mobility in samples with artificially induced defects [38]. It should be noted that inhomogeneous

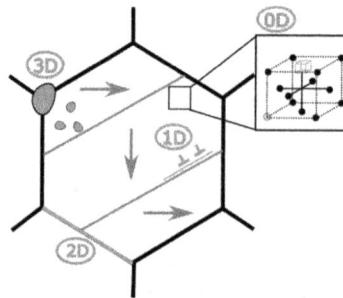

Figure 4.9. Schematic representation of common defects in polycrystalline ferroic materials: point defects (0D), dislocations (1D), grain boundaries and other domain walls (2D), and secondary phases and precipitates (3D).

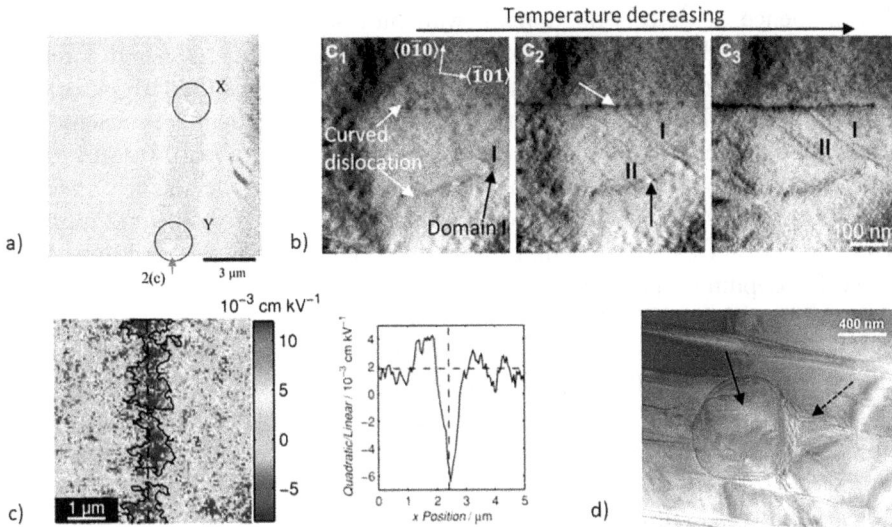

Figure 4.10. (a) Bending of a ferroelectric domain wall in LiTaO$_3$ at +2 kV/mm, which is being pinned by two point defects (X, Y). (b) Ferroelectric/ferroelastic domains (black arrow) nucleate and are being pinned at a dislocation (white arrow) in BaTiO$_3$ upon cooling the material below the Curie temperature. (c) Decreased piezoelectric non-linear response due to reduced domain wall mobility at a grain boundary (dashed vertical line) in a Pb(Zr$_{0.45}$Ti$_{0.55}$)O$_3$ thin film. (d) Interactions between ferroelectric/ferroelastic domains (dashed arrow) and second phase CaTiO$_3$ precipitates (solid arrow) in (Ba,Ca)TiO$_3$. ((a) Reproduced with permission from [35]. Copyright 1999 the American Physical Society. (b) Reproduced with permission from [66]. Copyright 2021 AAAS. (c) Reproduced with permission from [71]. Copyright 2013 John Wiley and Sons. (d) Reproduced from [72]. CC BY 4.0.)

distributions of point defects can also lead to space charge layers, which additionally influence domain wall movement [39].

The exact mechanism by which the movement of ferroic domain walls is hindered depends on the material and defect type [40]. For example, electronic charge trapping was shown to lock domain walls and suppress the switchable polarization in PZT thin films [41]. On the other hand, ionic point defects are successfully used for material engineering in hardened ferroelectrics, which are widely used for piezo-electric applications. Hereafter, intentionally introduced acceptor dopants and the charge-compensating oxygen vacancies hinder the movement of ferroelectric domain walls by one of the following mechanisms: stabilization of domains by defect complexes distributed within domains (volume effect), defects that have diffused and accumulated near domain walls (domain wall effect), or defects agglomerated in the vicinity of grain boundaries (grain boundary effect) [42–44]. This results in lower losses and enables the use of these materials at high frequencies, e.g. for ultrasound generation or voltage transformation. However, one major drawback of the use of point defects in domain wall stabilization is their inherent temperature dependence [45, 46], as their mobility is thermally activated and increases upon increasing the temperature, thus decreasing their domain wall pinning strength [43].

The influence of defects on domain wall motion has also been considered by different theoretical models. Using density-functional theory, He and Vanderbilt demonstrated that oxygen vacancies tend to migrate towards the ferroelectric 180° domain walls in $PbTiO_3$ and pin them [47]. Note that subsequent experimental work showed that curved 90° domain walls are even more absorptive for oxygen vacancies, compared to 180° domain walls [48]. Following the classical theory for polarization switching considering domain nucleation and growth, Jo *et al.* showed that the defects act as strong pinning sites for domain walls and a critical field is required for depinning [49]. On a larger scale, the relaxation models, applicable to domain wall motion at low fields, considered the damped motion of a 90° domain wall, whereby crystal defects can be introduced to the equation of motion via the periodic potential term [50, 51]. Incorporating distributed switchable dipolar defects into the Preisach approach enabled the description of pinched ferroelectric polarization loops [52] often observed in acceptor-doped ferroelectrics. The ability of point defects to reduce the driving force on the domain wall has also been demonstrated using continuum mechanics and numerical simulations [53].

Imperfections with sizes extending beyond a point are collectively referred to as extended defects. Among these, the line defects (one-dimensional) have the lowest dimensionality and are represented by various types of dislocations. These can naturally form during material processing, for example during sintering of ceramics or thin film growth [54], or can be introduced on purpose using mechanical deformation [54]. Although natural dislocation densities in oxides are relatively low ($\sim 10^8$ to 10^{10} m^{-2}), their potential for interaction with ferroic domains is large and opens up new opportunities for dislocation-based functionalities if the densities can be increased up to $\sim 10^{13}$ m^{-2}. The interactions are based on the dislocations' strain field, as well as their charged dislocation cores and the surrounding space charge layers [55], described by the Debye–Hückel cloud. The importance of electrostatic interactions is demonstrated by the observation that even the movement of purely ferroelectric (i.e. non-ferroelastic) domain walls in triglycine sulfate is influenced by dislocations [56]. Moreover, certain dislocations with Ti-rich cores were demonstrated to be magnetic, although the surrounding matrix material was non-magnetic [57]. It should be noted that a great advantage of dislocations as domain-stabilizing defects over point defects is their inherent thermal stability (up to 500°C–700°C [58]). In addition to domain wall pinning, dislocations also represent preferential sites for domain nucleation [59] and, interestingly, domains too can act as sources for dislocations [60].

Examples of dislocation–domain wall interactions have been reported for different ferroics. Early works on magnetic materials linked the observed changes in the macroscopic magnetic behavior to the dislocation-related local stress [61, 62]. Magnetic domain wall pinning was later experimentally confirmed and was shown to depend on dislocation density, arrangement, and length [63, 64]. It was demonstrated that ferroelastic/ferroelectric domains in thin films can be effectively and permanently stabilized by misfit dislocations, which are formed at the substrate interface to reduce the lattice mismatch [65]. On the other hand, imprinted dislocation networks have been used for targeted local strain engineering, which resulted in strong mechanical restoring forces and a giant dielectric and piezoelectric response in $BaTiO_3$ (figure 4.10(b)) [66]. Moreover, increased dislocation densities were found to decrease the ferroelectric domain size [67].

Phase-field models are frequently applied to study dislocation–ferroelectric domain wall interactions, evaluate the pinning strength, and determine the influence on the macroscopic properties [68, 69]. Dislocation arrays are reported to induce kinking of the domain walls, whereby higher pinning strengths were reported for dislocations with larger Burger's vectors. The pinning strength is dependent on orientation, dislocation spacing, and domain wall type. Besides, the inhomogeneous strain field of the dislocation locally increased the Curie temperature and facilitated ferroelectric domain nucleation [70].

Another type of extended defect is two-dimensional defects or interfaces, which include grain boundaries, stacking faults, and also domain walls. As ferroics are often utilized in their polycrystalline form, grain boundaries represent the dominant two-dimensional defects that interfere with domain walls. Neighboring grains are mechanically coupled via rigid grain boundaries which, in combination with random grain orientations and anisotropy, impose mechanical stress on the microstructure. Intergranular stress in ferroelectrics can reach up to 100 MPa [73] and was found to play a key role in domain switching processes [74]. The elastic strain and the corresponding changes in the ferroelectric/ferroelastic domains [75] are also one of the main causes for the scaling behavior and grain size effects in ferroics, which have been reviewed in numerous publications (see for example [76, 77]).

The interaction of domain walls with grain boundaries has also been confirmed at the microscopic level. Regions of grains in the vicinity of a grain boundary were found to have strains twice as high as the grain interior, which originates from the combination of random grain orientation, thermal expansion anisotropy, or spontaneous strain formation below transition temperatures. Ferroelastic materials can to some extent compensate for these strains by increasing the local domain wall density, which in turn reduces the domain wall mobility. For example, in a PZT bicrystal the domain wall movement was reported to be impacted up to a distance of 450 nm from the grain boundary (figure 4.10(c)) [71], while pinning and an increased local coercive field were observed in the grain boundary vicinity in polycrystalline PZT ceramics [78, 79]. However, note that this effect is strongly influenced by boundary type and orientation angle, which may enable continuity and motion of domains/domain walls across boundaries [80, 81].

The mobility of domain walls additionally depends strongly on the presence of other domain walls, which can thus also be considered as inhibiting interfaces. This interaction can have an electrostatic origin in the case of charged domain walls or a mechanical one in the case of ferroelastic walls. For non-cubic materials, the angle between two adjacent domains deviates from the ideal 90° strain free configuration [82] and the mechanical stress in the domain wall vicinity scales with the lattice distortion. At junctions these stresses can reach values of GPa [83] and the respective elastic fields can extend up to a few micrometers into the domain interior [84], considerably influencing the domain wall movement. Direct evidence has been provided by Li and Alexe, showing that the presence of 90° domain walls inhibits the movement of 180° domain walls in epitaxial PZT films [85]. It should be noted that in thin films, a strong clamping effect on domain walls is also imposed by the substrate interface [86], which could be decreased by optimizing the seed layer, removing the substrate, or patterning the ferroelectric layer [87, 88]. Domain pinning effects have also been reported at the electrode interfaces and at electrode imperfections [89, 90].

The final group of extended defects are three-dimensional objects, which include second phase inclusions, precipitates, and pores. While these were traditionally considered as culprits that should be avoided, examples of targeted implementation of these defects have emerged over the past decade. These studies have additionally improved our understanding of their interaction with ferroic domain walls. For example, charges induced in semiconducting ZnO inclusions upon electric field application stabilized the domain structure of Bi-based lead-free ferroelectric relaxors and thus increased their depolarization temperature [91]. In addition, the strain incompatibility between ZnO inclusions and the matrix established a long-range ferroelectric order and pinned the domain walls, resulting in piezoelectric hardening [92]. Further modifications of this approach include the use of metallic second phase particles for ferroelectric/ferroelastic domain stabilization [93] and precipitation of non-ferroelectric secondary phase particles within a ferroelectric matrix (figure 4.10(d)) [72]. The additional benefit of precipitates is their position within the matrix grains, increasing their interaction with domain walls compared to the second phase inclusions located at grain boundaries, and the ability to control their size and number density by adjusting the precipitation annealing conditions. Particularly strong domain wall pinning has been observed upon orienting the anisometric precipitates in specific crystallographic directions [94]. Note that a similar hardening concept is also well established in magnetic materials, where precipitates are used to pin Bloch-type domain walls [95, 96].

An undesired influence of second phase inclusions has been reported in textured piezoceramics, where plate- or needle-like template particles are used during processing to induce a crystallographic texture in the piezoceramic and thus enhance the electromechanical performance [97]. In the case when non-reactive templates are used, e.g. $BaTiO_3$ for textured Pb-based ferroelectrics [98], the templates remain within the grains after sintering and result in an increased coercive field and a decreased remanent polarization. This was related to the influence of the local stress fields on the domain structure and domain wall mobilities.

Another traditional three-dimensional microstructural elements are pores, which were found to have a manifold effect on the domain wall movement [77]. The large difference in dielectric permittivities between the pore and the ferroic matrix broadens the distribution of the local electric fields, resulting in differences in driving fields for domain wall movement. While some regions of the material exhibit local fields below the threshold for domain wall movement, others will experience field concentrations up to two-times higher than the externally applied field. Due to the relaxed mechanical constraints in the vicinity of the pores, the mechanical driving force for domain wall movement is also altered. Reduction of residual stress from 70 MPa to 40 MPa was reported for $BaTiO_3$ upon increasing the pore fraction [99].

These effects resulted in increased domain switching fractions and enhanced piezo-electric response in porous ferroelectric films [100, 101] and bulk piezoceramics with oriented anisometric porosity [102].

4.5 Electrical characterization methods for domain wall movement

The movement of domain walls can be characterized directly, by imaging the polarization of the domains using piezoresponse force microscopy for example, or indirectly, by monitoring a variation of a property that depends on the number of domain walls or their mobilities, such as the dielectric permittivity or the piezo-electric coefficient. Characterization based on dielectric or piezoelectric measure-ments methods, although they provide indirect information, may be easier to perform in some cases.

4.5.1 Small signal perturbations

This section describes the Rayleigh law and its consequence on the dielectric/piezoelectric responses.

4.5.1.1 Linear response measurement

To measure the properties of a material, a common approach consists of applying a periodic sinusoidal excitation, $F(t) = F_0 \sin(\omega t)$, where F_0 is the magnitude of the excitation and ω its angular frequency. In the case of a linear material, the response is

$$R(t) = R_0 \sin(\omega t + \phi) = m F_0 \sin(\omega t + \phi), \qquad (4.2)$$

where m is the material coefficient and ϕ is a phase shift of the response, which corresponds to the lag. Both quantities can vary with the frequency of the excitation but, for a linear response, they are constant for any magnitude of the driving field F. The material response is thus derived using the ratio between the response and the excitation.

For example, if we consider a dielectric material, the excitation is the electric field E and the response is the polarization P. For a linear material, the ratio gives the dielectric susceptibility χ and the phase shift is represented by the dielectric loss $\tan\delta$. For ferroelectric materials, the susceptibility is large, and the approximation $\varepsilon_r = 1 + \chi \approx \chi$ is commonly done. Nevertheless, in the general case, the response is not purely linear and the material coefficient m depends on the magnitude and sense of variation of F (and may be hysteretic). The $R(F)$ expression thus contained additional terms as in a series expansion.

In the materials described in this section, the response can be the polarization P or the strain/deformation x and the driving field can be the electric field E or the pressure σ. Depending on the coupled excitation/response, the material coefficient can be the dielectric permittivity ε, converse or direct piezoelectric coefficient d, or the reciprocal of the Young modulus.

4.5.1.2 Rayleigh law

Domain walls contribute to the macroscopic response of materials (polarization or deformation). Their contribution is strongly influenced by the magnitude of the excitation field. For a random distribution of the depth of pinning centers, the potential energy as function of the position is non-regular (figure 4.11(a)) [16].

Depending on the magnitude of the perturbation, there are two possibilities:
1. The stimulus is small compared to the energy required to unpin the domain wall (figure 4.11(a)): this latter can only vibrate around an equilibrium position (figure 4.11(b)). In this case, the contribution of the domain wall to the response is said to be reversible and does not depend on the magnitude of the perturbation.
2. The stimulus is large compared to the energy required to unpin the domain wall: this can move from a pinning center to another. In this case, the contribution of the domain wall is irreversible.

This irreversible contribution increases when the stimulus increases, which gives a strong sensitivity of the response to the magnitude of the perturbation, even for small signals. Therefore, by measuring the response of the material for multiple magnitudes, it is possible to obtain information regarding the domain dynamics and to link it to the energy profile. In the case of materials with domain walls, and for a homogeneous distribution of pinning centers, the response R of the material can be described using the Rayleigh law over a limited field range [17, 103, 104]:

$$R(F) = (m_{\mathrm{init}} + \alpha_m F_0)F \pm (\alpha/2) * (F_0^2 - F^2). \qquad (4.3)$$

The signs $+$ and $-$ stand for the decreasing and increasing parts of the driving field, respectively. Figure 4.12 shows the response for different magnitudes of the driving field. This relation has been experimentally discovered by Lord Rayleigh for permeability and magnetization in ferromagnetic materials [105] and theorized by Néel [106, 107]. Initially, it has been successfully applied to describe the ferroelectric

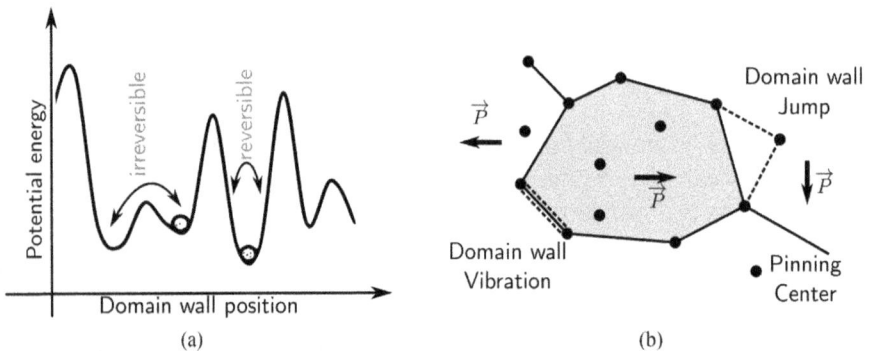

Figure 4.11. (a) Potential energy of a domain wall, as a function of its position, in a sample with randomly distributed pinning centers and (b) representation of ferroelectric domains and associated domain wall motion possibilities.

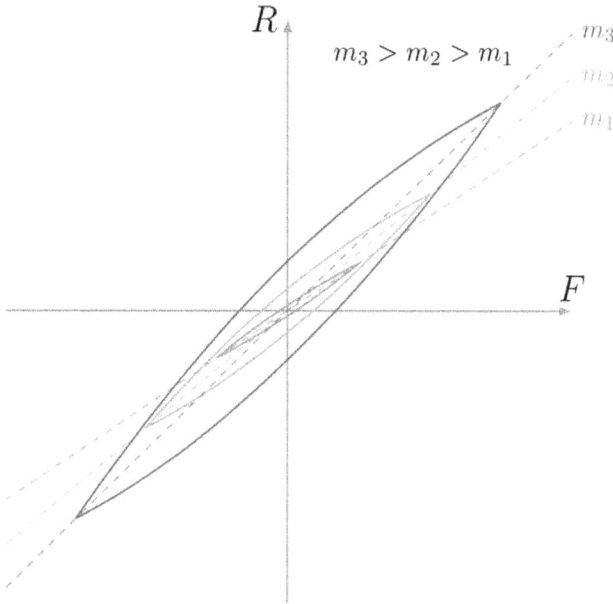

Figure 4.12. Response of a material with a contribution from domain wall motion, and a homogeneous distribution of pinning centers in the Rayleigh range, for various driving field magnitudes.

response [108], then the piezoelectric response of ferroelectric materials such as Pb(Zr,Ti)O$_3$ [104, 109] or BaTiO$_3$ [110].

The response of a material can thus be split in two terms: (i) $m_{\text{init}} + \alpha_m F_0$ corresponding to the first-harmonic response (the slope in figure 4.12) and (ii) $(\alpha/2) *$ $(F_0^2 - F^2)$ corresponding to the hysteretic contribution of the domain wall motion (the opening of the loop in figure 4.12). The following sections describe those two terms separately since in practice they are, most of the time, measured using different set-ups.

4.5.1.3 First harmonic response

The detection of the response of a material to a small signal perturbation is performed only at the fundamental frequency ω, also called the first harmonic. The experimental set-up will depend on the material coefficient measured: (i) it may be an impedance analyser for the dielectric permittivity, (ii) a laser interferometer and a voltage source for the converse piezoelectric coefficient or (iii) a charge amplifier coupled with an actuator for the direct piezoelectric coefficient. For very small magnitudes of the signal, lock-in amplifiers (e.g. SR830 Stanford Research, MFLI Zurich Instruments) are widely used since they allow the measurement of weak signals in the presence of substantial noise.

When domain walls are present, the material coefficient corresponds to the first term of equation (4.4), and this coefficient increases linearly with the magnitude of the driving field (for a purely linear material, this coefficient is constant):

$$m = m_{\text{init}} + \alpha_m F_0. \tag{4.4}$$

This value corresponds to the slope of the response in figure 4.12. This generic expression can be written explicitly for permittivity and the piezoelectric coefficient or reciprocal Young's modulus, depending on the context of measurement:

$$\text{Dielectric permittivity} \qquad \varepsilon_r = \varepsilon_{\text{init}} + \alpha_{\varepsilon/E} E_0 \qquad (4.5)$$

$$\text{Inverse piezoelectric coefficient} \qquad d_{33} = d_{\text{init}} + \alpha_{d/E} E_0 \qquad (4.6)$$

$$\text{Direct piezoelectric coefficient} \qquad d_{33} = d_{\text{init}} + \alpha_{d/\sigma} \sigma_0 \qquad (4.7)$$

$$\text{Reciprocal Young's modulus} \qquad \frac{1}{Y} = \frac{1}{Y_{\text{init}}} + \alpha_{Y/\sigma} \sigma_0. \qquad (4.8)$$

If we consider the first equation, E_0 is the magnitude of the measuring electric field, $\varepsilon_{\text{init}}$ is the zero field permittivity and corresponds to the reversible contribution to the permittivity (lattice contribution and domain wall vibrations), and $\alpha_{\varepsilon/E}$ quantifies the irreversible contribution from the motion of domain walls (domain wall pinning/unpinning), polar cluster boundaries, or phase boundaries and corresponds to the slope of the permittivity versus electric field curve. Most of the time, the subscript for α is omitted since the presented measurements deal with only one response and one driving field.

This first-harmonic response measurement has been applied to many ferroelectric and related materials, including relaxor ferroelectrics and antiferroelectrics [111–113]. The value of the Young's modulus has been studied as a function of the magnitude of the stress amplitude in only one article [114]. In this study, the authors did not find a linear relation. Nevertheless, the Young's modulus instead of its reciprocal value has been considered, which may explain the negative value of the Rayleigh coefficient α.

The linear increase of the material coefficient is valid for a limited range of driving fields. The upper limit is often approximately $E_C/2$ [115], where E_C is the coercive field. Close to this upper limit, additional contributions appear such as domain switching [116], field-induced phase transitions [112, 113], saturation of the polarization [117–119] or a limitation of the domain wall movement due to finite size of the ferroelectric medium [120]. The lower limit depends on the distribution of the depth of the pinning centers and depends on the studied material.

In a real material, the distribution of pinning centers is not homogeneous and for low driving fields the material coefficient is almost constant, corresponding to reversible domain wall contributions, also called domain wall vibration [121, 122]. A generalized expression can then be used to describe the permittivity evolution, called the hyperbolic law [112, 123, 124]:

$$\varepsilon_r = \varepsilon_{(r-l)} + \sqrt{\left(\varepsilon_{(r-\text{rev})}^2 + (\alpha_r E_{\text{AC}})^2\right)} \qquad (4.9)$$

with $\varepsilon_{(r-\text{rev})}$ the reversible domain wall motion contribution, which is proportional to the domain wall density [115, 125, 126]. $\varepsilon_{(r-l)}$, $\varepsilon_{(r-\text{rev})}$ and α_r can be obtained by measuring the relative permittivity as a function of the driving field and by fitting the data using equation (4.9) instead of a linear regression. The major difference to the

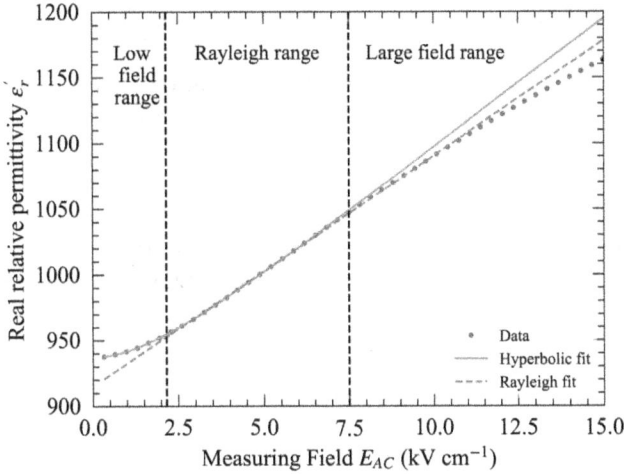

Figure 4.13. Permittivity as a function of the driving field for a BCTZ thin film, at 10 kHz and 230 K. Data from [117].

original Rayleigh law is that it avoids the arbitrary definition of a threshold field to perform the linear regression. It also proposes the decorrelation of the lattice contribution and the reversible domain contribution since they are in the same term ε_{init} in the original law. Figure 4.13 shows the relative permittivity as a function of the driving field. From 2.1 to 7.5 kV cm^{-1}, the linear increase corresponds to the Rayleigh range. Below 2.1 kV cm^{-1} the relative permittivity dependency with the electric field is low and it corresponds to the low field region where the wall motion is mainly reversible. This range is also visible for piezoelectric coefficient measurements [127]. For fields above 7.5 kV cm^{-1}, the increase is sub-linear, which means the pinning center distribution is not homogeneous at large fields or the field is sufficiently high to perturb the domain wall density.

The evolution of the Rayleigh coefficient α as a function of other parameters gives information on the mobility of domain walls. The evolution of α can be obtained as a function of the driving frequency [109, 118, 128–132], temperature [117, 122, 133], DC electric field [134, 135], pressure [110] or as a function of previous states, residual ferroelectricity [133], contributions to tunability [128, 135], the effect of dopants [115, 131, 136, 137] and annealing/cycling effects [120, 125]. These studies indicate that the contribution of domain walls is reduced when frequency increases. Dopants can facilitate domain wall motion, such as in Nb-doped PZT [138], and they can also act as a pinning centers. When the dopant acts as a pinning center for domain walls, the reciprocal of the Rayleigh coefficient increases with dopant concentration [115, 131, 136, 137].

The frequency evolution of the Rayleigh law can be described using the following expressions [139, 140]:

$$\alpha'(\omega) = \alpha' \left(\ln \left(\frac{1}{\omega \tau_{dw}} \right) \right)^{\theta} \quad \text{and} \quad \alpha''(\omega) = \alpha'' \left(\ln \left(\frac{1}{\omega \tau_{dw}} \right) \right)^{\theta-1}. \quad (4.10)$$

The parameter θ is related to the interaction between domain walls and defects/pinning centers, and τ_{dw} is the characteristic relaxation time. A very broad peak for the imaginary part indicates a very large distribution of the relaxation times, which is common for the contribution of domain walls to the permittivity [141]. The real part can be expressed as [122, 128, 142]

$$\alpha'(\omega) = \alpha_{r-0} - \alpha_{r-\omega} \ln(\omega). \tag{4.11}$$

α_{r-0} is the value at $\omega = 1$ and $\alpha_{r-\omega}$ represents the frequency decay.

According to the Kramers–Kronig relations, the imaginary part can be expressed as [143]

$$\alpha''(\omega) = -\frac{\pi}{2} \frac{\partial \alpha'(\omega)}{\partial \ln \omega} = \frac{\pi}{2} \alpha_{r-\omega}, \tag{4.12}$$

giving a constant imaginary part of the irreversible contribution.

In some cases, the imaginary part is not constant but decreases slightly with frequency. In this case, a power law can be used to describe the frequency decay [131, 132, 141]:

$$\alpha_r(\omega) = \alpha_{r-\omega} \omega^s. \tag{4.13}$$

$\alpha_{r-\omega}$ corresponds to the value at $\omega = 1 \text{ rad s}^{-1}$ and s is conditioning the decay rate.

A maximum in α at a defined temperature has been observed in many materials: $NaNbO_3$ [144], $Pb_{0.92}La_{0.08}Zr_{0.52}Ti_{0.48}O_3$ [145], Nb-doped $PbZrO_3$ [146] or $PbMn_{1/3}Nb_{2/3}O_3$ [147].

4.5.1.4 Hysteretic response description

In this section, the hysteretic contribution term is described. By using the same sinusoidal stimulus mentioned previously, the response (4.14) can be decomposed as a Fourier series expansion [17, 148–150]:

$$\begin{aligned} R(t, F_0) = (m_{\text{init}} + \alpha_m F_0)F_0 \sin(\omega t) &- \frac{4\alpha_m F_0^2}{3\pi} \cos(\omega t) \\ &- \frac{4\alpha_m F_0^2}{3\pi}\left(\frac{1}{5}\cos(3\omega t) - \frac{1}{35}\cos(5\omega t) + \dots \right). \end{aligned} \tag{4.14}$$

In the first line of the equation, an additional term at the fundamental frequency, but in quadratic form is visible. This indicates that the contribution of the domain wall motion to losses is visible also in the first-harmonic response.

In addition to the first term, which is at the first harmonic, the response contains additional terms which are only odd harmonics and out of phase. The irreversible domain wall motion contribution is thus out-of-phase with the measuring electric field in the case of an ideal material. To describe a real material, equation (4.14) can contain additional terms, reflecting the degree of randomness of the energy profile, [151] and in that case, harmonics may not be purely out-of-phase.

Even if the measurement of only the first-harmonic term should be enough to study the motion of domain walls in ferroelectric materials (by extracting the

Rayleigh coefficient α), the hysteretic or non-hysteretic character of the motion of domain wall can also be studied by measuring higher order harmonics. For this reason, the non-linear response of a relaxor or a ferroelectric material can be investigated by extracting the phase angle of the higher order harmonics and their evolution with the measuring field amplitude.

The measurement is performed by applying a sinusoidal excitation and collecting the response with a lock-in amplifier [149, 151–153] (which can analyze any multiple of the fundamental frequency), or by analyzing the time domain response [120, 154, 155]. The measured response is

$$R(t) = R^{(0)} + \sum_{n=1}^{\infty} R^{(n)} \sin{(n\omega t + \delta_n)}, \tag{4.15}$$

where $R^{(0)}$ is a DC term, $R^{(n)}$ is the magnitude of the n-th harmonic and δ_n is its phase angle. Even harmonics are in principle forbidden in a centrosymmetric material [148]. Nevertheless, they have been measured in many materials, even in the paraelectric and relaxor phases [151, 156, 157]. Due to its higher magnitude compared to the other harmonics, the third-order harmonics is the most studied. For the third-order harmonics, δ_3 is close to $-90°$ for an ideal response. In real materials, a deviation from this ideal value of $-90°$ can be observed (as described later on). Odd harmonics with a higher order (fifth to ninth) are sometimes studied but for large electric fields only, close to or above the coercive field [153, 155]. Figure 4.14 shows the hysteresis deformation due to the third-harmonic contribution in four cases:

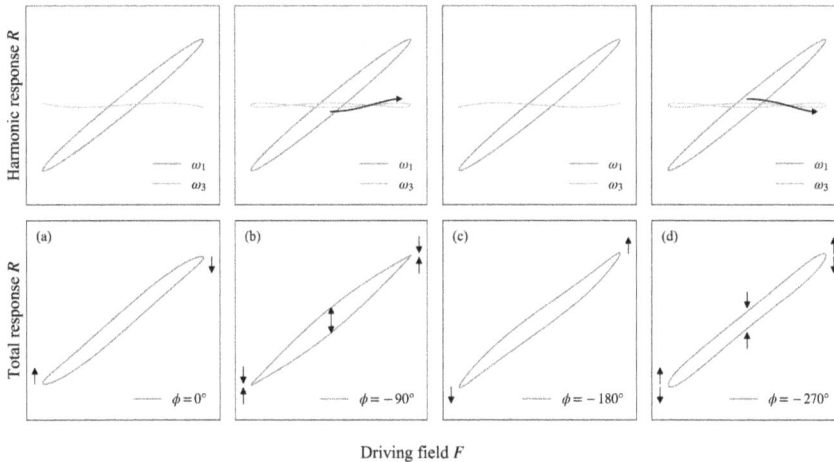

Figure 4.14. Hysteresis deformation due to the third-harmonic contribution shown for four cases: (a) $\delta_3 = 0°$, $\delta_3 = -90°$ (b), $\delta_3 = -180°$ (c) and $\delta_3 = -270°$ (d). The upper line shows the individual contribution of the first and the third harmonics and the lower line shows the total hysteresis loop. The black arrows in the upper line show the rotational sense of the third harmonic and the black arrows in the lower line show the deformation of the hysteresis. (Adapted with permission from [148]. Copyright 2021 AIP Publishing.)

- $\delta_3 = 0°$, the third harmonic is minimum when the field is maximum, which decreases the total response. This corresponds to a saturation-like response (figure 4.14(a)).
- $\delta_3 = -90°$, the third harmonic opens the center of the loop. This corresponds to the ideal Rayleigh behavior (figure 4.11(b)).
- $\delta_3 = -180°$, the third harmonic is maximum when the field is maximum, which increases the total response. This corresponds to a divergent-like response (figure 4.11(c)).
- $\delta_3 = -270°$, the this harmonic constricts the center of the loop. This corresponds to the pinching of the loop (figure 4.11(d)).

4.5.1.5 First-harmonic and third-harmonic responses

The characterization of the domain wall motion using only the first harmonic is commonly used and gives important information. Nevertheless, higher harmonic measurements are useful to understand the hysteretic behavior and classify the non-linear responses, in particular when working with relaxor ferroelectrics. In this section, typical first-harmonic and third-harmonic responses are given.

Figure 4.15 shows the typical relative permittivity and third-harmonic phase angle as a function of the AC measuring field normalized according to E_C for undoped, soft and hard PZT. For undoped PZT, in the range 0.1 to 0.3 the phase angle is close to $-90°$ and the permittivity increases linearly, corresponding to the ideal Rayleigh behavior. For low fields, as for other materials, a phase angle of the third harmonic close to $-180°$ is observed, which is the case for many ferroelectric materials, including relaxors [117, 149, 151]. A phase angle of $-180°$ indicates mainly hysteretic contributions and in that range ($\frac{E_{AC}}{E_C} \leqslant 0.05$), the permittivity is almost constant, indicating mainly reversible contributions (domain wall vibration). For hard PZT, the permittivity increase is very small and the phase angle stays around $-180°$, corresponding to weak and hysteretic irreversible domain wall contribution. A similar response has also been found for weak ferroelectricity in $PbZrO_3$ [120]. For soft PZT, the increase in permittivity is very high, indicating a large irreversible domain wall contribution. In addition, the constant range expected

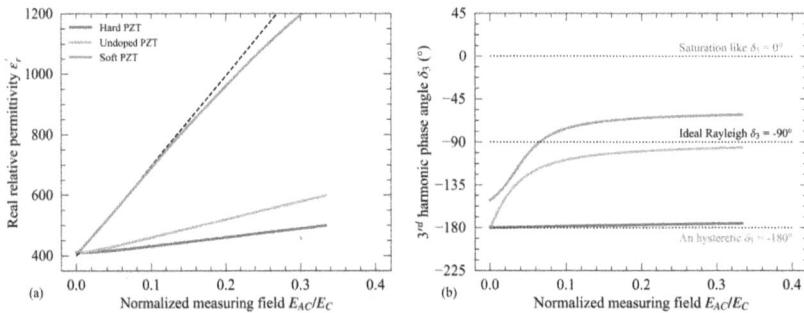

Figure 4.15. Typical (a) relative permittivity and (b) third-harmonic phase angle as a function of the AC measuring field for undoped, soft and hard PZT. (Adapted from [149]. CC BY 4.0.)

at low fields is almost not visible, indicating that the threshold field is small, and thus the mobility of the domain walls is high [128]. For normalized fields in the range from 0.02 to 0.1, the increase of the relative permittivity is linear and the phase angle is around $-90°$. When the field increases, the relative permittivity deviates from the linear increase predicted by the Rayleigh law and exhibits instead a sub-linear increase. The phase angle in this range is around $-60°$, which corresponds to an intermediate regime between irreversible domain wall contribution and saturation, this latter corresponding to $\delta = 0°$. The origin of this saturation can be complex to determine, however, it indicates that the depth of the pinning center is not homogeneous in the measured field range [149] or that a change of the domain wall density took place.

For relaxors, the phase-angle response exhibits an 'S' shape similar to the one observed for soft ferroelectrics. The major difference is that for low fields the value is $-180°$ and for high field it is $0°$ [117, 152]. This response depends on the temperature for relaxors—at low temperatures they behave more like conventional ferroelectrics and thus exhibit a permittivity and phase-angle response like ferroelectrics [117, 158, 159].

4.5.2 Large signal perturbations

(Parts of this section have been reproduced with permission from [167]. Copyright 2023 Elsevier.)

A classic way to characterize the motion of domain walls in ferroelectrics is to perform the measurement of the polarization versus electric field loop, also called the hysteresis loop (for the dielectric response) and the measurement of deformation versus electric field loop, also called the butterfly loop (for the piezoelectric response). In this case, the goal is to apply a field that is sufficiently high—three times the coercive field is considered enough—to switch the domain. It is also possible to measure the deformation versus pressure $S(\sigma)$ loop [15], but it is less common.

This section presents the measurement techniques that are commonly used for large signal perturbation measurements of ferroelectric materials.

4.5.2.1 Hysteresis loop measurement

For the polarization, most methods are based on the measurement of a current flowing through a capacitor containing the ferroelectric material. Then polarization is obtained by numerical or analytical integration of the current over time. This method assumes that the leakage current and the dielectric losses are low. If this is not the case, wrong interpretations of the results occur [160, 161].

The first circuit used for measurment is called the modified Sawyer–Tower bridge [162], presented in figure 4.16(a)[1]. When the ferroelectric capacitor is loss-less, the charge stored on its negative electrode is the same as the charge stored on the positive electrode of the reference capacitor $Q_{ref} = C_{ref} V$. The polarization of the ferroelectric can thus be obtained:

---

[1] The original schematics are slightly different since the results directly on a cathode-ray tube are illustrated.

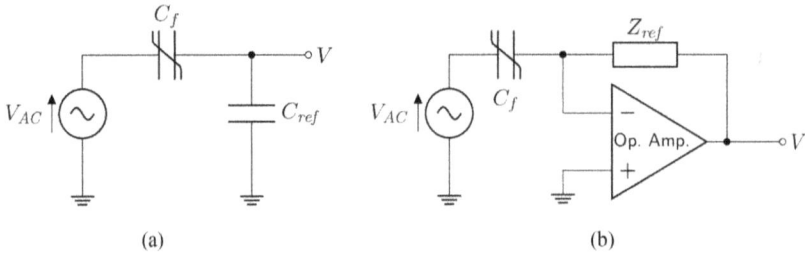

Figure 4.16. Electrical schematics for the polarization measurement: (a) modified Sawyer–Tower circuit and (b) virtual ground circuit.

$$P = \frac{C_{\text{ref}} V}{A}, \qquad (4.16)$$

where A is the area of the ferroelectric capacitor.

The applied voltage, to determine the electric field using the thickness of the material, can be computed using $V_f = V_{\text{AC}} - V \simeq V_{\text{AC}}$ if $C_{\text{ref}} \gg C_f$.

The second possibility is to use a virtual ground circuit, whose principle is shown schematically in figure 4.16(b). The operational amplifier forces the second electrode of the capacitor containing the ferroelectric material to a null potential. The two main advantages are that the voltage applied on the sample is directly V_{AC}, without any requirement on the reference impedance Z_{ref}, and the parasitic capacitance of the connection cables (between the electrode and the measurement circuit) has almost no influence. Depending of the type of reference impedance, an image of the current (for a resistor) or of the charge (for a capacitor) is available at the output of the circuit. This circuit is used in many commercial ferroelectric measurement systems.

The main drawback of those two methods is that they assume that the dielectric losses and the conductivity of the sample are negligible. When these contributions to the current are high, misinterpretation of the results can occur [161]. Other measurement techniques have been proposed in order to reduce the influence of this leakage contribution.

4.5.2.2 Positive up negative down
The switching in ferroelectric materials using pulses was initially studied by Merz in the fifties, in particular to study the stability after partial switching [30]. The method described in this section corresponds to an extension of this pulse method, by using multiple consecutive pulses with specific polarity. During the application of the electric field, the resulting current (most of the time measured using the virtual ground method) is the sum of multiple contributions:

$$I(t) = I_S(t) + I_C(t) + I_L(t), \qquad (4.17)$$

where I_C is the capacitive current, corresponding to the charge stored in the ferroelectric capacitor and the parasitic capacitance, I_L is the leakage current, corresponding to the dielectric losses and I_S to the domain switching, and thus to the movement of domain walls.

Capacitive and leakage contributions are always present, irrespective of the state of the material[2]. The switching contribution depends on the previous state and thus on the polarity of the previous pulses. This method is called PUND for 'positive up negative down' [125, 164–167], corresponding to the polarity of the applied pulses.

An initial pulse (here of negative polarity, rectangular, trapezoidal or triangular) is applied in order to switch the net ferroelectric polarization in a given direction (figure 4.17(a)). Then, two pulses of opposite polarity (P and U, here positive) are successively applied and finally, the application of the pulses N and D, again of opposite polarity (here negative) allow observation of the full hysteresis cycle. The current signal of the respective pulses is recorded.

In the case of the pulses P and N (the so-called switching pulses), the measured current is the sum of the contributions of the (i) leakage current, (ii) the capacitive charging and (iii) the possible ferroelectric phase switching if the previous pulse was of opposite polarity. In the case of a purely capacitive behavior, a triangular applied voltage should result in a constant current: positive if the voltage is increasing and

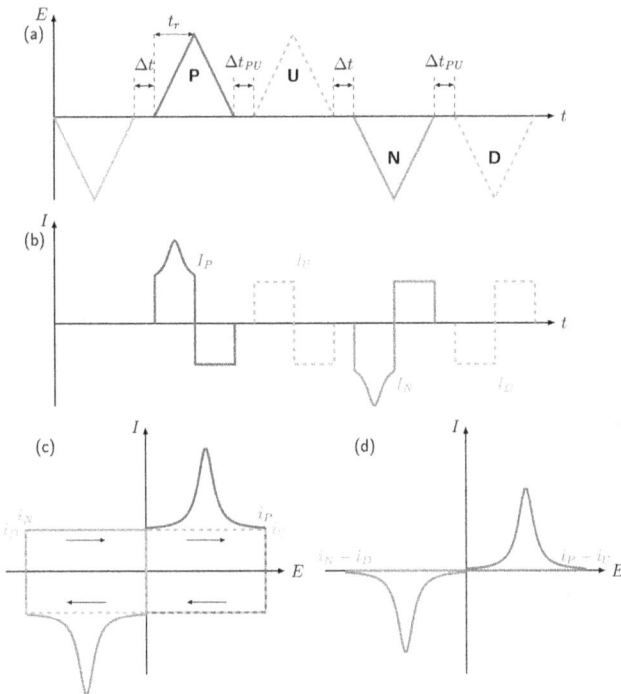

Figure 4.17. (a) Applied PUND signal and (b) measured current as functions of time. (c) Global $I(E)$ cycle and (d) $I(E)$ cycle of only the ferroelectric switching contribution. t_r is the raise time of the pulse, Δt the delay between the orientation pulse and the PUND measurement, and Δt_{PU} the delay between the pulses P and U as well as N and D. The arrows indicate the direction of the cycle.

---

[2] Except in the case of a very thin layer, to form ferroelectric tunnel junctions, in which the direct current can be modulated by the polarization direction [163].

negative if the voltage is decreasing. A deviation from this behavior indicates a polarization change at this particular field, which may include ferroelectric switching. The example of the current signals obtained is shown in figure 4.17(b), which can be used for representing the obtained global $I(E)$ cycle (figure 4.17(c)). For the pulses U and D (non-switching pulses), no ferroelectric phase switching should occur as domains have already switched with the previous pulses of the same polarity, and hence the current is the sum of only the three other contributions (assuming of course, that the written polarization is stable over the time scale of the measurement). By subtracting the non-switching currents from the switching currents ($i_P - i_U$ and $i_N - i_D$), it is possible to extract the pure switching contribution for both polarities, to compute the respective $I(E)$ loop (figure 4.17(d)).

In order to obtain a sufficient signal-to-noise ratio, the switching contribution needs to be high compared to the capacitive and the leakage contributions. Otherwise, current subtraction prior to the amplification and acquisition can be used [168]. This method has been widely used for ferroelectric materials in order to remove excessive capacitive and leakage contributions and to study the switching times of the different contributions to the polarization [125, 164–166, 169].

In order to study the kinetics of domain switching and ascertain the limiting factor (domain expansion or nucleation), pulses with different amplitude can be used [21–24].

4.5.2.3 The Preisach plane and first order reversal curves

In a real material, ceramic or thin film, the apparent coercive field is not the same from one volume to another, due to different size, orientation, local defect concentration or strain in the structure. To access this distribution, it is possible to use first order reversal curves (FORCs).

The Preisach model

The Preisach model describes a hysteresis cycle as a sum of elementary cycles, called hysterons. The elementary cycle can be seen as a two state commutator ($+1$ and -1), and the switching from one state to another occurs for two values of the input α and β (figure 4.18(a)). α corresponds to the input going from the state -1 to $+1$, and β to the opposite. In the materials of interest in this book, $\alpha \geqslant \beta$.

By summing a large number of elementary cycles, with different values of α and β and different weights, it is possible to represent the complete cycle of a material [17]. To represent this distribution of the hysterons the Preisach plane is widely used (figure 4.18). The horizontal coordinate indicates the α value (the input required to switch from -1 to $+1$) and the vertical coordinate gives the β coordinate. Each point of the plane corresponds to a hysteron density having the switching couple (α; β). Since $\alpha \geqslant \beta$, only the lower right triangle is occupied. In this plane, four zones can be defined, corresponding to particular cases of α and β (indicated by different colors in figure 4.18).

The Preisach model was initially developed for the description of the ferromagnetic hysteresis loop [170, 171] was later extended to ferroelectric materials [119, 172–174]. The knowledge of the hysteron decomposition of the $P(E)$ cycle gives information on the fatigue and polarization of the material [119, 136, 175, 176]. It is

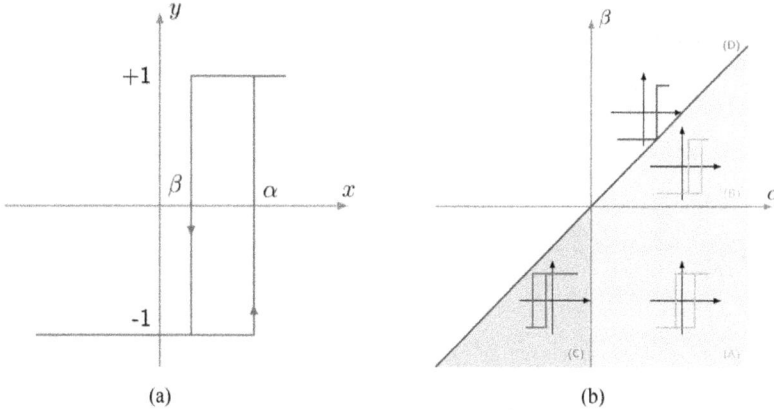

(a)   (b)

Figure 4.18. (a) Representation of an elementary cycle, also called a hysteron. (b) Preisach distribution. (A) (the square on the bottom right) corresponds to the threshold with opposite signs. (B) (the triangle on the upper right) corresponds to the case in which both threshold fields have positive values and (C) (the triangle in the lower left) corresponds to the case in which both threshold fields have negative values. (D) (axis $\beta = \alpha$) corresponds to reversible hysterons.

also helpful to identify different phases of a material [177, 178] or a different behavior of a portion of the material [112, 179].

Measurement of the Preisach density

The Preisach density can be obtained using the FORC method, which is equivalent under some conditions [180]. The measurement method is presented in figure 4.19(a). In the case of a ferroelectric material, it consists of applying a large electric field to saturate the material, i.e. switch all the domains to a given orientation (figure 4.19(a)). In the case of an initial field E_{max}, the field is decreased to E_r, before it increases again to E_{max}. The polarization variation is measured (by current integration) and this curve is noted $P^-(E_r, E)$. The field is then decreased to another value E_r, and the procedure is repeated until negative saturation is reached.[3] Once the measurement is done, a set of $P(E)$ loops is obtained (figure 4.19(b)).

The FORC distribution is calculated using the mixed second derivative:

$$\rho^-(E_r, E) = \frac{1}{2} \frac{\partial^2 P^-(E_r, E)}{\partial E_r \partial E},$$ (4.18)

which also corresponds to the variation of the differential susceptibility when E_r changes:

$$\rho^-(E_r, E) = \frac{1}{2} \frac{\partial \chi^-(E_r, E)}{\partial E_r}.$$ (4.19)

A large variation of the susceptibility when the initial field goes from E_r to $E_r + \Delta E_r$ in a given range of field $[E, E + \Delta E]$ indicates a large number of cells which have

---

[3] Fields E_r and E are sometimes denoted as β and α, respectively, to correspond to the mathematical model [119, 181, 182].

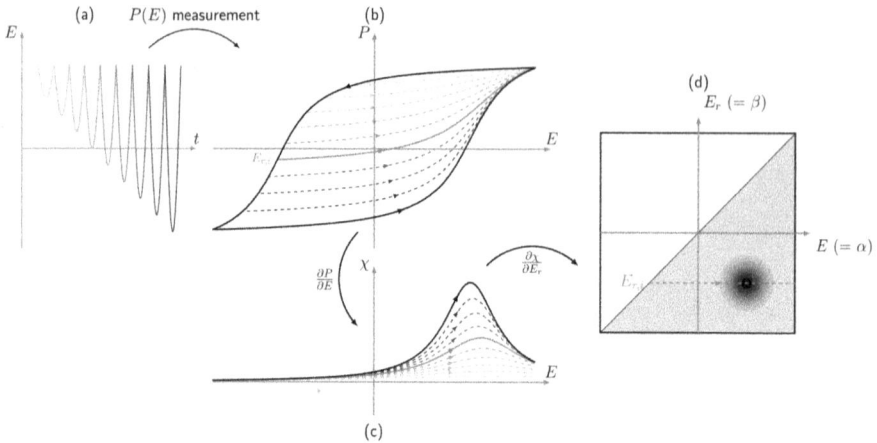

Figure 4.19. Principle of the FORC measurement. (a) Applied electric field. (b) Polarization as a function of the applied field. (c) Successive derivatives and (d) FORC distribution. (Adapted with permission from [183]. Copyright 2015 American Chemical Society.)

back-switching fields between E_r and $E_r + \Delta E_r$ and switching fields in the interval $[E, E + \Delta E]$ (figure 4.19(c)). A large value of the FORC distribution is thus obtained in this area. A similar procedure is used in order to fill the right triangle of the FORC diagram using the decreasing part of the $P(E)$ loops. In theory, both distributions are identical and give the Preisach distribution.

In the given example, the maximum of density is on the axis $E_r = E$, indicating that the cycle is symmetric: positive and negative coercive fields have the same absolute value but opposite signs. The difficulty when measuring a FORC distribution is twofold: it requires a large number of cycles and the mean value of the signal is not null (which can allow charge movement). This kind of measurement in nonetheless a useful tool to study phenomena in ferroelectrics, such as wake-up in HfO_2 and its derivatives [183], coercive field control [179], modeling the response [119] and identifying different contributions [112].

Data handling

The FORC distribution extraction is performed by calculating the mixed order derivative of the polarization. Even if in the illustration the differentiations are made successively, a more robust method is to fit $P(E, E_r)$ with a two-variable surface of the second order:

$$P(E_r, E) = a_{00} + a_{10}E + a_{20}E^2 + a_{01}E_r + a_{02}E_r^2 + a_{11}EE_r. \quad (4.20)$$

After the fitting of the polarization, the mixed second order derivative is simply a_{11}. The advantage of this method is its robustness, since the number of points used to make the fitting can be adjusted [170, 171, 183] and different weighting of the points used in the fitting can also be set. This procedure is repeated for all points of the FORC distribution.

4.6 X-ray diffraction methods

Diffraction methods are well-established non-destructive characterization techniques, which have become widespread and accessible over the past few decades. In addition to their use in crystal structure determination, phase identification, texture analysis, and strain/microstrain analysis, they also enable the examination of static domain configurations and domain dynamics. Although in principle such measurements can be conducted using electron, neutron, or x-ray diffraction, the latter has been most widely used for technical reasons and availability. This section is therefore focused on methods for studying the response of ferroic domains to various external stimuli using x-ray diffraction, while electron diffraction is described later in this chapter. In terms of domain analysis, these techniques can be divided into approaches that average the domain behavior across a large sample volume, and imaging methods that provide a detailed local view of a specific region/volume of the sample.

Pioneering work on methods that average the domain behavior over a given sample volume has been done by Subbarao et al. [184], who related the differences in relative intensities of x-ray reflections to the changes in polarization direction. For example, polarization switching in tetragonal $BaTiO_3$ changes the intensities of the (200) and (002) peaks, which is directly related to the different lengths of the unit cell's a and the polar c axes. This approach was later developed into a robust method for the determination of domain switching fractions and lattice strains (based on peak position), i.e. extrinsic and intrinsic contributions to electromechanical response in ferroic materials, even in situ under applied electric fields or mechanical stress [185–188]. Note that due to the low penetration depth of typical laboratory x-rays into ferroelectric ceramics (up to a few micrometers below the surface, which is comparable to typical grain sizes), many researchers have instead employed high-energy synchrotron radiation that penetrates millimeter-sized samples using a transmission diffraction geometry.

Li et al. [189] have used the above methodology to determine that polarization switching in polycrystalline PZT predominantly takes place through two successive 90° domain walls processes. This model was further refined by Schultheiß et al. using time-resolved in situ high-energy diffraction and data collection with a 4 ms time resolution, separating the switching behavior into three regimes [74]. In addition to new mechanistic insights into switching, this measurement principle has also been extensively applied for quantifying the various contributions to a material's macroscopic electromechanical response [190]. Using stroboscopic data collection, this method can be utilized to determine these contributions at high frequencies, e.g. at piezoelectric resonance [191]. Note that valuable information on domain switching can also be obtained from neutron diffraction [192].

The other group of diffraction methods provides detailed information on a local scale. One example is Bragg coherent x-ray diffractive ptychography, which is a reciprocal-space based imaging technique using coherent x-ray nanodiffraction and enables mapping of ferroic domain structures in thin films [193]. An example of its use is the imaging of local polarization in $PbTiO_3$ thin films with a spatial resolution of 5.7 nm [194].

Another powerful tool for analysing polycrystalline materials is the 3D maps obtained by focused x-ray beams [195]. Majkut *et al.* have used 3D x-ray diffraction to reconstruct the microstructure and track the domain switching strains in individual grains of $BaTiO_3$ [196] and NBT–KBT [197] under *in situ* electric field loading. In analogy to dark-field electron microscopy, Simons *et al.* developed dark-field x-ray microscopy, that enables fast imaging of deeply embedded grains or other features within the samples with a spatial resolution of 100 nm [198]. This method provided a detailed view of domain walls in their natural environment, revealing the complex strain fields in their surrounding [84]. The approach was later adopted for *in situ* measurements and used for domain analysis in ferroelectrics [199] and antiferroelectrics [200].

In situ synchrotron x-ray total scattering with applied electric field has been utilized to detect local structural changes at different directions relative to the applied electric field. High scattering angles are used to extract a normalized structure factor which is then converted through a sine Fourier transform to a pair distribution function (PDF). This allows one to determine field-induced lattice changes at the atomic scale by detecting interatomic distances in polycrystalline ceramics. For example, in relaxor-like ferroelectric NBT it was found that with applied electric field a monoclinic to rhombohedral phase transition is accompanied by switching of nanoscale domains [201]. In $BaTiO_3$, however, piezoelectric lattice strain is dominant. Increasing structural ordering with increasing field is observed in the case of morphotropic phase boundary NBT–KBT compositions through decreased diffuse scattering and ordering of Bi–Ti distances [202]. Local redistribution of Bi–Ti distances and off-centered shifts of Bi indicate a piezoelectric lattice strain. In the case of PZT and PMN-PT, the Pb-B (B = Zr/Ti) distances change significantly with field, proving a polarization rotation mechanism [203]. PDF studies on the BZT–BCT system showed non-linear lattice strains with applied electric field, with lattice distortion minimized in compositions with the largest piezoelectric effect [204]. *In situ* PDF measurements with applied pressure were also performed on $BaTiO_3$ and it was shown that while on average a phase transition from tetragonal to cubic happens with applied pressure, the lattice remains tetragonally distorted on the local scale [205].

4.7 Microscopy methods

Various microscopy techniques have been used to monitor microstructural changes in ferroelastic (and ferroelectric) materials, specifically the movement of domain walls, strained regions and accumulated charges, in response to applied external fields. Optical microscopy appears to be most reliable for *in situ* studies of domain wall movement as the photons do not modify the materials' properties [206, 207]. However, optical methods are strongly limited in their spatial resolution and are often not sufficient to detect domains below the micrometer range. Piezoresponse force microscopy techniques are beneficial as they detect surface charges and strains, determine the orientation of dipoles, and directly measure the domain switching dynamics [208]. However, reading and writing is done through the same source (the tip) which can interfere with the material and change its initial state. Next are electron microscopy techniques which offer significantly higher resolution, even down to the sub-atomic scale, allowing the identification of individual dipoles and the ability to monitor their switching, thus, looking into the domain wall. Most experiments are focused on applying voltage to the samples *in situ*, while only a few studies entailed applying a compressive stress and following domain wall movement [209]. The biggest drawback with electron microscopy methods when voltage is brought to the sample *in situ* is the possible interference with the 5–300 kV electron beam that hits the sample, with electrons backscattered from the surface (detected by scanning electron microscope) or chanelling through the sample (detected by the transmission electron microscope). Nevertheless, continuous progress in the development of all microscopy techniques in the last two decades has significantly contributed to understanding of ferroic materials' responses to external fields at the micro-to-nano-scales.

4.7.1 Transmission electron microscopy

Employing transmission electron microscopy, combined with electron diffraction measurements, for investigating the ferroelastic/ferroelectric domain structures started in the 1980s with studies of domain walls in classical ferroelectrics such as $BaTiO_3$ and PZT [210–213], and local structural modulations and perturbations from oxygen octahedral tilting [214] and cationic disorders [215], respectively. Previously unseen nanoscale polar regions and short-range polar ordering could be identified by diffuse scattering in the electron diffraction and contrast-dominated platelets and nanofeatures in dark-field images [216–218]. At the same time the dependence of domain structure on grain size and film thickness was studied [219, 220], and the importance of domain wall mobility in response to external fields was put to the forefront. The latter strongly depends not just on the type of defects in materials, their mobility and concentration, but also on the domain wall thickness and structure [221], as well as electromechanical cycling history [83]. With atomic resolution imaging, better understanding of domain walls was established [222]; at the atomic level, the thickness of ferroelectric DWs was shown to be one-to-many unit cells thick [223, 224], extended in a step-like manner or an almost random wave through the lattice. Charged and strained domain walls were studied using TEM [225, 226], and linked to their mobility and macroscopic electromechanical response to external fields. Along with the discoveries of the domain structure, *in situ* TEM experiments became increasingly important, with the aim to visualize structural changes in the materials when an external field is applied, to provide direct proof of the mechanisms behind the measured responses. However, for ferroic materials, the majority of *in situ* TEM experiments were performed with the electric field applied to the thin TEM samples. Dedicated TEM holders for mechanical tests are mainly designed for tensile stress analysis (not suitable for oxide materials), while compression tests could be envisaged using diamond nanoindenter tips pressing into the wedge TEM lamella [227], while causing dislocation formation and plastic deformation of the material rather than inducing strain redistribution through domain wall movement.

The first *in situ* biasing TEM studies were performed by Snoeck *et al.* using a modified *in situ* heating TEM holder with Cu-electrodes put on both sides of the $BaTiO_3$ TEM lamella and wired to an external voltage source [228]. With electric field applied, wedge-domain tip motion occurred first, followed by a lateral shift of the 90° DWs. For the 180° ferroelectric domain walls, it was postulated that they have a much smaller threshold field for switching due to no elastic energy and could already be switched by the TEM electron beam.

Tan *et al.* also used modified heating TEM holders, by putting electrodes onto the top surface of the disc-shaped TEM lamellas (see figure 4.20), to ensure electric field in the direction perpendicular to the electron beam (and viewing direction) [229]. While the applied electric field on the sample was calculated from the distance between the electrodes and corresponded to large-signal voltage, it was observed that the electric field concentrated around the central perforation, causing cracking along the domain walls and grain boundaries even at fields far below the dielectric breakdown fields of bulk ferroelectrics. Another problem associated with applying voltage in the TEM is that the electron beam can deflect due to its interference with externally applied voltage, and can also cause bending of the sample. Nevertheless, it was observed in the case of antiferroelectric $Pb_{0.99}Nb_{0.02}[(Zr_{1-x}Sn_x)_{1-y}Ti_y]_{0.98}O_3$

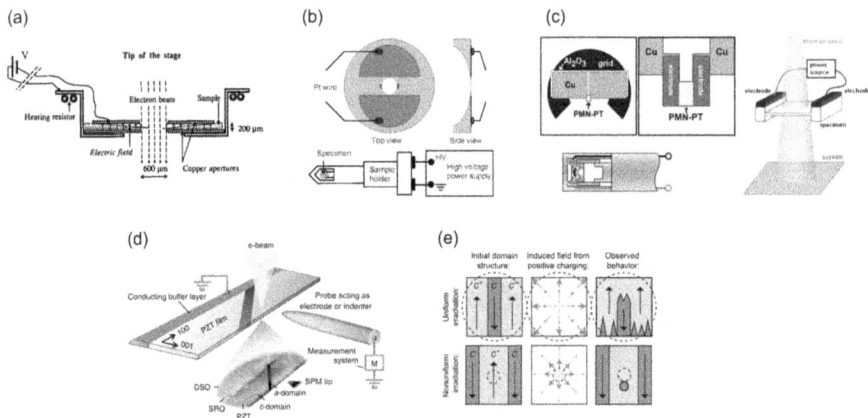

Figure 4.20. TEM sample and holder set-ups for *in situ* biasing experiments. (a) Perforated disc-shaped TEM sample with top and bottom copper apertures as electrodes, wired to an exterior voltage supply. (b) Perforated disc-shaped TEM sample with gold electrodes sputtered onto the bottom side of the sample for E applied perpendicular to the viewing direction. (c) Thinned TEM lamella, glued using tungsten onto copper plates and grid, serving as electrodes. The orientation to the electron beam in the TEM is presented. (d) Set-up for biasing and mechanical excitation with an STM probe as a part of the TEM holder. (e) Electrical excitation of the TEM sample by source electron beam irradiation. ((a) Reproduced with permission from [228]. Copyright 2006 Taylor and Francis. (b) Reproduced with permission from [229]. Copyright 2001 Elsevier. (c) Reproduced with permission from [236]. Copyright 2011 by the American Physical Society. (d) Reproduced with permission from [65]. Copyright 2014 Springer Nature. (e) Reproduced with permission from [237]. Copyright 2016 the American Physical Society.)

polycrystalline ceramics, from following the electron diffraction patterns, that the incommensurate modulation had disappeared, rendering an irreversible transformation to a ferroelectric state [230, 231]. Superlattice reflections corresponding to the anti-phase oxygen octahedral tilting, however, remained, rendering the octahedra less sensitive to external disturbance. In another experiment, switching of 90° ferroelastic domains into preferred 90° orientations was observed, maintaining the head-to-tail domain configuration [232]. Guo *et al.* [233] studied the microstructural origins of electric fatigue upon cycling (10^3 cycles) of a PMN-30PT relaxor ferroelectric. They observed irreversible changes in domain configuration and concluded that progressive degradation of polarization, dielectric constant and piezoelectric coefficient is caused by frozen-in domain states upon electrical cycling. When alternating electric field was applied to a lead-free ergodic relaxor ferroelectric material (NBT-6BT-3KNN) *in situ* in the TEM, lamellar domains emerged, indicating a reversible *E*-induced phase transition from a nonpolar to a polar phase, contributing to an abrupt increase in polarization and strain [234]. In the study of switching behavior of $Ba(Zr_{0.2}Ti_{0.8})O_3$–$0.3(Ba_{0.7}Ca_{0.3})TiO_3$ piezoelectric material, a field-induced transformation from a multi- to a single-domain state was observed *in situ*, pointing to the source of high extrinsic contribution to the piezoelectric properties of this material [235].

A somewhat different sample set-up was used by Sato *et al.* to perform *in situ* TEM experiments, i.e. a thin lamella prepared by focused ion beam (FIB) was fixed onto a Cu-grid by depositing parallel tungsten lines, serving as electrodes [236, 238]. A dedicated TEM holder was used. DC bias was applied to PMN-PT single-crystal samples and changes in domain structure were recorded at different magnitudes of the field, times and after biasing (figure 4.21). Irreversible changes to low-angle nanodomains, characteristic for lead-based relaxor ferroelectrics, were observed upon the first exposure to electric field, corresponding to a poling process. The initial change was confirmed by *ex situ* poling of TEM samples of PMN-PT, showing switching and redistribution of nanodomain variants, changing the microscale domain configuration [239]. With electric field applied to the poled sample, redistribution of nanodomain variants and movement of microscale wedge domains was observed, and the changes were largely reversible [236, 238]. These experiments elucidate the importance of hierarchical domain structures with high-density low-angle nanodomains, their switching and their mobility (influenced by polar nano-clusters) [240], crucially affecting the large dielectric and piezoelectric responses in relaxor ferroelectrics [152].

Domain switching and domain wall pinning at defects were also studied using a modified *in situ* TEM technique with a scanning tunneling microscope (STM) probe integrated into the TEM holder [65, 241–244]. Such experiments were done on thin films with deposited bottom electrodes, thinned to obtain a TEM lamella. DC electric field from an external source was applied to the sample via the STM tungsten tip *in situ* when in contact with the tip of the lamella. For examining ferroelastic domain switching under compressive stresses, a tungsten tip was used to apply in-plane pressure to the sample, while a nanoindentation system with a diamond punch was incorporated into the TEM holder for the evaluation of mechanical properties. The dynamics of

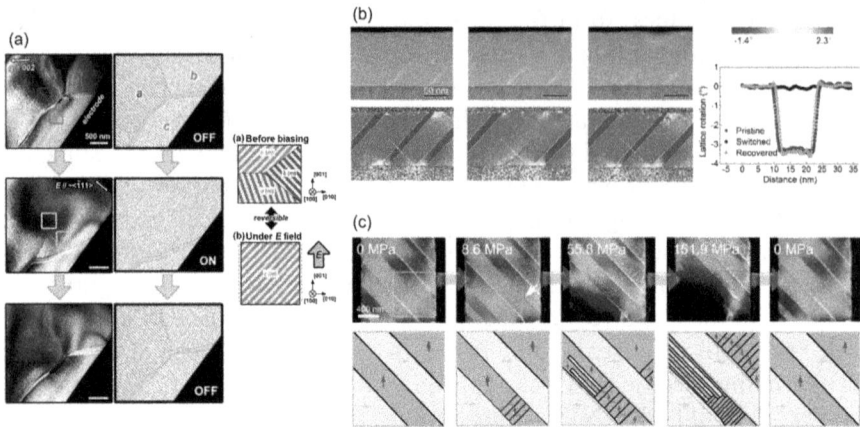

Figure 4.21. *In situ* TEM of reversible domain changes by electrical or mechanical excitations. (a) Polarization switching and reversal of hierarchically arranged domains in PMN-30PT crystal upon application and removal of the electric field, respectively. (b) HAADF STEM images of *in situ* switching and recovery of 90° ferroelastic domains in a PTO film. The same evolution was observed with electrical and mechanical excitations via contact with a tungsten tip. (c) Compression test using a probe tip inside the TEM on the PMN-38PT crystal, showing fragmentation of lamellar domains by stress-induced switching and recovery of the initial domain state upon release. ((a) Reproduced with permission from [236]. Copyright 2011 the American Physical Society. (b) Reproduced with permission from [243]. Copyright 2019 Elsevier. (c) Reproduced with permission from [245]. Copyright 2017 the American Physical Society.)

domain switching was observed from the diffraction contrast in bright-field TEM or using STEM and evaluated with the help of phase-field modeling.

In the ferroelectric BFO thin film the applied external field showed nucleation of domains at the electrode interface and DW pinning to point defects, impeding polarization switching [241]. The influence of electrical and mechanical boundary conditions in epitaxial thin films on non-switchable polarization under external fields was studied on PZT thin film, showing the formation of energetically unfavorable head-to-head DWs [246]. It was further found that ferroelastic 90° domain walls hinder the motion of ferroelectric 180° domain walls by forming charged ferroelastic domain boundaries, leading to incomplete switching [242]. Under electrical and mechanical excitations, ferroelastic domains could be effectively stabilized by dislocations at the substrate interface, while in the absence of dislocations, domains near the surface could be permanently removed by external fields [65]. It was further shown that in the case of defect-free PbTiO$_3$ films with strongly distorted lattice at the interface, 90° domain walls can be switched easily due to weakened interface clamping; the switching is reversible, yielding a high control of field-induced polarization switching (figure 4.21) [243]. Reversible switching of charged domain walls in BFO thin films that showed strong conductance at such DWs, yielded high resistance changes upon switching, which showed its high potential for non-destructive readout [244].

Further progress in following the domain dynamics using *in situ* TEM marks the development of dedicated *in situ* biasing and heating holders. A focused ion beam is used to mount the thinned samples onto specifically designed MEMS chips, and platinum electrodes are deposited on both sides of a TEM lamella to glue it onto the chip. Contacts are provided by tips that are an integral part of the TEM holder, and the holder is connected to an external voltage source. It was shown on a lead-free nonergodic relaxor ferroelectric material at its morphotropic phase boundary that domains can be permanently created, moved and irreversibly switched and these changes are followed by symmetry changes [247]. Anti-phase tilting of the octahedra, corresponding to the rhombohedral symmetry of perovskite, was observed to be strongly coupled to the induced polar order, while in-phase tilting of the octahedra from the tetragonal short-range symmetry appeared unaffected by the applied external field and decoupled from the lattice polarity. Investigation of the voltage-driven response in KNN single crystal showed highly mobile needle-like domains that can be pinned to immobile lamellar domain walls or other discontinuities [248]. Upon further voltage increase and domain wall pinning, the domains tend to fragment into thinner domains, retract, and new nucleation sites for domain growth can emerge, such as at the lamella surface. The *in situ* observed electric field response of a compositionally graded KNN crystal showed reversible switching of non-180° domain walls. The authors assign the reversible domain wall motion to the built-in flexoelectricity, arising from compositional fluctuation, resulting in high piezo-electric performance of the material [249]. In the case for which a subcoercive field is applied to a BFO single crystal, changes of zig-zag domains that correspond to 180° domain walls were observed by *in situ* STEM at the atomic scale [250]. Domain wall movement was observed based on 180° switching of Fe-displacements, which was observed to be decoupled from the lattice strain field, identified from bright-field contrast that did not change upon field application. This was assigned to defect segregation at the initial DW position (e.g. clusters of oxygen vacancies), prohibiting the change in strain states. A cyclic electric loading experiment on tetragonal PMN-PT showed ferroelectric degradation due to charges accumulated at the domain walls, which served as compensation charges to stabilize the c domains that were subsequently unresponsive to applied field [251]. In the case of ferroelectric Aurivillius structured $SrBi_2Ta_2O_9$, *in situ* switching of 180° domains and (anti) vortices was observed, leaving the 90° domain walls in the electrically poled state [252]. An *in situ* biasing experiment was also performed on a ferroelectric $Hf_{0.5}Zr_{0.5}O_2$ thin film with top and bottom electrode layers, monitoring oxygen migration using an applied electric field at the atomic scale by differential phase contrast (DPC) imaging [253]. If was found that the dielectric layer can act as a fast conduit for oxygen, and can also represent the source and sink for the oxygen itself.

As an alternative way to bias a material inside the TEM and directly observe the ferroelastic/ferroelectric changes in the material, the high-energy electron beam for sample irradiation can be used in a controlled fashion to serve as a source of applied electric field. Beam-induced changes in the materials have been observed before, but this was first demonstrated in a controlled fashion on a thinned BTO single crystal with striped 180° nanodomains. The latter under strong beam irradiation partially switched to striped nanodomains at 90° orientation to the original domain walls, forming a zig-zag microdomain pattern [254]. Using phase-field modeling, the authors suggested that the domain arrangement was caused by an anisotropic interaction of the incident beam with the ferroelectric. Similarly, Hart *et al.* [237] suggested that a focused beam irradiation in TEM causes ferroelectric nanodomain nucleation and motion, which they demonstrated on $KTiOPO_4$ material that has only two ferroelectric domain variants; the growth of domains from the nucleation site to the edge of the sample is a slow process, taking an hour. The authors identified positive sample charging (common for insulating samples if uncoated under the electron beam) as the driving force for domain nucleation and movement upon beam irradiation. More complex nanostructures have also been formed in BTO under the electron beam, resembling vortex-like structures, or the so-called nanodots that show distinct quadrants of 90° ferroelastic domains with radial net polarization [255]. These were suggested to be caused by a radial electric field arising from the electron beam charging. An extreme example of mechanical displacement of a BTO lamella (a free-standing BTO film) under the electron beam was controllably stimulated with focusing and spreading the electron beam to stimulate a 'sushi-rolling-like' effect, proving giant superelastic piezoelectricity of the thin membrane [256]. Again, a strong tendency for a mechanical strain response to the electron beam proved beneficial to study the material's response.

4.8 Summary

The properties of ferroelastic materials, such as the elastic stiffness, the relative permittivity, and the piezoelectric coefficients, are partly governed by the motion of ferroelastic domain walls. This motion is rarely smooth in real materials where defects (0D, 1D, 2D and 3D) act as pinning centers that impede wall motion and induce hysteresis in the presence of oscillating fields. Different electrical measurements have been developed to probe these movements: hysteresis loops, the Rayleigh law, third-harmonic responses, and first order reversal curves. With the support of structural and chemical characterization techniques, such as x-ray diffraction and transmission electron microscopy, they provide a comprehensive overview of the physics at play in ferroelastic materials.

Acknowledgments

MO would like to acknowledge the Slovenian Research and Innovation Agency (ARIS) core funding P2-0105, projects J2-50077 and N2-0366. STM is grateful for financial support from the National Science Foundation through the Pennsylvania State University Materials Research Science and Engineering Center DMR-201183, through DMR 2025439, and through the Center for Dielectrics and Piezoelectrics.

References

[1] Tagantsev A K, Eric Cross L and Fousek J 2010 *Domains in Ferroic Crystals and Thin Films* (Berlin: Springer)
[2] Aizu K 1972 Electrical, mechanical and electromechanical orders of state shifts in nonmagnetic ferroic crystals *J. Phys. Soc. Japan* **32** 1287–301
[3] Amin A and Newnham R E 1980 Tertiary ferroics *Phys. Stat. Solidi* a **61** 215–20

[4] Newnham R E and Cross L E 1974 Symmetry of secondary ferroics. I *Mater. Res. Bull.* **9** 927–33

[5] Aizu K 1970 Possible species of ferromagnetic, ferroelectric, and ferroelastic crystals *Phys. Rev.* B **2** 754–72

[6] Mitsui T 2018 Ferroelectrics and antiferroelectrics *Springer Handbook of Condensed Matter and Materials Data* (Berlin: Springer) pp 901–34

[7] Li W, Wang Z, Deschler F, Gao S, Friend R H and Cheetham A K 2017 Chemically diverse and multifunctional hybrid organic–inorganic perovskites *Nat. Rev. Mater.* **2** 16099

[8] Bierlein J D and Sleight A W 1975 Ferroelasticity in BiVO$_4$ *Solid State Commun.* **16** 69–70

[9] Baker T L, Faber K T and Readey D W 1991 Ferroelastic toughening in bismuth vanadate *J. Am. Ceram. Soc.* **74** 1619–23

[10] Zhou D, Pang L-X, Wang H, Guo J, Yao X and Randall C A 2011 Phase transition, Raman spectra, infrared spectra, band gap and microwave dielectric properties of low temperature firing (Na$_{0.5x}$Bi$_{1-0.5x}$)(Mo$_x$V$_{1-x}$)O$_4$ solid solution ceramics with scheelite structures *J. Mater. Chem.* **21** 18412

[11] Sleight A W, Chen H-y, Ferretti A and Cox D E 1979 Crystal growth and structure of BiVO$_4$ *Mater. Res. Bull.* **14** 1571–81

[12] David W I F and Wood I G 1983 Ferroelastic phase transition in BiVO$_4$: V. Temperature dependence of Bi^{3+} displacement and spontaneous strains *J. Phys. C: Solid State Phys.* **16** 5127–48

[13] Tullis J 1983 Deformation of feldspars *Feldspar Mineralogy* (Berlin: De Gruyter) ch 13 pp 297–323

[14] Nylund I-E, Tsoutsouva M, Grande T and Meier D 2022 Observation of cation-specific critical behavior at the improper ferroelectric phase transition in Gd$_2$(MoO$_4$)$_3$ *Phys. Rev. Mater.* **6** 034402

[15] Salje E K H 2000 Ferroelasticity *Contemp. Phys.* **41** 79–91

[16] Damjanovic D 1998 Ferroelectric, dielectric and piezoelectric properties of ferroelectric thin films and ceramics *Rep. Prog. Phys.* **61** 1267–324

[17] Damjanovic D 2006 Hysteresis in piezoelectric and ferroelectric materials *The Science of Hysteresis* vol 3 (Amsterdam: Elsevier) pp 337–465

[18] Shur V 1996 Fast polarization reversal process: evolution of ferroelectric domain structure in thin films *Ferroelectric Thin Films: Synthesis and Basic Properties* (London: Gordon and Breach Science) pp 154–92

[19] Webber K G, Vögler M, Khansur N H, Kaeswurm B, Daniels J E and Schader F H 2017 Review of the mechanical and fracture behavior of perovskite lead-free ferroelectrics for actuator applications *Smart Mater. Struct.* **26** 063001

[20] Schäufele A B and Heinz Härdtl K 1996 Ferroelastic properties of lead zirconate titanate ceramics *J. Am. Ceram. Soc.* **79** 2637–40

[21] Tagantsev A K, Stolichnov I, Setter N, Cross J S and Tsukada M 2002 Non-Kolmogorov–Avrami switching kinetics in ferroelectric thin films *Phys. Rev.* B **66** 214109

[22] Guido R, Wang X, Xu B, Alcala R, Mikolajick T, Schroeder U and Lomenzo P D 2024 Ferroelectric Al$_{0.85}$Sc$_{0.15}$N and Hf$_{0.5}$Zr$_{0.5}$O$_2$ domain switching dynamics *ACS Appl. Mater. Interfaces* **16** 42415–25

[23] So Y W, Kim D J, Noh T W, Yoon J-G and Song T K 2005 Polarization switching kinetics of epitaxial Pb(Zr$_{0.4}$Ti$_{0.6}$)O$_3$ thin films *Appl. Phys. Lett.* **86** 1–3

[24] Gong N, Sun X, Jiang H, Chang-Liao K S, Xia Q and Ma T P 2018 Nucleation limited switching (NLS) model for HfO_2-based metal–ferroelectric–metal (MFM) capacitors: switching kinetics and retention characteristics *Appl. Phys. Lett.* **112** 262903

[25] Chen Z, Zhang Y, Li S, Lu X-M and Cao W 2017 Frequency dependence of the coercive field of $0.71Pb(Mg_{1/3}Nb_{2/3})O_3$-$0.29PbTiO_3$ single crystal from 0.01 Hz to 5 MHz *Appl. Phys. Lett.* **110** 1–5

[26] Chen X, Dong X, Zhang H, Cao F, Wang G, Gu Y, He H and Liu Y 2011 Frequency dependence of coercive field in soft $Pb(Zr_{1-x}Ti_x)O_3$ ($0.20 < x < 0.60$) bulk ceramics *J. Am. Ceram. Soc.* **94** 4165–8

[27] Scott J F 1996 Models for the frequency dependence of coercive field and the size dependence of remanent polarization in ferroelectric thin films *Integr. Ferroelectr.* **12** 71–81

[28] Yang S M, Jo J Y, Kim T H, Yoon J-G, Song T K, Lee H N, Marton Z, Park S, Jo Y and Noh T W 2010 ac dynamics of ferroelectric domains from an investigation of the frequency dependence of hysteresis loops *Phys. Rev.* B **82** 174125

[29] Lente M H, Picinin A, Rino J P and Eiras J A 2004 90° domain wall relaxation and frequency dependence of the coercive field in the ferroelectric switching process *J. Appl. Phys.* **95** 2646–653

[30] Merz W J 1954 Domain formation and domain wall motions in ferroelectric $BaTiO_3$ single crystals *Phys. Rev.* **95** 690–8

[31] Maier J 2004 *Physical Chemistry of Ionic Materials: Ions and Electrons in Solids* (New York: Wiley)

[32] Arlt G and Neumann H 1988 Internal bias in ferroelectric ceramics: origin and time dependence *Ferroelectrics* **87** 109–20

[33] Nowick A S and Heller W R 1963 Anelasticity and stress-induced ordering of point defects in crystals *Adv. Phys.* **12** 251–98

[34] Postnikov V S, Pavlov V S and Turkov S K 1970 Internal friction in ferroelectrics due to interaction of domain boundaries and point defects *J. Phys. Chem. Solids* **31** 1785–91

[35] Yang T J, Gopalan V, Swart P J and Mohideen U 1999 Direct observation of pinning and bowing of a single ferroelectric domain wall *Phys. Rev. Lett.* **82** 4106–9

[36] Salje E K H and Ishibashi Y 1996 Mesoscopic structures in ferroelastic crystals: needle twins and right-angled domains *J. Phys.: Condens. Matter.* **8** 8477–95

[37] Kronmüller H and Hilzinger H R 1973 The coercive field of hard magnetic materials *Int. J. Magn* **5** 27–30

[38] Paruch P, Giamarchi T, Tybell T and Triscone J-M 2006 Nanoscale studies of domain wall motion in epitaxial ferroelectric thin films *J. Appl. Phys.* **100** 051608

[39] Li K T and Lo V C 2005 Simulation of oxygen vacancy induced phenomena in ferroelectric thin films *J. Appl. Phys.* **97** 034107

[40] Pramanick A, Prewitt A D, Forrester J S and Jones J L 2012 Domains, domain walls and defects in perovskite ferroelectric oxides: a review of present understanding and recent contributions *Crit. Rev. Solid State Mater. Sci.* **37** 243–75

[41] Warren W L, Dimos D, Tuttle B A, Nasby R D and Pike G E 1994 Electronic domain pinning in Pb(Zr,Ti)O_3 thin films and its role in fatigue *Appl. Phys. Lett.* **65** 1018–20

[42] Jonker G H 1972 Nature of aging in ferroelectric ceramics *J. Am. Ceram. Soc.* **55** 57–8

[43] Carl K and Hardtl K H 1977 Electrical after-effects in Pb(Ti,Zr)O_3 ceramics *Ferroelectrics* **17** 473–86

[44] Genenko Y A, Glaum J, Hirsch O, Kungl H, Hoffmann M J and Granzow T 2009 Aging of poled ferroelectric ceramics due to relaxation of random depolarization fields by space-charge accumulation near grain boundaries *Phys. Rev.* B **80** 224109

[45] Warren W L, Vanheusden K, Dimos D, Pike G E and Tuttle B A 1996 Oxygen vacancy motion in perovskite oxides *J. Am. Ceram. Soc.* **79** 536–8

[46] Morozov M I and Damjanovic D 2010 Charge migration in Pb(Zr,Ti)O$_3$ ceramics and its relation to ageing, hardening, and softening *J. Appl. Phys.* **107** 1–20

[47] He L and Vanderbilt D 2003 First-principles study of oxygen-vacancy pinning of domain walls in PbTiO$_3$ *Phys. Rev.* B **68** 134103

[48] Li W, Ma J, Chen K, Su D and Zhu J S 2005 Absorption of 90° domain walls to oxygen vacancies investigated through internal friction technique *Europhys. Lett.* **72** 131–6

[49] Jo J Y, Yang S M, Kim T H, Lee H N, Yoon J G, Park S, Jo Y, Jung M H and Noh T W 2009 Nonlinear dynamics of domain-wall propagation in epitaxial ferroelectric thin films *Phys. Rev. Lett.* **102** 045701

[50] Fousek J and Brezina B 1964 Relaxation of 90° domain walls of BaTiO$_3$ and their equation of motion *J. Phys. Soc. Japan* **19** 830–8

[51] Arlt G and Dederichs H 1980 Complex elastic, dielectric and piezoelectric constants by domain wall damping in ferroelectric ceramics *Ferroelectrics* **29** 51–4

[52] Robert G, Damjanovic D and Setter N 2000 Preisach modeling of ferroelectric pinched loops *Appl. Phys. Lett.* **77** 4413–5

[53] Mueller R, Gross D and Lupascu D C 2006 Driving forces on domain walls in ferroelectric materials and interaction with defects *Comput. Mater. Sci.* **35** 42–52

[54] Chu M-W, Szafraniak I, Scholz R, Harnagea C, Hesse D, Alexe M and Gösele U 2004 Impact of misfit dislocations on the polarization instability of epitaxial nanostructured ferroelectric perovskites *Nat. Mater.* **3** 87–90

[55] Gao P, Ishikawa R, Feng B, Kumamoto A, Shibata N and Ikuhara Y 2018 Atomic-scale structure relaxation, chemistry and charge distribution of dislocation cores in SrTiO$_3$ *Ultramicroscopy* **184** 217–24

[56] Nakamura T and Nakamura H 1962 Domain wall caught in dislocations in ferroelectric glycine sulfate crystals *Japan. J. Appl. Phys.* **1** 253

[57] Shimada T, Xu T, Araki Y, Wang J and Kitamura T 2017 Multiferroic dislocations in ferroelectric PbTiO$_3$ *Nano Lett.* **17** 2674–80

[58] Adepalli K K, Yang J, Maier J, Tuller H L and Yildiz B 2017 Tunable oxygen diffusion and electronic conduction in SrTiO$_3$ by dislocation-induced space charge fields *Adv. Funct. Mater.* **27** 6

[59] Bradt R C and Ansell G S 1967 Dislocations and 90° domains in barium titanate *J. Appl. Phys.* **38** 5407–8

[60] Tan X and Shang J K 2004 Partial dislocations at domain intersections in a tetragonal ferroelectric crystal *J. Phys.: Condens. Matter.* **16** 1455–66

[61] Brown W F 1940 Theory of the approach to magnetic saturation *Phys. Rev.* **58** 736–43

[62] Seeger A and Kronmüller H 1960 Die Einmündung in die ferromagnetische Sättigung—I *J. Phys. Chem. Solids* **12** 298–313

[63] Taylor R A, Jakubovics J P, Astié B and Degauque J 1983 Direct observation of the interaction between magnetic domain walls and dislocations in iron *J. Magn. Magn. Mater.* **31-34** 970–2

[64] Lindquist A K, Feinberg J M, Harrison R J, Loudon J C and Newell A J 2015 Domain wall pinning and dislocations: investigating magnetite deformed under conditions analogous to nature using transmission electron microscopy *J. Geophys. Res.: Solid Earth* **120** 1415–30

[65] Gao P *et al* 2014 Ferroelastic domain switching dynamics under electrical and mechanical excitations *Nat. Commun.* **5** 3801

[66] Höfling M *et al* 2021 Control of polarization in bulk ferroelectrics by mechanical dislocation imprint *Science* **372** 961–4

[67] Höfling M, Trapp M, Porz L, Uršič H, Bruder E, Kleebe H, Rödel J and Koruza J 2021 Large plastic deformability of bulk ferroelectric KNbO$_3$ single crystals *J. Eur. Ceram. Soc.* **41** 4098–107

[68] Li Y L, Hu S Y, Choudhury S, Baskes M I, Saxena A, Lookman T, Jia Q X, Schlom D G and Chen L Q 2008 Influence of interfacial dislocations on hysteresis loops of ferroelectric films *J. Appl. Phys.* **104** 104110

[69] Kontsos A and Landis C M 2009 Computational modeling of domain wall interactions with dislocations in ferroelectric crystals *Int. J. Solids Struct.* **46** 1491–8

[70] Hu S Y, Li Y L and Chen L Q 2003 Effect of interfacial dislocations on ferroelectric phase stability and domain morphology in a thin film—a phase-field model *J. Appl. Phys.* **94** 2542–7

[71] Marincel D M, Zhang H, Kumar A, Jesse S, Kalinin S V, Rainforth W M, Reaney I M, Randall C A and Trolier-McKinstry S 2014 Influence of a single grain boundary on domain wall motion in ferroelectrics *Adv. Funct. Mater.* **24** 1409–17

[72] Zhao C *et al* 2021 Precipitation hardening in ferroelectric ceramics *Adv. Mater.* **33** 2102421

[73] Daniel L, Hall D A and Withers P J 2014 A multiscale modelling analysis of the contribution of crystalline elastic anisotropy to intergranular stresses in ferroelectric materials *J. Phys. D: Appl. Phys.* **47** 325303

[74] Schultheiß J, Liu L, Kungl H, Weber M, Kodumudi Venkataraman L, Checchia S, Damjanovic D, Daniels J E and Koruza J 2018 Revealing the sequence of switching mechanisms in polycrystalline ferroelectric/ferroelastic materials *Acta Mater.* **157** 355–63

[75] Arlt G 1990 Twinning in ferroelectric and ferroelastic ceramics: stress relief *J. Mater. Sci.* **25** 2655–66

[76] Ihlefeld J F, Harris D T, Keech R, Jones J L, Maria J-P and Trolier-McKinstry S 2016 Scaling effects in perovskite ferroelectrics: fundamental limits and process-structure-property relations *J. Am. Ceram. Soc.* **99** 2537–57

[77] Schultheiß J, Picht G, Wang J, Genenko Y A, Chen L Q, Daniels J E and Koruza J 2023 Ferroelectric polycrystals: structural and microstructural levers for property-engineering via domain-wall dynamics *Prog. Mater. Sci.* **136** 101101

[78] Schultheiß J, Checchia S, Uršič H, Frömling T, Daniels J E, Malič B, Rojac T and Koruza J 2020 Domain wall–grain boundary interactions in polycrystalline Pb(Zr$_{0.7}$Ti$_{0.3}$)O$_3$ piezoceramics *J. Eur. Ceram. Soc.* **40** 3965–73

[79] Jesse S, Baddorf A P and Kalinin S V 2006 Switching spectroscopy piezoresponse force microscopy of ferroelectric materials *Appl. Phys. Lett.* **88** 2004–7

[80] Marincel D M, Zhang H, Jesse S, Belianinov A, Okatan M B, Kalinin S V, Mark Rainforth W, Reaney I M, Randall C A and Trolier-McKinstry S 2015 Domain wall motion across various grain boundaries in ferroelectric thin films *J. Am. Ceram. Soc.* **98** 1848–57

[81] Mantri S, Oddershede J, Damjanovic D and Daniels J E 2017 Ferroelectric domain continuity over grain boundaries *Acta Mater.* **128** 400–5

[82] Arlt G and Sasko P 1980 Domain configuration and equilibrium size of domains in BaTiO$_3$ ceramics *J. Appl. Phys.* **51** 4956–60

[83] MacLaren I, Schmitt L A, Fuess H, Kungl H and Hoffmann M J 2005 Experimental measurement of stress at a four-domain junction in lead zirconate titanate *J. Appl. Phys.* **97** 094102

[84] Simons H, Haugen A B, Jakobsen A C, Schmidt S, Stöhr F, Majkut M, Detlefs C, Daniels J E, Damjanovic D and Poulsen H F 2018 Long-range symmetry breaking in embedded ferroelectrics *Nat. Mater.* **17** 814–9

[85] Li W and Alexe M 2007 Investigation on switching kinetics in epitaxial Pb(Zr$_{0.2}$Ti$_{0.8}$)O$_3$ ferroelectric thin films: role of the 90° domain walls *Appl. Phys. Lett.* **91** 2005–8

[86] Wallace M, Johnson-Wilke R L, Esteves G, Fancher C M, Wilke R H T, Jones J L and Trolier-McKinstry S 2015 In situ measurement of increased ferroelectric/ferroelastic domain wall motion in declamped tetragonal lead zirconate titanate thin films *J. Appl. Phys.* **117** 054103

[87] Nagarajan V, Roytburd A, Stanishevsky A, Prasertchoung S, Zhao T, Chen L, Melngailis J, Auciello O and Ramesh R 2003 Dynamics of ferroelastic domains in ferroelectric thin films *Nat. Mater.* **2** 43–7

[88] Denis L M, Esteves G, Walker J, Jones J L and Trolier-McKinstry S 2018 Thickness dependent response of domain wall motion in declamped 001 Pb(Zr$_{0.3}$Ti$_{0.7}$)O$_3$ thin films *Acta Mater.* **151** 243–52

[89] Alexe M, Harnagea C, Hesse D and Gösele U 2001 Polarization imprint and size effects in mesoscopic ferroelectric structures *Appl. Phys. Lett.* **79** 242–4

[90] Schrade D, Mueller R, Gross D, Utschig T, Ya Shur V and Lupascu D C 2007 Interaction of domain walls with defects in ferroelectric materials *Mech. Mater.* **39** 161–74

[91] Zhang J *et al* 2015 Semiconductor/relaxor 0–3 type composites without thermal depolarization in Bi$_{0.5}$Na$_{0.5}$TiO$_3$-based lead-free piezoceramics *Nat. Commun.* **6** 6615

[92] Lalitha K V, Riemer L M, Koruza J and Rödel J 2017 Hardening of electromechanical properties in piezoceramics using a composite approach *Appl. Phys. Lett.* **111** 1–6

[93] Zheng M, Zhao C, Yan X, Khachaturyan R, Zhuo F, Hou Y and Koruza J 2023 Metal particle composite hardening in Ba$_{0.85}$Ca$_{0.15}$Ti$_{0.90}$Zr$_{0.10}$O$_3$ piezoceramics *Adv. Funct. Mater.* **33** 2301356

[94] Zhao C, Benčan A, Bohnen M, Zhuo F, Ma X, Dražić G, Müller R, Li S, Koruza J and Rödel J 2024 Impact of stress-induced precipitate variant selection on anisotropic electrical properties of piezoceramics *Nat. Commun.* **15** 10327

[95] Shilling J W and Soffa W A 1978 Magnetic precipitation hardening in a semihard permanent magnet alloy *Acta Metall.* **26** 413–27

[96] Fidler J and Kronmüller H 1979 Nucleation and pinning of magnetic domains at Co$_7$Sm$_2$ precipitates in Co$_5$Sm crystals *Phys. Status Solidi* a **56** 545–56

[97] Messing G L *et al* 2017 Texture-engineered ceramics—property enhancements through crystallographic tailoring *J. Mater. Res.* **32** 3219–41

[98] Chang Y, Wu J, Sun Y, Zhang S, Wang X, Yang B, Messing G L and Cao W 2015 Enhanced electromechanical properties and phase transition temperatures in [001] textured Pb(In$_{1/2}$Nb$_{1/2}$)O$_3$-Pb(Mg$_{1/3}$Nb$_{2/3}$)O$_3$-PbTiO$_3$ ternary ceramics *Appl. Phys. Lett.* **107** 082902

[99] Roscow J I, Li Y and Hall D A 2022 Residual stress and domain switching in freeze cast porous barium titanate *J. Eur. Ceram. Soc.* **42** 1434–44

[100] Johnson-Wilke R L, Wilke R H T, Wallace M, Rajashekhar A, Esteves G, Merritt Z, Jones J L and Trolier-McKinstry S 2015 Ferroelectric/ferroelastic domain wall motion in dense and porous tetragonal lead zirconate titanate films *IEEE Trans. Ultrason. Ferroelectr. Freq. Control* **62** 46–55

[101] Matavž A, Bradeško A, Rojac T, Malič B and Bobnar V 2019 Self-assembled porous ferroelectric thin films with a greatly enhanced piezoelectric response *Appl. Mater. Today* **16** 83–9

[102] Schultheiß J, Roscow J I and Koruza J 2019 Orienting anisometric pores in ferroelectrics: piezoelectric property engineering through local electric field distributions *Phys. Rev. Mater.* **3** 084408

[103] Taylor D V and Damjanovic D 1997 Evidence of domain wall contribution to the dielectric permittivity in PZT thin films at sub-switching fields *J. Appl. Phys.* **82** 1973

[104] Damjanovic D and Demartin M 1996 The Rayleigh law in piezoelectric ceramics *J. Phys. D: Appl. Phys.* **29** 2057–60

[105] Rayleigh L 1887 XXV. Notes on electricity and magnetism—III. On the behaviour of iron and steel under the operation of feeble magnetic forces *Phil. Mag. J. Sci.* **23** 225–45

[106] Néel L 1942 Théorie des lois d'aimantation de Lord Rayleigh et les déplacements d'une paroi isolée *Cahiers de Physique* **12** 1–20

[107] Néel L 1943 Théorie des lois daaimantation de Lord Rayleigh: II. Multiples domaines et Champ coercitif *Cahiers Phys.* **13** 18–30

[108] Turik A V 1963 Theory of polarization and hysteresis of ferroelectrics *Sov. Phys. Solid State* **5** 885–6

[109] Damjanovic D 1997 Stress and frequency dependence of the direct piezoelectric effect in ferroelectric ceramics *J. Appl. Phys.* **82** 1788–97

[110] Damjanovic D and Demartin M 1997 Contribution of the irreversible displacement of domain walls to the piezoelectric effect in barium titanate and lead zirconate titanate ceramics *J. Phys.: Condens. Matter.* **9** 4943–53

[111] Bharadwaja S S N, Laha A, Halder S and Krupanidhi S B 2002 Reversible and irreversible switching processes in pure and lanthanum modified lead zirconate thin films *Mater. Sci. Eng. B* **94** 218–22

[112] Nadaud K, Borderon C, Renoud R, Bah M, Ginestar S and Gundel H W 2021 Evidence of residual ferroelectric contribution in antiferroelectric lead-zirconate thin films by first-order reversal curves *Appl. Phys. Lett.* **118** 042902

[113] Luo Z *et al* 2014 Rayleigh-like nonlinear dielectric response and its evolution during electrical fatigue in antiferroelectric (Pb,La)(Zr,Ti)O$_3$ thin film *Appl. Phys. Lett.* **104** 142904

[114] Alguero M, Jiménez B and Pardo L 2003 Rayleigh type behavior of the Young's modulus of unpoled ferroelectric ceramics and its dependence on temperature *Appl. Phys. Lett.* **83** 2641–3

[115] Boser O 1987 Statistical theory of hysteresis in ferroelectric materials *J. Appl. Phys.* **62** 1344–8

[116] Lee Y *et al* 2022 The influence of crystallographic texture on structural and electrical properties in ferroelectric Hf$_{0.5}$Zr$_{0.5}$O$_2$ *J. Appl. Phys.* **132** 244103

[117] Nadaud K, Nataf G F, Jaber N, Bah M, Negulescu B, Andreazza P, Birnal P and Wolfman J 2024 Subcoercive field dielectric response of 0.5(Ba$_{0.7}$Ca$_{0.3}$TiO$_3$)-0.5(BaZr$_{0.2}$Ti$_{0.8}$O$_3$) thin film: peculiar third harmonic signature of phase transitions and residual ferroelectricity *Appl. Phys. Lett.* **124** 042901

[118] Nadaud K, Borderon C, Renoud R, Ghalem A, Crunteanu A, Huitema L, Dumas-Bouchiat F, Marchet P, Champeaux C and Gundel H W 2018 Diffuse phase transition of BST thin films in the microwave domain *Appl. Phys. Lett.* **112** 262901

[119] Fujii I F, Hong E H and Trolier-McKinstry S 2010 Thickness dependence of dielectric nonlinearity of lead zirconate titanate films *IEEE Trans. Ultrason. Ferroelectr. Freq. Control* **57** 1717–23

[120] Nadaud K, Borderon C, Renoud R, Bah M, Ginestar S and Gundel H W 2023 Metastable and field-induced ferroelectric response in antiferroelectric lead zirconate thin film studied by the hyperbolic law and third harmonic response *J. Appl. Phys.* **133** 174102

[121] Schenk T, Hoffmann M, Pešić M, Park M H, Richter C, Schroeder U and Mikolajick T 2018 Physical approach to ferroelectric impedance spectroscopy: the Rayleigh element *Phys. Rev. Appl.* **10** 064004

[122] Bassiri-Gharb N, Fujii I, Hong E, Trolier-McKinstry S, Taylor D V and Damjanovic D 2007 Domain wall contributions to the properties of piezoelectric thin films *J. Electroceram.* **19** 49–67

[123] Borderon C, Renoud R, Ragheb M and Gundel H W 2011 Description of the low field nonlinear dielectric properties of ferroelectric and multiferroic materials *Appl. Phys. Lett.* **98** 112903

[124] Bai J W, Yang J, Zhang Y Y, Bai W, Lv Z F, Tang K, Sun J L, Meng X J, Tang X D and Chu J H 2017 The ac sub-coercive-field dielectric resonses of (Pb, Sr)TiO$_3$ films at low temperature *Ceram. Int.* **43** S516–9

[125] Nadaud K, Sadl M, Bah M, Levassort F and Ursic H 2022 Effect of thermal annealing on dielectric and ferroelectric properties of aerosol-deposited 0.65Pb(Mg$_{1/3}$Nb$_{2/3}$)O$_3$-0.35PbTiO$_3$ thick films *Appl. Phys. Lett.* **120** 112902

[126] Borderon C, Brunier A E, Nadaud K, Renoud R, Alexe M and Gundel H W 2017 Domain wall motion in Pb(Zr$_{0.20}$Ti$_{0.80}$)O$_3$ epitaxial thin films *Sci. Rep.* **7** 3444

[127] Robert G, Damjanovic D and Setter N 2001 Preisach distribution function approach to piezoelectric nonlinearity and hysteresis *J. Appl. Phys.* **90** 2459–64

[128] Bassiri Gharb N and Trolier-Mckinstry S 2005 Dielectric nonlinearity of Pb(Yb$_{1/2}$Nb$_{1/2}$)O$_3$-PbTiO$_3$ thin films with {100} and {111} crystallographic orientation *J. Appl. Phys.* **97** 064106

[129] Becker M T, Burkhardt C J, Kleiner R and Koelle D 2022 Impedance spectroscopy of ferroelectrics: the domain wall pinning element *J. Appl. Phys.* **132** 044104

[130] Marquardt R, Petersen D, Gronenberg O, Zahari F, Lamprecht R, Popkirov G, Carstensen J, Kienle L and Kohlstedt H 2023 Domain wall movement in undoped ferroelectric HfO$_2$: a Rayleigh analysis *ACS Appl. Electron. Mater.* **5** 3251–60

[131] Nadaud K, Borderon C, Renoud R and Gundel H W 2015 Effect of manganese doping of BaSrTiO$_3$ on diffusion and domain wall pinning *J. Appl. Phys.* **117** 084104

[132] Coulibaly M D, Borderon C, Renoud R and Gundel H W 2020 Effect of ferroelectric domain walls on the dielectric properties of PbZrO$_3$ thin films *Appl. Phys. Lett.* **117** 142905

[133] Garten L M, Lam P, Harris D, Maria J-P and Trolier-McKinstry S 2014 Residual ferroelectricity in barium strontium titanate thin film tunable dielectrics *J. Appl. Phys.* **116** 044104

[134] Nadaud K, Borderon C, Renoud R, Ghalem A, Crunteanu A, Huitema L, Dumas-Bouchiat F, Marchet P, Champeaux C and Gundel H W 2017 Effect of the incident power on permittivity, losses and tunability of BaSrTiO$_3$ thin films in the microwave frequency range *Appl. Phys. Lett.* **110** 212902

[135] Nadaud K, Borderon C, Renoud R and Gundel H W 2016 Decomposition of the different contributions to permittivity, losses, and tunability in BaSrTiO$_3$ thin films using the hyperbolic law *J. Appl. Phys.* **119** 114101

[136] Zhu W, Fujii I, Ren W and Trolier-McKinstry S 2011 Influence of Mn doping on domain wall motion in $Pb(Zr_{0.52}Ti_{0.48})O_3$ films *J. Appl. Phys.* **109** 064105

[137] Bolten D, Böttger U, Schneller T, Grossmann M, Lohse O and Waser R 2000 Reversible and irreversible processes in donor-doped $Pb(Zr,Ti)O_3$ *Appl. Phys. Lett.* **77** 3830–2

[138] Zhu W, Fujii I, Ren W and Trolier-McKinstry S 2012 Domain wall motion in A and B site donor-doped $Pb(Zr_{0.52}Ti_{0.48})O_3$ films *J. Am. Ceram. Soc.* **95** 2906–13

[139] Fedorenko A A, Mueller V and Stepanow S 2004 Dielectric response due to stochastic motion of pinned domain walls *Phys. Rev.* B **70** 224104

[140] Nattermann T, Shapir Y and Vilfan I 1990 Interface pinning and dynamics in random systems *Phys. Rev.* B **42** 8577–86

[141] Damjanovic D, Bharadwaja S S N and Setter N 2005 Toward a unified description of nonlinearity and frequency dispersion of piezoelectric and dielectric responses in $Pb(Zr,Ti)O_3$ *Mater. Sci. Eng.* B **120** 170–4

[142] Bassiri Gharb N, Trolier-McKinstry S and Damjanovic D 2006 Piezoelectric nonlinearity in ferroelectric thin films *J. Appl. Phys.* **100** 044107

[143] Colla E V, Furman E L, Gupta S M, Yushin N K and Viehland D 1999 Dependence of dielectric relaxation on ac drive in $[Pb(Mg_{1/3}Nb_{2/3})O_3]_{(1-x)}$-$(PbTiO_3)_x$ single crystals *J. Appl. Phys.* **85** 1693–7

[144] Cai B, Schwarzkopf J, Hollmann E, Braun D, Schmidbauer M, Grellmann T and Wördenweber R 2016 Electronic characterization of polar nanoregions in relaxor-type ferroelectric $NaNbO_3$ films *Phys. Rev.* B **93** 224107

[145] Ma B, Hu Z, Liu S, Tong S, Narayanan M, Koritala R E and Balachandran U 2013 Temperature-dependent dielectric nonlinearity of relaxor ferroelectric $Pb_{0.92}La_{0.08}Zr_{0.52}Ti_{0.48}O_3$ thin films *Appl. Phys. Lett.* **102** 202901

[146] Ye M, Huang H, Li T, Ke S, Lin P, Peng B, Mai M, Sun Q, Peng X and Zeng X 2015 Temperature-dependent reversible and irreversible processes in Nb-doped $PbZrO_3$ relaxor ferroelectric thin films *Appl. Phys. Lett.* **107** 202902

[147] Shetty S, Damodaran A, Wang K, Yuan Y, Gopalan V, Martin L and Trolier-McKinstry S 2019 Relaxor behavior in ordered lead magnesium niobate $(PbMg_{1/3}Nb_{2/3}O_3)$ thin films *Adv. Funct. Mater.* **29** 1804258

[148] Riemer L M, Jin L, Uršič H, Otonicar M, Rojac T and Damjanovic D 2021 Dielectric and electro-mechanic nonlinearities in perovskite oxide ferroelectrics, relaxors, and relaxor ferroelectrics *J. Appl. Phys.* **129** 054101

[149] Otonicar M, Dragomir M and Rojac T 2022 Dynamics of domain walls in ferroelectrics and relaxors *J. Am. Ceram. Soc.* **105** 6479–507

[150] Morozov M, Damjanovic D and Setter N 2005 The nonlinearity and subswitching hysteresis in hard and soft PZT *J. Eur. Ceram. Soc.* **25** 2483–6

[151] Hashemizadeh S and Damjanovic D 2017 Nonlinear dynamics of polar regions in paraelectric phase of $(Ba_{1-x},Sr_x)TiO_3$ ceramics *Appl. Phys. Lett.* **110** 192905

[152] Otoničar M *et al* 2020 Connecting the multiscale structure with macroscopic response of relaxor ferroelectrics *Adv. Funct. Mater.* **30** 2006823

[153] Morozov M I and Damjanovic D 2008 Hardening-softening transition in Fe-doped $Pb(Zr,Ti)O_3$ ceramics and evolution of the third harmonic of the polarization response *J. Appl. Phys.* **104** 034107

[154] Miga S, Dec J and Kleemann W 2007 Computer-controlled susceptometer for investigating the linear and nonlinear dielectric response *Rev. Sci. Instrum.* **78** 033902

[155] Schenk T, Schroeder U, Pešić M, Popovici M, Pershin Y V and Mikolajick T 2014 Electric field cycling behavior of ferroelectric hafnium oxide *ACS Appl. Mater. Interfaces* **6** 19744–51

[156] Bharadwaja S S N, Hong E, Zhang S J, Cross L E and Trolier-McKinstry S 2007 Nonlinear dielectric response in $(1-x)Pb(Zn_{1/3}Nb_{2/3})O_{3-x}$ PbTiO$_3$ ($x = 0.045$ and 0.08) single crystals *J. Appl. Phys.* **101** 104102

[157] Dec J, Kleemann W, Miga S, Filipic C, Levstik A, Pirc R, Granzow T and Pankrath R 2003 Probing polar nanoregions in $Sr_{0.61}Ba_{0.39}Nb_2O_6$ *Phys. Rev.* B **68** 092105

[158] Li F, Zhang S, Damjanovic D, Chen L-Q and Shrout T R 2018 Local structural heterogeneity and electromechanical responses of ferroelectrics: learning from relaxor ferroelectrics *Adv. Funct. Mater.* **28** 1801504

[159] Shetty S, Kim J, Martin L W and Trolier-McKinstry S 2020 Non-linearity in engineered lead magnesium niobate $(PbMg_{1/3}Nb_{2/3}O_3)$ thin films *J. Appl. Phys.* **128** 194102

[160] Dawber M, Rabe K M and Scott J F 2005 Physics of thin-film ferroelectric oxides *Rev. Mod. Phys.* **77** 1083–130

[161] Scott J F 2008 Ferroelectrics go bananas *J. Phys.: Condens. Matter.* **20** 021001

[162] Sawyer C B and Tower C H 1930 Rochelle salt as a dielectric *Phys. Rev.* **35** 269–73

[163] Boyn S *et al* 2017 Learning through ferroelectric domain dynamics in solid-state synapses *Nat. Commun.* **8** 14736

[164] Wang C-Y, Wang C-I, Yi S-H, Chang T-J, Chou C-Y, Yin Y-T, Shiojiri M and Chen M-J 2020 Paraelectric/antiferroelectric/ferroelectric phase transformation in As-deposited ZrO$_2$ thin films by the TiN capping engineering *Mater. Des.* **195** 109020

[165] Liu C, Lin S X, Qin M H, Lu X B, Gao X S, Zeng M, Li Q L and Liu J M 2016 Energy storage and polarization switching kinetics of (001)-oriented $Pb_{0.97}La_{0.02}(Zr_{0.95}Ti_{0.05})O_3$ antiferroelectric thick films *Appl. Phys. Lett.* **108** 112903

[166] Mart C, Kühnel K, Kämpfe T, Zybell S and Weinreich W 2019 Ferroelectric and pyroelectric properties of polycrystalline La-doped HfO$_2$ thin films *Appl. Phys. Lett.* **114** 102903

[167] Nadaud K, Borderon C, Renoud R, Bah M, Ginestar S and Gundel H W 2023 Study of the long time relaxation of the weak ferroelectricity in PbZrO$_3$ antiferroelectric thin film using positive up negative down and first order reversal curves measurements *Thin Solid Films* **773** 139817

[168] Martin S, Baboux N, Albertini D and Gautier B 2017 A new technique based on current measurement for nanoscale ferroelectricity assessment: nano-positive up negative down *Rev. Sci. Instrum.* **88** 023901

[169] Si M, Lyu X, Shrestha P R, Sun X, Wang H, Cheung K P and Ye P D 2019 Ultrafast measurements of polarization switching dynamics on ferroelectric and anti-ferroelectric hafnium zirconium oxide *Appl. Phys. Lett.* **115** 072107

[170] Pike C R 2003 First-order reversal-curve diagrams and reversible magnetization *Phys. Rev.* B **68** 104424

[171] Harrison R J and Feinberg J M 2008 FORCinel: an improved algorithm for calculating first-order reversal curve distributions using locally weighted regression smoothing *Geochem. Geophys. Geosyst.* **9** Q05016

[172] Mitoseriu L, Ciomaga C E, Buscaglia V, Stoleriu L, Piazza D, Galassi C, Stancu A and Nanni P 2007 Hysteresis and tunability characteristics of $Ba(Zr,Ti)O_3$ ceramics described by first order reversal curves diagrams *J. Eur. Ceram. Soc.* **27** 3723–6

[173] Mitoseriu L, Stoleriu L, Stancu A, Galassi C and Buscaglia V 2009 First order reversal curves diagrams for describing ferroelectric switching characteristics *Process. Appl. Ceram.* **3** 3–7

[174] Ricinschi D, Stancu A, Mitoseriu L, Postolache P and Okuyama M 2004 First order reversal curves diagrams applied for the ferroelectric systems *J. Optoelectro. Adv. Mater.* **6** 623–7

[175] Stoleriu L, Stancu A, Mitoseriu L, Piazza D and Galassi C 2006 Analysis of switching properties of porous ferroelectric ceramics by means of first-order reversal curve diagrams *Phys. Rev.* B **74** 174107

[176] Hoffmann M, Schenk T, Pešić M, Schroeder U and Mikolajick T 2017 Insights into antiferroelectrics from first-order reversal curves *Appl. Phys. Lett.* **111** 182902

[177] Roberts A P, Liu Q, Rowan C J, Chang L, Carvallo C, Torrent J and Horng C-S 2006 Characterization of hematite (α -Fe$_2$O$_3$), goethite (α -FeOOH), greigite (Fe$_3$S$_4$), and pyrrhotite (Fe$_7$S$_8$) using first-order reversal curve diagrams *J. Geophys. Res.: Solid Earth* **111** B12S35

[178] Muxworthy A R, King J G and Heslop D 2005 Assessing the ability of first-order reversal curve (FORC) diagrams to unravel complex magnetic signals *J. Geophys. Research: Solid Earth* **110** B01105

[179] Saremi S, Xu R, Allen F I, Maher J, Agar J C, Gao R, Hosemann P and Martin L W 2018 Local control of defects and switching properties in ferroelectric thin films *Phys. Rev. Mater.* **2** 084414

[180] Stancu A, Pike C, Stoleriu L, Postolache P and Cimpoesu D 2003 Micromagnetic and Preisach analysis of the first order reversal curves (FORC) diagram *J. Appl. Phys.* **93** 6620–2

[181] Zhu W, Fujii I, Ren W and Trolier-McKinstry S 2014 Influence of Li doping on domain wall motion in Pb(Zr$_{0.52}$Ti$_{0.48}$)O$_3$ films *J. Mater. Sci.* **49** 7883–9

[182] Zhu W, Fujii I, Ren W and Trolier-McKinstry S 2011 Influence of Mn doping on domain wall motion in Pb(Zr$_{0.52}$Ti$_{0.48}$)O$_3$ films *J. Appl. Phys.* **109** 064105

[183] Schenk T, Hoffmann M, Ocker J, Pešić M, Mikolajick T and Schroeder U 2015 Complex internal bias fields in ferroelectric hafnium oxide *ACS Appl. Mater. Interfaces* **7** 20224–33

[184] Subbarao E C, McQuarrie M C and Buessem W R 1957 Domain effects in polycrystalline barium titanate *J. Appl. Phys.* **28** 1194–200

[185] Hammer M, Monty C, Endriss A and Hoffmann M J 1998 Correlation between surface texture and chemical composition in undoped, hard, and soft piezoelectric PZT ceramics *J. Am. Ceram. Soc.* **81** 721–4

[186] Hall D A, Steuwer A, Cherdhirunkorn B, Mori T and Withers P J 2004 A high energy synchrotron x-ray study of crystallographic texture and lattice strain in soft lead zirconate titanate ceramics *J. Appl. Phys.* **96** 4245–52

[187] Guo R, Cross L E, Park S-E, Noheda B, Cox D E and Shirane G 2000 Origin of the high piezoelectric response in PbZr$_{12x}$Ti$_x$O$_3$ *Phys. Rev. Lett.* **84** 5423–6

[188] Jones J L, Daniels J E and Üstündag E 2007 Advances in the characterisation of domain switching in ferroelectric ceramics *Tenth European Powder Diffraction Conference* (München: Oldenbourg Wissenschaftsverlag) pp 441–6

[189] Li S, Bhalla A S, Newnham R E, Cross L E and Huang C Y 1994 90‡ domain reversal in Pb (Zr$_x$Ti$_{11-x}$)O$_3$ ceramics *J. Mater. Sci.* **29** 1290–4

[190] Pramanick A, Damjanovic D, Daniels J E, Nino J C and Jones J L 2011 Origins of electromechanical coupling in polycrystalline ferroelectrics during subcoercive electrical loading *J. Am. Ceram. Soc.* **94** 293–309

[191] Slabki M, Kodumudi Venkataraman L, Checchia S, Fulanović L, Daniels J and Koruza J 2021 Direct observation of domain wall motion and lattice strain dynamics in ferroelectrics under high-power resonance *Phys. Rev.* B **103** 174113

[192] Jones J L, Hoffman M, Daniels J E and Studer A J 2006 Direct measurement of the domain switching contribution to the dynamic piezoelectric response in ferroelectric ceramics *Appl. Phys. Lett.* **89** 092901

[193] Shi X, Shi J and Fohtung E 2022 Applicability of coherent x-ray diffractive imaging to ferroelectric, ferromagnetic, and phase change materials *J. Appl. Phys.* **131** 040901

[194] Hruszkewycz S O, Highland M J, Holt M V, Kim D, Folkman C M, Thompson C, Tripathi A, Stephenson G B, Hong S and Fuoss P H 2013 Imaging local polarization in ferroelectric thin films by coherent x-ray Bragg projection ptychography *Phys. Rev. Lett.* **110** 177601

[195] Poulsen H F, Nielsen S F, Lauridsen E M, Schmidt S, Suter R M, Lienert U, Margulies L, Lorentzen T and Juul Jensen D 2001 Three-dimensional maps of grain boundaries and the stress state of individual grains in polycrystals and powders *J. Appl. Crystallogr.* **34** 751–6

[196] Majkut M, Daniels J E, Wright J P, Schmidt S and Oddershede J 2017 Electromechanical response of polycrystalline barium titanate resolved at the grain scale *J. Am. Ceram. Soc.* **100** 393–402

[197] Daniels J E, Majkut M, Cao Q, Schmidt S, Wright J, Jo W and Oddershede J 2016 Heterogeneous grain-scale response in ferroic polycrystals under electric field *Sci. Rep.* **6** 22820

[198] Simons H, King A, Ludwig W, Detlefs C, Pantleon W, Schmidt S, Stöhr F, Snigireva I, Snigirev A and Poulsen H F 2015 Dark-field x-ray microscopy for multiscale structural characterization *Nat. Commun.* **6** 6098

[199] Schultheiß J, Porz L, Kodumudi Venkataraman L, Höfling M, Yildirim C, Cook P, Detlefs C, Gorfman S, Rödel J and Simons H 2021 Quantitative mapping of nanotwin variants in the bulk *Scr. Mater.* **199** 113878

[200] Oliveira L, Zhang M-H, Koruza J, Rodiquez-Lamas R, Yildirim C and Simons H 2024 Heterogeneous antiferroelectric ordering in $NaNbO_3$–$SrSnO_3$ ceramics revealed by direct superstructure imaging *ACS Mater. Lett.* **6** 3745–9

[201] Usher T-M, Levin I, Daniels J E and Jones J L 2015 Electric-field-induced local and mesoscale structural changes in polycrystalline dielectrics and ferroelectrics *Sci. Rep.* **5** 14678

[202] Goetzee-Barral A J, Usher T M, Stevenson T J, Jones J L, Levin I, Brown A P and Bell A J 2017 Electric field dependent local structure of $(K_xNa_{1-x})_{0.5}Bi_{0.5}TiO_3$ *Phys. Rev.* B **96** 014118

[203] Zhao C, Hou D, Chung C-C, Yu Y, Liu W, Li S and Jones J L 2017 Local structural behavior of $PbZr_{0.5}Ti_{0.5}O_3$ during electric field application via *in situ* pair distribution function study *J. Appl. Phys.* **122** 174102

[204] Manjón-Sanz A, Culbertson C M, Hou D, Jones J L and Dolgos M R 2019 Total scattering and diffraction studies of lead-free piezoelectric $(1-x)Ba(Zr_{0.2}Ti_{0.8})O_3$-$x(Ba_{0.7}Ca_{0.3})TiO_3$ deconvolute intrinsic and extrinsic contributions to electromechanical strain *Acta Mater.* **171** 79–91

[205] Chapman K W, Chupas P J, Halder G J, Hriljac J A, Kurtz C, Greve B K, Ruschman C J and Wilkinson A P 2010 Optimizing high-pressure pair distribution function measurements in diamond anvil cells *J. Appl. Crystallogr.* **43** 297–307

[206] Lazar I, Majchrowski A, Kajewski D, Soszyński A and Roleder K 2021 Strong piezoelectric properties and electric-field-driven changes in domain structures in a $PbZr_{0.87}Ti_{0.13}O_3$ single crystal *Acta Mater.* **216** 117129

[207] Nataf G F and Guennou M 2020 Optical studies of ferroelectric and ferroelastic domain walls *J. Phys.: Condens. Matter.* **32** 183001

[208] Lei S, Eliseev E A, Morozovska A N, Haislmaier R C, Lummen T T A, Cao W, Kalinin S V and Gopalan V 2012 Origin of piezoelectric response under a biased scanning probe microscopy tip across a 180° ferroelectric domain wall *Phys. Rev.* B **86** 134115

[209] Reichmann A, Mitsche S, Zankel A, Poelt P and Reichmann K 2014 *In situ* mechanical compression of polycrystalline $BaTiO_3$ in the ESEM *J. Eur. Ceram. Soc.* **34** 2211–5

[210] Arlt G, Hennings D and de With G 1985 Dielectric properties of fine-grained barium titanate ceramics *J. Appl. Phys.* **58** 1619–25

[211] Randall C A, Barber D J and Whatmore R W 1987 Ferroelectric domain configurations in a modified-PZT ceramic *J. Mater. Sci.* **22** 925–31

[212] Asada T and Koyama Y 2007 Ferroelectric domain structures around the morphotropic phase boundary of the piezoelectric material $PbZr_{1-x}$ *Phys. Rev.* B **75** 214111

[213] Tsai F, Khiznichenko V and Cowley J M 1992 High-resolution electron microscopy of 90° ferroelectric domain boundaries in $BaTiO_3$ and $Pb(Zr_{0.52}Ti_{0.48})O_3$ *Ultramicroscopy* **45** 55–63

[214] Woodward D, Knudsen J and Reaney I 2005 Review of crystal and domain structures in the $PbZr_xTi_{1-x}O_3$ solid solution *Phys. Rev.* B **72** 104110

[215] Randall C, Barber D, Whatmore R and Groves P 1987 Short-range order phenomena in lead-based perovskites *Ferroelectrics* **76** 277–82

[216] Dorcet V and Trolliard G 2008 A transmission electron microscopy study of the A-site disordered perovskite $Na_{0.5}Bi_{0.5}TiO_3$ *Acta Mater.* **56** 1753–61

[217] Levin I, Stennett M C, Miles G C, Woodward D I, West A R and Reaney I M 2006 Coupling between octahedral tilting and ferroelectric order in tetragonal tungsten bronze-structured dielectrics *Appl. Phys. Lett.* **89** 2006–8

[218] Levin I, Reaney I M, Anton E-M, Jo W, Rödel J, Pokorny J, Schmitt L A, Kleebe H-J, Hinterstein M and Jones J L 2013 Local structure, pseudosymmetry, and phase transitions in $Na_{1/2}Bi_{1/2}TiO_3$-$K_{1/2}Bi_{1/2}TiO_3$ ceramics *Phys. Rev.* B **87** 024113

[219] Cao W and Randall C A 1996 Grain size and domain size relations in bulk ceramic ferroelectric materials *J. Phys. Chem. Solids* **57** 1499–505

[220] Li S, Zhu Y L, Tang Y L, Liu Y, Zhang S R, Wang Y J and Ma X L 2017 Thickness-dependent a_1/a_2 domain evolution in ferroelectric $PbTiO_3$ films *Acta Mater.* **131** 123–30

[221] Schmitt L A, Schönau K A, Theissmann R, Fuess H, Kungl H and Hoffmann M J 2007 Composition dependence of the domain configuration and size in $Pb(Zr_{1-x}Ti_x)O_3$ ceramics *J. Appl. Phys.* **101** 074107

[222] Foeth M, Sfera A, Stadelmann P and Buffat P A 1999 A comparison of HREM and weak beam transmission electron microscopy for the quantitative measurement of the thickness of ferroelectric domain walls *J. Electron Microsc.* **48** 717

[223] Stemmer S, Streiffer S K, Ernst F and Rüuhle M 1995 Atomistic structure of 90° domain walls in ferroelectric $PbTiO_3$ thin films *Phil. Mag.* **71** 713–24

[224] Rojac T *et al* 2017 Domain-wall conduction in ferroelectric BiFeO$_3$ controlled by accumulation of charged defects *Nat. Mater.* **16** 322–7

[225] Moore K, Conroy M, O'Connell E N, Cochard C, Mackel J, Harvey A, Hooper T E, Bell A J, Marty Gregg J and Bangert U 2020 Highly charged 180 degree head-to-head domain walls in lead titanate *Commun. Phys.* **3** 231

[226] Ge W, Beanland R, Alexe M, Ramasse Q and Sanchez A M 2023 180° head-to-head flat domain walls in single crystal BiFeO$_3$ *Microstructures* **3** 2023026

[227] Legros M 2014 *In situ* mechanical TEM: seeing and measuring under stress with electrons *Comptes Rendus. Physique* **15** 224–40

[228] Snoeck E, Normand L, Thorel A and Roucau C 1994 Electron microscopy study of ferroelastic and ferroelectric domain wall motions induced by the *in situ* application of an electric field in BaTiO$_3$ *Phase Transit.* **46** 77–88

[229] Tan X, Xu Z and Shang J K 2001 *In situ* transmission electron microscopy observations of electric-field-induced domain switching and microcracking in ferroelectric ceramics *Mater. Sci. Eng.* A **314** 157–61

[230] He H and Tan X 2004 *In situ* transmission electron microscopy study of the electric field-induced transformation of incommensurate modulations in a Sn-modified lead zirconate titanate ceramic *Appl. Phys. Lett.* **85** 3187–9

[231] Tan X, He H and Shang J-K 2005 *In situ* transmission electron microscopy studies of electric-field-induced phenomena in ferroelectrics *J. Mater. Res.* **20** 1641–53

[232] Qi X Y, Liu H H and Duan X F 2006 *In situ* transmission electron microscopy study of electric-field-induced 90° domain switching in BaTiO$_3$ single crystals *Appl. Phys. Lett.* **89** 2004–7

[233] Guo H, Tan X and Zhang S 2015 *In situ* TEM study on the microstructural evolution during electric fatigue in 0.7Pb(Mg$_{1/3}$Nb$_{2/3}$)O$_3$–0.3PbTiO$_3$ ceramic *J. Mater. Res.* **30** 364–72

[234] Kling J, Tan X, Jo W, Kleebe H-J, Fuess H and Rödel J 2010 *In situ* transmission electron microscopy of electric field-triggered reversible domain formation in Bi-based lead-free piezoceramics *J. Am. Ceram. Soc.* **93** 2452–5

[235] Zakhozheva M, Schmitt L A, Acosta M, Jo W, Rödel J and Kleebe H-J 2014 *In situ* electric field induced domain evolution in Ba(Zr$_{0.2}$Ti$_{0.8}$)O$_3$-0.3(Ba$_{0.7}$Ca$_{0.3}$)TiO$_3$ ferroelectrics *Appl. Phys. Lett.* **105** 112904

[236] Sato Y, Hirayama T and Ikuhara Y 2011 Real-time direct observations of polarization reversal in a piezoelectric crystal: Pb(Mg$_{1/3}$Nb$_2$) *Phys. Rev. Lett.* **107** 187601

[237] Hart J L, Liu S, Lang A C, Hubert A, Zukauskas A, Canalias C, Beanland R, Rappe A M, Arredondo M and Taheri M L 2016 Electron-beam-induced ferroelectric domain behavior in the transmission electron microscope: toward deterministic domain patterning *Phys. Rev.* B **94** 174104

[238] Sato Y, Hirayama T and Ikuhara Y 2012 Evolution of nanodomains under DC electrical bias in Pb(Mg$_{1/3}$Nb$_{2/3}$)O$_3$-PbTiO$_3$: an in-situ transmission electron microscopy study *Appl. Phys. Lett.* **100** 86–90

[239] Otonicar M, Ursic H, Dragomir M, Bradesko A, Esteves G, Jones J L, Bencan A, Malic B and Rojac T 2018 Multiscale field-induced structure of $(1 - x)$Pb(Mg$_{1/3}$Nb$_{2/3}$)O$_{3-x}$PbTiO$_3$ ceramics from combined techniques *Acta Mater.* **154** 14–24

[240] Takenaka H, Grinberg I, Liu S and Rappe A M 2017 Slush-like polar structures in single-crystal relaxors *Nature* **546** 391–5

[241] Nelson C T *et al* 2011 Domain dynamics during ferroelectric switching *Science* **334** 968–71

[242] Gao P, Britson J, Jokisaari J R, Nelson C T, Baek S-H, Wang Y, Eom C-B, Chen L-Q and Pan X 2013 Atomic-scale mechanisms of ferroelastic domain-wall-mediated ferroelectric switching *Nat. Commun.* **4** 2791

[243] Li M *et al* 2019 Direct observation of weakened interface clamping effect enabled ferroelastic domain switching *Acta Mater.* **171** 184–9

[244] Li L, Britson J, Jokisaari J R, Zhang Y, Adamo C, Melville A, Schlom D G, Chen L Q and Pan X 2016 Giant resistive switching via control of ferroelectric charged domain walls *Adv. Mater.* **28** 6574–80

[245] Chen Z, Hong L, Wang F, Ringer S P, Chen L-Q, Luo H and Liao X 2017 Facilitation of ferroelectric switching via mechanical manipulation of hierarchical nanoscale domain structures *Phys. Rev. Lett.* **118** 017601

[246] Han M-G, Marshall M S J, Wu L, Schofield M A, Aoki T, Twesten R, Hoffman J, Walker F J, Ahn C H and Zhu Y 2014 Interface-induced nonswitchable domains in ferroelectric thin films *Nat. Commun.* **5** 4693

[247] Otonicar M, Park J, Logar M, Esteves G, Jones J L and Jancar B 2017 External-field-induced crystal structure and domain texture in $(1-x)Na_{0.5}Bi_{0.5}TiO_3-xK_{0.5}Bi_{0.5}TiO_3$ piezoceramics *Acta Mater.* **127** 319–31

[248] Condurache O, Dražić G and Benčan A 2023 Voltage-driven ferroelectric domain dynamics in $(K,Na)NbO_3$ investigated by *in situ* transmission electron microscopy *Appl. Phys. Lett.* **122** 202902

[249] Guo Q *et al* 2023 *In situ* TEM analysis of reversible non-180° domain switching in (K,Na)NbO$_3$ single crystals *J. Mater. Chem.* A **11** 10828–33

[250] Condurache O, Dražić G, Rojac T, Uršič H, Dkhil B, Bradeško A, Damjanovic D and Benčan A 2023 Atomic-level response of the domain walls in bismuth ferrite in a subcoercive-field regime *Nano Lett.* **23** 750–6

[251] Huang Q *et al* 2021 Direct observation of nanoscale dynamics of ferroelectric degradation *Nat. Commun.* **12** 2095

[252] Mun J, Huang F-T, Pivak Y, Fang X, Camino F, Cheong S-W, Zhu Y and Han M-G 2024 Probing ferroelectric domain structures and their switching dynamics in $SrBi_2Ta_2O_9$ by *in-situ* electric biasing in transmission electron microscopy *Commun. Mater.* **5** 145

[253] Nukala P *et al* 2021 Reversible oxygen migration and phase transitions in hafnia-based ferroelectric devices *Science* **372** 630–5

[254] Matsumoto T and Okamoto M 2011 Effects of electron irradiation on the ferroelectric 180° in-plane nanostripe domain structure in a thin film prepared from a bulk single crystal of $BaTiO_3$ by focused ion beam *J. Appl. Phys.* **109** 014104

[255] Ahluwalia R, Ng N, Schilling A, McQuaid R G P, Evans D M, Gregg J M, Srolovitz D J and Scott J F 2013 Manipulating ferroelectric domains in nanostructures under electron beams *Phys. Rev. Lett.* **111** 165702

[256] Elangovan H, Barzilay M, Seremi S, Cohen N, Jiang Y, Martin L W and Ivry Y 2020 Giant superelastic piezoelectricity in flexible ferroelectric $BaTiO_3$ membranes *ACS Nano* **14** 5053–60

IOP Publishing

Ferroelastic Materials

Guillaume F Nataf, Blai Casals and Ekhard K H Salje

Chapter 5

Avalanches in ferroelastic materials

Jordi Baró, Guillaume F Nataf and Blai Casals

5.1 Introduction

Avalanches are discrete events that occur when an external force is applied to a material. These events occur intermittently, and their magnitudes are statistically distributed according to a power law. Avalanches can originate from various physical mechanisms; for instance, in ferroelastic materials, they are associated with changes in the order parameter of the phase transition, which manifests itself as variations in the spontaneous strain (see chapter 2). Historically, avalanches in solids were first observed indirectly through the study of ferromagnetic materials. The Barkhausen noise [1]—the sudden, discrete changes in magnetization under an applied magnetic field—was the first phenomenon to provide indirect evidence for the existence of magnetic domains. However, at that time, these events were not referred to as avalanches but simply as Barkhausen noise. The terminology of 'avalanches' was introduced later, following the work of Babcock and Westervelt [2], and Cote and Meisel [3], who demonstrated that the magnitudes of the individual Barkhausen jumps followed a power-law distribution. Initially, theoretical descriptions connected these avalanche phenomena to the concept of self-organized criticality (SOC) [4], a framework in which a system naturally evolves into a critical state without the need to fine-tune external parameters, resulting in scale-free, avalanche-like behavior. In contrast, the Ising model approach, particularly when driven quasistatically with disorder, provides a controlled way to study criticality tuned by external parameters, offering more direct insights into the scaling behavior and hysteresis associated with first-order phase transitions. Thus, while SOC emphasizes spontaneous criticality without fine-tuning, the Ising model highlights critical phenomena emerging from the interplay between external driving and internal disorder. In this context, the seminal works by Sethna *et al.* [5] explored the hysteresis behavior of first-order phase transitions using the Ising model, offering a new theoretical framework for understanding avalanche dynamics. These seminal studies are particularly relevant for ferroic systems, including ferroelastic materials,

doi:10.1088/978-0-7503-6089-0ch5

which exhibit hysteresis under an applied strain. As a result, avalanches in ferroelastic materials occur during both branches of the hysteresis loop.

Although avalanche theories were developed mainly later within the field of materials science in the context of ferromagnets [6, 7], avalanche phenomena have also been observed in a wide variety of other systems. In particular, from the beginning, avalanches were detected in ferroelastic materials, particularly during martensitic phase transitions, as demonstrated by the pioneering works of Eduard Vives, Antoni Planes, and collaborators [8], a topic that is discussed in greater detail in chapter 8. Here, we first review the dynamics of avalanches in ferroelastic materials in section 5.2, with a focus on their origin in general terms and in ferroelastics, their propagation mechanisms and micromechanical models. In section 5.3 we highlight experimental techniques capable of probing avalanches on different spatial and temporal scales. Finally, in section 5.4, we propose statistical tools to model avalanches as stochastic point events.

5.2 Avalanche dynamics

Avalanches are out-of-equilibrium phenomena responsible for the large heterogeneity observed in the dynamics of many complex systems and materials exhibiting discrete symmetry breaking [4], such as metastability in ferroelastic switching [9–12] and structural phase transitions [8, 13–15], and are prominent in mechanical processes from earthquakes [16], microfracturing [17–19] and the plastic deformation of single crystals [20, 21], bulk metallic [22] and colloidal [23] glasses, high entropy alloys [24], and slips in granular assemblies [25, 26].

Consider the temporal evolution of some extensive physical measure $v(t)$ representative of an internal order parameter, such as magnetization, polarization or spontaneous strain in ferromagnetic, ferroelectric and ferroelastic materials. In many processes, instead of the characteristic smooth response with self-averaging Gaussian or exponential noise typically expected in thermal systems, the observed power spectrum is inversely proportional to the frequency, i.e. $1/f$, pink, or fractal noise [27].

In extreme cases where all external stimuli are halted except for a quasistatically slow driving of a control parameter, $v(t)$ can exhibit intermittency. This means that the signal is neatly separated between prolonged periods of quiescence and fast punctual scale-free excursions away from quiescence well delimited in time. By labeling such excursions as point events i in time with a defined time origin t_i, we can mark their magnitude or size,

$$S_i = \int_{t_i}^{t_i + T_i} v(t)dt, \tag{5.1}$$

and by the duration of the excursion T_i away from quiescence. Early studies on this phenomenon [4] suggested that $1/f$-noise can often be associated with the statistical frequencies of both magnitudes, which are often well modeled by truncated or pure power-law distributions and, therefore, scale free for at least a few decades:

$$\rho(S_i = S)dS = S^{-\kappa_S}\Phi_S\left(\frac{S}{S_c}\right); \quad \rho(T_i = T)dT = T^{-\kappa_T}\Phi_T\left(\frac{T}{T_c}\right); \quad E(S|T) \sim T^{\frac{\kappa_S - 1}{\kappa_T - 1}}, \quad (5.2)$$

κ_S and κ_T being the so-called critical, i.e. Fisher, exponents, and S_c and T_c the characteristic scales truncating the scale-free behavior due to system correlation lengths or system sizes [28].

Both $1/f$-noise and scale-free intermittency have been regarded as trademarks of criticality, i.e. an underlying divergence of correlation lengths. However, $1/f$-noise exists in regimes that cannot be associated with thermal critical points. In the late 1980s, Per Bak and co-authors introduced the theory of SOC [4] which explains the ubiquity of $1/f$-noise and power-law intermittency as a signal that the system is undergoing a critical avalanche process. We use the term *avalanche dynamics* to describe and study such non-linear dynamics consisting in a sequence of fast transformation between marginally stable or metastable arrested states exhibited in complex systems and materials dominated by quenched disorder rather than thermal noise.

Avalanches in ferroelastic materials can occur under different scenarios, which can be classified according to the external force acting on (or coupled to) the order parameter. In this context, three distinct cases can be considered: first, avalanches occurring during a phase transition as a function of temperature; second, avalanches driven by external stress while exploring the hysteresis of the ferroelastic response (and thus keeping the same overall symmetry); and third, avalanches as a function of electric field. Although ferroelastic materials intrinsically respond to external strain, many ferroelectric materials are also ferroelastic, exhibiting a strong coupling between polarization and local strain.

5.2.1 Avalanches during ferroelastic phase transitions

First, we consider the temperature as the external driving force. Temperature changes induce both thermal expansion and thermal fluctuations, which can eventually trigger a phase transition in the material. Avalanches appear in first-order phase transitions, where the order parameter undergoes an abrupt change to a different state, in contrast to second-order phase transitions, where the order parameter evolves smoothly between states. An important example of a first-order phase transition is the martensitic transformation, described in chapter 8, where avalanches are observed during the transition. In contrast, a typical second-order phase transition, such as the cubic-to-tetragonal structural transition of $SrTiO_3$ [29], does not exhibit avalanches.

5.2.2 Avalanches during ferroelastic switching

Under an external stress, ferroelastic materials change the preferred orientation of their spontaneous strain, exhibiting a hysteretic behavior between two distinct strain states. When an external stress is applied gradually and quasistatically, avalanches appear as abrupt changes in the configuration of ferroelastic domains [30].

5.2.3 Avalanches during ferroelectric/ferroelastic switching

It is essential to consider ferroelectric materials, as the majority of them exhibit ferroelastic properties. This dual behavior, in which both electric and mechanical interactions influence the response of the material, plays a critical role in the performance and functionality of these materials in various applications. In particular, the domain structure of these materials can be altered under the influence of an electric field, resulting in a coupling between the ferroelectric and ferroelastic behaviors.

The switching mechanism that occurs in these materials involves both ferro-electric and ferroelastic phenomena, with the domain configurations being modified not only by the applied electric field but also by the stress or strain within the material. This coupling between the electric and mechanical domains is fundamental to understand the material's response to external stimuli and is a key aspect in the design of devices that exploit both the electric and mechanical properties, such as actuators, sensors, and transducers. During switching under an electric field, avalanches can be observed as jumps in polarization and strain [11, 31], reflecting the reconfiguration of domains [32].

5.2.4 Origin of avalanches

Avalanches are non-linear collective activation of multistable elements caused by the interplay between positive feedback interactions, e.g. ferroic or autocatalityc interactions, and disorder or heterogeneity in adiabatic and athermal conditions. In the case of ferroic switching, the origin of avalanche dynamics can be addressed considering condensed matter arguments. Consider a system of multistable interact-ing elements, e.g. materials, which, in the absence of disorder, typically exhibit a first-order and continuous phase transition. In terms of ferroelasticity these can be different crystallographic structures (phases) or different crystallographic orienta-tions (domains). According to the so-called Harris criterion, the presence of quenched disorder introduced in such systems, e.g. defects as described in chapter 4 (point defects, dislocations, grain boundaries and other domain walls, secondary phases and precipitates), dominates correlation lengths if the hyper-scaling relation $\nu < 2/d$ [33]. This is the typical scenario in bistable systems, such as the Ising model in 2D and 3D lattices, and the mean-field approximation which we take here as a prototype model to introduce the micromechanics of avalanches in ferroelastic materials [34]. Quenched disorder hinders spontaneous ordering and prevents thermal criticality at the thermodynamic limit by damping long-range correlations. Below the Curie temperature thermal fluctuations play a minor role and the system effectively behaves as athermal at the thermodynamic limit, as can be shown through renormalization group arguments [35]. A new fixed point appears under renormalization at $T = 0$ with properties of a critical point given the right amount of heterogeneity. Truncated correlations in such athermal conditions generate rare regions that transform collectively. The system is intermittently trapped in metastable states associated with rare regions causing the observe quiescence in $v(t)$. The punctual excursions of $v(t)$ mark the stability limit triggering the fast transformation of such regions between the metastable states in short periods of time. These events are what we properly refer here as avalanches. Note that, according to this definition, avalanches correspond to the transformation of spatially correlated regions. Consequently, the power-law distributions of avalanche magnitudes correspond to scale-free correlation lengths. The match between avalanches at the conceptual level and intermittencies in measures is a delicate matter when spatio-temporal correlations are observed between individual avalanches (section 5.4).

5.2.5 Micromechanics of avalanches

Avalanche dynamics is a collective phenomenon described in atomistic models constituted by interactive elements or regions with multiple competing stable states. The transformation of elements or regions favors the transformation of other elements linked by positive feedback interactions. Consider an Ising model where an initial element, e.g. a crystallographic cell, rapidly transitions between configurations due to external driving. This event will change the internal stress, electric field or magnetic field of all interactive elements, e.g. neighboring cells. This will trigger the transition of some of them, constituting what we call the first shell, or time-step of the avalanche. The avalanche propagates from each of the transformed elements through the same triggering mechanism. In idealized conditions, we consider that the transformation profile, i.e. the number of elements in each shell, is proportional to the external measure $v(t)$. Considering this atomistic Ising representation, the avalanche size (S) and duration (T) are proportional to the total number of transformations triggered within the avalanche and the number of shells, respectively (figure 5.1).

The avalanche process represented in such terms is, by definition, a stochastic branching process. In particular avalanches in the mean-field approximation, i.e. within high dimensional spaces or all-to-all interactions, correspond to the case where the reproduction number, i.e. the probability of propagating a number of

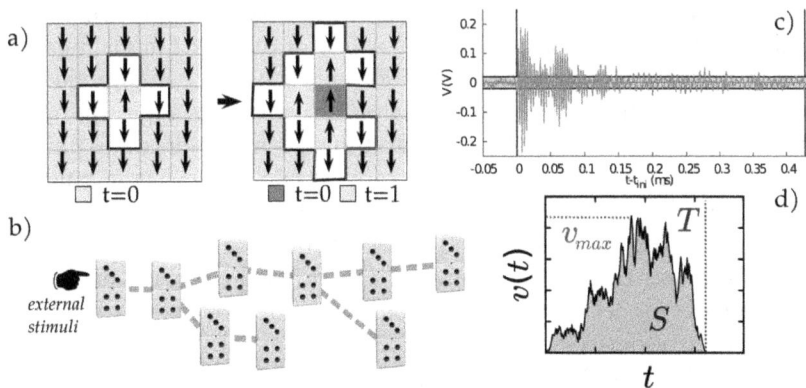

Figure 5.1. Schematic representation of avalanche profiles. Propagation in systems, even (a) lattice models can be interpreted as (b) branching processes. Ferroelastic avalanches can be detected as well as (c) delimited acoustic events or (d) through other channels $v(t)$ such that duration (T), size (S) and amplitude (v_{max}) can be assigned.

elements, is invariant. This is a memoryless, i.e. Galton–Watson, branching process, which renders a universality class with exponents $\kappa_S = 3/2$ and $\kappa_T = 2$ [36]. These statistical laws can also be linked to return-time random walks [37]. Note, however, that the avalanche profiles scale as a multiplicative process. Rather than a random $v(t)$ profile, the connection between mean-field (MF) avalanches and random walks is better achieved through alternative representations such as the Harris path, i.e. depth-first exploration, of the shells [38], where avalanche sizes are proportional to return times and durations to drifts.

Many natural avalanche processes fit well within the mean-field universality class. This might occur due to interactions that are long-range interactions, small-world networks, or remote [25, 39–42], or even self-organized pruning of loops in directed networks [43]. In contrast, short-range interactions will assist or inhibit avalanche propagation and hinder such a direct mapping to branching processes by introducing non-trivial dependencies in the reproduction numbers. Alternative stochastic models are used to study avalanches in reduced dimensions, such as percolation theory in regular lattices [44] or more complex constructs such as quenched Edwards–Wilkinson models or the so-called ABBM [45] or Hébraud Lequeux models [46], which allow us to describe interface depinning such as the propagation of domain walls, imbibition fronts, or growth of cracks. Each theoretical framework provide specific critical exponents beyond the so-called mean-field universality class. Note that natural avalanche phenomena might elude any classification within universality classes, but rather depend on external conditions such as the anisotropic external stress (see chapter 8), the internal properties such as the deformability of the materials [42], or the formation of power-law pseudo-gaps [47]. Section 5.4.2 provides further details on how to estimate the critical exponents defining marginal and bivariate distributions.

5.2.6 Molecular dynamics simulations of ferroelastics

Molecular dynamics (MD) simulations provide a powerful approach for studying ferroelastic materials at the atomic scale, allowing for the direct modeling of domain structures, wall dynamics, and transformation mechanisms. Interatomic potentials are constructed to capture the key features of ferroelasticity, typically by incorporating multiwell energy landscapes that reflect the symmetry-breaking nature of the phase transition and the associated spontaneous strain states. These potentials can be derived from phenomenological models, such as Landau theory, or calibrated against first-principles calculations and experimental data to reproduce the correct elastic and structural responses. Figure 5.2 (a) shows the scheme of the neighbor interactions of the model. These interactions are designed so that the equilibrium distances between atoms correspond to local minima of the total potential and preserve the underlying lattice symmetry. In particular, in figure 5.2 (a), the arrangement of the equilibrium positions corresponds to the plane group *c2mm*.

This model can be used to study phase transitions and elastic properties [48], but also to study the avalanches produced by the twin reconfigurations. A representative implementation of such a model is presented in [48, 49], where large-scale molecular

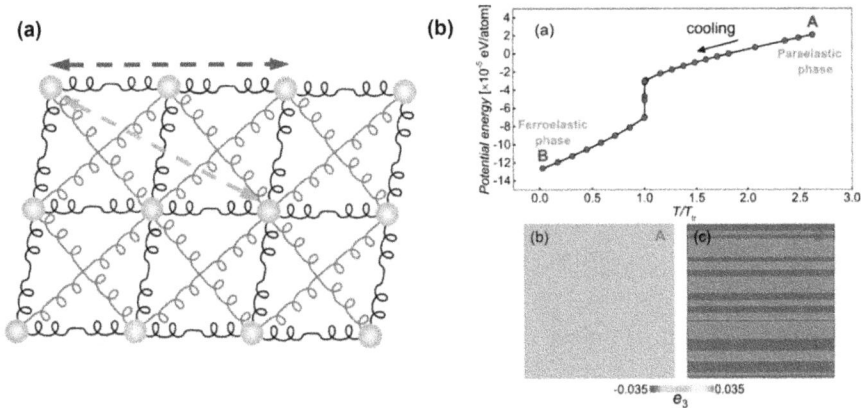

Figure 5.2. (a) Schematic of the neighbor interactions. (b) Ferroelastic phase transition simulated with the model where the transition is observed as a change of the potential energy and the twin domain structure is formed at the ferroelastic state. (Reproduced from [48]. CC BY 4.0.)

dynamics simulations using the LAMMPS package are used to investigate the kinetics of ferroelastic domain wall motion. The study demonstrates how kink nucleation and lateral propagation along the domain walls give rise to avalanche-like dynamics, in particular at low temperatures. The model reproduces key features such as Vogel–Fulcher behavior and power-law statistics of jerky wall motion without invoking extrinsic disorder [49–51].

Further insight into the kinetics of ferroelastic switching is provided in [9], where molecular dynamics simulations reveal that the strain rate has little influence on avalanche statistics at low driving, preserving robust power-law scaling. At high strain rates, however, the system deviates from critical-like behavior, with a suppression of large avalanches and a breakdown of scale invariance due to limited relaxation time.

The influence of lattice defects on ferroelastic switching dynamics is investigated in [52], where molecular dynamics simulations show that immobile vacancies interacting with twin walls (see figure 5.3), junctions and kinks lead to enhanced pinning and a suppression of large avalanches. This is reflected in an increase of the power-law exponent governing the avalanche size distribution, indicating a shift toward more localized, less correlated switching. Similarly, [53] shows that fixed point defects serve as preferential nucleation centers for domain wall segments, leading to a modification of the critical scaling behavior.

Twin interfaces can host emergent properties such as polarity, conductivity, or magnetism, even in the absence of magnetic constituents. This opens the way for domain wall-based multiferroic device concepts [10]. Furthermore, these models have also been extended to explore the potential of ferroelastic domain structures in neuromorphic applications, as shown in [54], where strain-mediated interactions and avalanche-like switching are proposed as mechanisms for memory and logic functionalities in nanoscale devices.

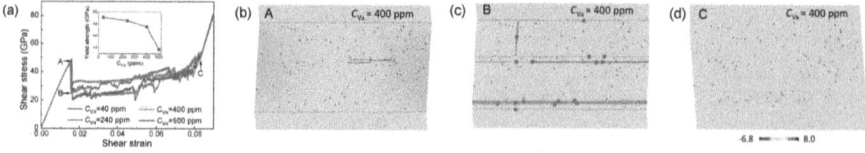

Figure 5.3. (a) The shear stress–strain response at various vacancy concentrations, highlighting yield and plastic regimes. (b) Atomistic snapshots at key points illustrate how vacancies (A) nucleate needle twins, (B) develop complex twin structures, and (C) lead to complete de-twinning. The color map encodes local shear angles relative to the bulk structure. (Reproduced with permission from [52]. Copyright 2019 Elsevier.)

Molecular dynamics models show that avalanche-like switching in ferroelastic materials can occur without quenched disorder, driven purely by intrinsic lattice interactions. However, topological defects such as kinks—localized steps along domain walls—play a crucial role in this dynamics. It has been shown that kinks mediate long-range elastic interactions between twin walls, with a crossover from dipolar to monopolar behavior depending on film thickness [55]. In addition, kinks can propagate at supersonic speeds along the walls, enabling rapid, inertia-driven switching [56]. These findings highlight the importance of kink dynamics in governing ferroelastic avalanches, even in structurally ideal crystals.

Moreover, molecular dynamics simulations have been used to directly compute the acoustic emission signal associated with ferroelastic switching events, as demonstrated in [57], where yield- and kink-driven domain collapses produce distinct acoustic emision (AE) energy bursts on the order of meV per atom.

5.2.7 Crackling noise in the random field Ising model

The zero-temperature random field Ising model (RFIM) with metastable dynamics was first introduced to study fluid invasion in porous media and front propagation in disordered systems [58]. Currently, the RFIM, usually implemented in regular lattices with nearest-neighbor interactions, acts as the prototype micromechanical model of avalanche dynamics in ferroic transitions. The Ising model with $J > 0$ (typically reduced to $J = 1$) within regular spin d-lattices with $d > 2$ is a paradigmatic case fulfilling the Harris criterion, and is hence dominated by the presence of quenched disorder, introduced here as the addition of internal magnetic fields (h_i) i.i.d. from a given distribution $\rho(h_i|R)$, where R is a measure of disorder inversely proportional to the density of field values $\nu(h)dh$. The Hamiltonian of the system reads as follows:

$$\mathcal{H} = -J\sum_{\langle i, j\rangle}S_i S_j - \sum_i^N h_i S_i - H(t)\sum_i^N S_i. \tag{5.3}$$

Under monotonous and quasistatically slow external driving $H(t)$ and starting from an original configuration $\{S_i\} = \{-1\}$, each spin flips according to the condition $h_i < H(t) + J\sum_{\langle i, j\rangle}S_j$. Therefore, a spin can flip due to the external driving, triggering an avalanche, or by the flip of an interacting spin, propagating the avalanche by building the aforementioned branching processes.

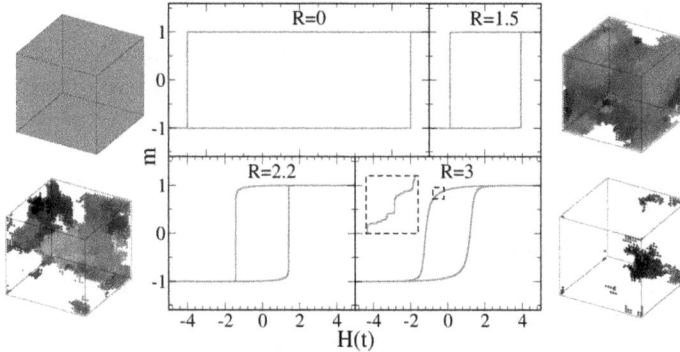

Figure 5.4. Magnetization loops in the athermal 3D-RFIM as a function of disorder. The largest avalanches in the loops are represented at the sides: Preisach hysteresis in the absence of disorder ($R = 0$) with a single avalanche; giant avalanche for low disorder ($R = 1.5$); scale-free avalanches around critical disorder ($R = 2.2$); and finite avalanches at high disorder ($R = 3.0$).

The propagation of avalanches in the RFIM is governed by the ratio between the interaction strength J and the magnitude of disorder R [59], giving rise to a phase transition around a critical degree of disorder (R_c) separating avalanche dynamics in three fixed points under renormalization (figure 5.4) [35]. For low disorder ($R < R_c$) the energy exchanged by the interaction overcomes the pinning due to disorder and ferroic switching is dominated by a massive and exponentially fast-growing avalanche, i.e. snapping noise [59], spanning a macroscopic fraction of the system, plus some small other avalanches. At high disorder ($R > R_c$) avalanches, also known as popping noise [59], are finite and disappear below noise at the disordered regime in the thermodynamic limit. Only close to critical disorder ($R \approx R_c$), i.e. the crackling noise regime [59], are avalanche dynamics relevant in the description of the switching, since their propagation is scale free and non-negligible at the thermodynamic limit. Close to R_c, avalanche statistics can be effectively used to characterize the state of the system. Rather than a fine-tuning coincidence, crackling noise is ubiquitous in nature [16–18, 28], which motivated the theory of SOC [4, 44], although scale-free statistics can be the result of integrating avalanche statistics around a critical point during the sweeping of instabilities [60]. In any case, scale-free avalanche distributions are characterized by critical exponents, which provides insight on the physical properties of the material and the avalanche propagation process.

5.3 Methods of avalanche recording

The investigation of avalanche dynamics in ferroelastic materials relies on experimental techniques capable of capturing the sudden, collective rearrangements that occur during domain wall motion or phase front propagation. These avalanches, which arise as a response to a slow, continuous external drive—such as applied stress, electric field, or temperature change—can manifest across a broad range of spatial and temporal scales. Each method used to detect these events probes a specific physical observable, offering insight into different facets of the phenomenon. In this section, we describe the principal experimental techniques employed to record avalanches, along with the physical quantities they measure and the interpretations they allow.

5.3.1 Acoustic emission

Among the earliest techniques employed to investigate avalanche phenomena in ferroelastic materials is acoustic emission (AE), which capitalizes on the detection of elastic waves generated by sudden microstructural events. When domain walls move abruptly or phase boundaries propagate discontinuously, they release stored elastic energy in the form of mechanical stress waves. These waves, typically in the ultrasonic frequency range, are detected by piezoelectric transducers coupled to the sample surface. The transducers convert mechanical vibrations into electrical signals, which can then be analysed in terms of amplitude, duration, energy content, and event rate. These parameters serve as indirect indicators of the underlying avalanche dynamics.

The strength of acoustic emission lies in its exceptional sensitivity to a broad range of event sizes and timescales. A single transducer can detect both minute localized rearrangements and large-scale system-spanning events, making AE particularly suited for statistical analyses over multiple orders of magnitude. Power-law distributions of event energies and interevent times are often observed, consistent with the critical behavior expected in driven disordered systems. Moreover, the technique is non-destructive and can be applied in real time under various experimental conditions, such as temperature ramps, mechanical loading or under an electric field.

However, this versatility comes with limitations. Acoustic emission does not provide spatial resolution; the exact location, geometry, and structural nature of the source event remain unknown unless triangulation is performed with an array of sensors, which is often impractical at small scales. Furthermore, AE signals are inherently indirect: they reflect the release of elastic energy rather than the domain wall movement itself. As a result, while AE can confirm the intermittent nature of the transformation and characterize the energy landscape statistically, it cannot, on its own, reveal the microscopic origins of avalanche behavior. An additional constraint is that the recorded AE signal is not a direct image of the event, but rather a convolution between the dynamic response of the sample and the transfer function of the sensor [61, 62]. As highlighted by Casals *et al.* [61], this convolution distorts the temporal shape and energy profile of the original avalanche, meaning that pulse duration, amplitude, and energy are shaped not only by the physical event but also by the acoustic transmission properties of the system and the detection sensors. With this motivation, recent work has proposed new experimental strategies designed to directly capture the avalanche source itself, using acceleration-based methods [63] that minimize filtering by the sensor and more accurately reflect the dynamics of the transformation front.

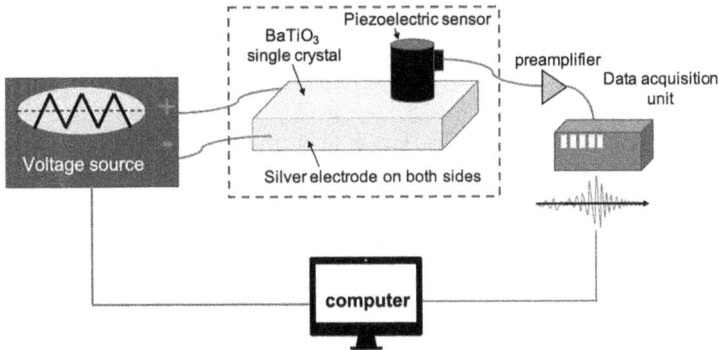

Figure 5.5. Schematic set-up for acoustic emission recording during the $BaTiO_3$ switching. (Reproduced with permission from [11]. Copyright 2019 the American Physical Society.)

Despite these constraints, acoustic emission has been successfully employed in a variety of ferroelastic systems. Thermally driven phase transitions, for instance, generate detectable AE signals as a function of temperature, as observed in materials such as $Pb_3(PO_4)_2$ [64], $BaTiO_3$, and PMN–PT [65], as well as in martensitic compounds [13, 14, 66]. Under applied electric fields, AE has also been used to investigate ferroelectric $BaTiO_3$ [11] (shown in figure 5.5), piezoelectric PZT [12], and materials such as gadolinium and terbium molybdates [67].

Beyond experimental observation, acoustic emission signals have also been the subject of theoretical modeling. Notably, statistical mechanical models have been developed to describe AE activity during martensitic transformations, treating the acoustic bursts as emergent phenomena arising from collective domain rearrangements under an external drive. For example, Planes, Castán, and Saxena [57] proposed a model that reproduces both the energy distribution and temporal correlations of AE events, capturing the key features of experimental AE data through a minimal description of the system's free energy landscape and dissipation mechanisms. Such approaches provide a crucial theoretical framework for interpreting AE signals not merely as empirical observations, but as the macroscopic manifestations of underlying thermodynamic and kinetic processes.

5.3.2 Dynamic mechanical analysis

Dynamic mechanical analysis (DMA) offers a direct way to probe ferroelastic materials by coupling to their fundamental order parameter—strain. Unlike external drives such as temperature or electric field, which act indirectly, DMA induces domain wall motion by applying mechanical stress—typically in a three-point bending or parallel-plate geometry—while recording the sample's deformation with high temporal resolution.

If the strain response were smooth, it would suggest a continuous reconfiguration of the domain structure. Instead, what is often observed is a jerky evolution: the deformation proceeds through a sequence of sudden jumps superimposed on a slowly varying background (see figure 5.6). These discontinuities in strain are

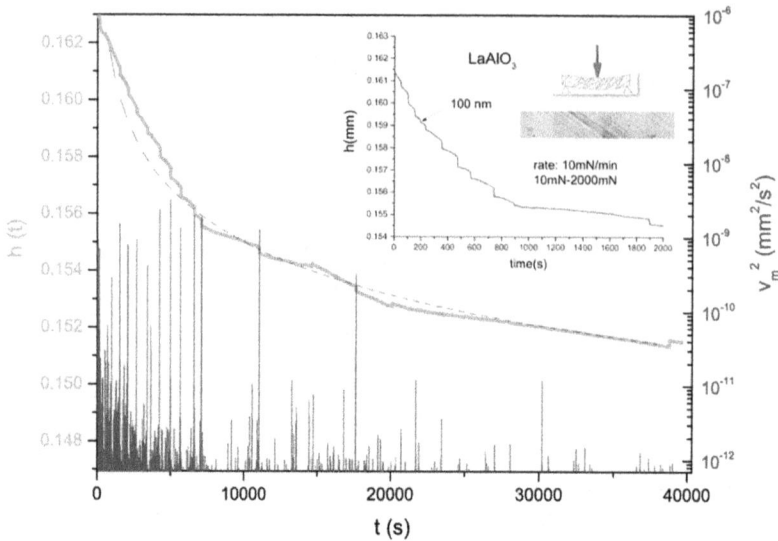

Figure 5.6. Compression experiment on LaAlO₃ at 295 K under a three-point bending set-up. The green line shows the sample height during loading at 3 mN/min. Blue lines represent the squared maximum velocity drops, indicating discrete deformation events. A stretched exponential fit is shown in red. (Reproduced from [30]. CC BY 4.0.)

interpreted as avalanches of ferroelastic domain walls, triggered when the external load exceeds local pinning thresholds. The time derivative of the strain signal, typically analysed in squared form, serves as a proxy for the energy released during each event.

Under slow loading, these jumps follow power-law statistics, consistent with critical avalanche behavior as observed in materials such as LaAlO₃ and PbZrO₃ [30, 68]. Like acoustic emission, however, DMA is a macroscopic technique. It captures the temporal signature of avalanches but does not resolve where in the material they occur. The measured signal reflects the collective response of the entire sample, and individual events cannot be spatially localized.

5.3.3 Optical microscopy

Optical microscopy provides a powerful means of observing ferroelastic avalanches with spatial resolution. Unlike macroscopic techniques such as acoustic emission or DMA, which only record the integrated response of the entire sample, optical imaging allows individual avalanches to be localized in space, offering a direct view of how domain walls move through specific regions of the material.

In particular, polarized light microscopy is widely used to visualize ferroelastic domain structures. The contrast arises from birefringence, a property of anisotropic crystals whereby different orientations of the crystallographic axes refract polarized light differently. During a ferroelastic phase transition, domains with different strain states and orientations form, and these differences manifest optically under crossed polarizers as regions of varying brightness or color. This makes it possible to track the evolution of domain patterns in real time, including sudden, discontinuous motions associated with avalanche events.

Like all optical techniques, this method is limited by diffraction-limited resolution, and therefore operates at the mesoscale. While it does not resolve individual domain walls, it does allow clear identification of domain shapes, positions, and interactions. This scale is particularly well suited for studying the collective behavior of domains, including how avalanches are initiated, propagate, and are influenced by nearby domains or microstructural features.

An illustrative example is $LaAlO_3$, where needle-shaped domains, observed under optical microscopy, advance in discrete steps under applied stress. When DMA is combined with optical imaging, these jerky movements can be tracked in both time and space, revealing localized avalanches at the mesoscopic scale [68].

Thanks to optical microscopy, different types of ferroelastic domains—distinguished by their twin symmetries—can be observed directly. In $SrTiO_3$, domain rearrangement under applied electric field exhibits distinct behaviors depending on the complexity of the domain pattern: simple, well-aligned twin structures give rise to jerky, avalanche-like motion, whereas more complex, interwoven configurations display smoother, glassy relaxation dynamics [29]. Similar behavior has also been observed in $LaAlO_3$, where optical imaging revealed that domain pattern complexity and sample geometry influence avalanche statistics, including transitions between distinct scaling regimes [69].

Optical microscopy has also been used to study the coupled ferroelectric and ferroelastic switching under an electric field in single crystals of $BaTiO_3$ and PMN–PT [32]. By imaging the switched regions during avalanches (see figure 5.7), the study revealed that the switching areas exhibit a fractal geometry, whose complexity varies with the applied electric field. At coercive field, the switching becomes

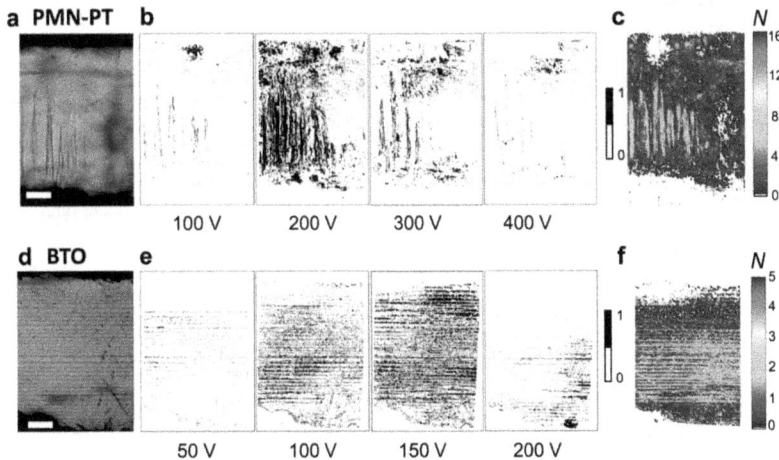

Figure 5.7. (a), (d) Initial domain configurations in PMN–PT and BTO, respectively. (b), (e) Binary maps showing switched (black) and unchanged (white) regions between consecutive frames under applied voltage. (c), (f) Cumulative maps over a full hysteresis loop, where the color scale represents the number of switching events per pixel (1 μ m^2). (Reproduced from [32]. CC BY 4.0.)

maximally fractal, and the size distributions follow a clearer power-law behavior, indicating that critical avalanche dynamics emerge most strongly near the coercive regime.

5.3.4 Transmission electron microscopy

Transmission electron microscopy (TEM) provides an exceptional window into the dynamics of ferroelastic avalanches at the nanometer scale. *In situ* TEM observations in $BaTiO_3$ single crystals have documented individual Barkhausen-like pulses during domain wall motion triggered by an electric field. Specifically, frame-by-frame tracking of single 90° needle domains revealed sudden, step-like displacements of their tips as shown in figure 5.8, corresponding to localized avalanche events and suggesting the influence of local pinning interactions [70]. These measurements, while limited in statistical breadth due to the small field of view, offer direct nanoscopic to microscopic evidence of domain wall avalanches and highlight the crucial role of nearby domain structures in triggering sudden jumps. In the future, this technique is expected to play a key role in resolving the dynamics of complex domain wall topologies—such as kinks, jamming points, steps, and curvature-driven rearrangements—which remain largely inaccessible to other experimental approaches.

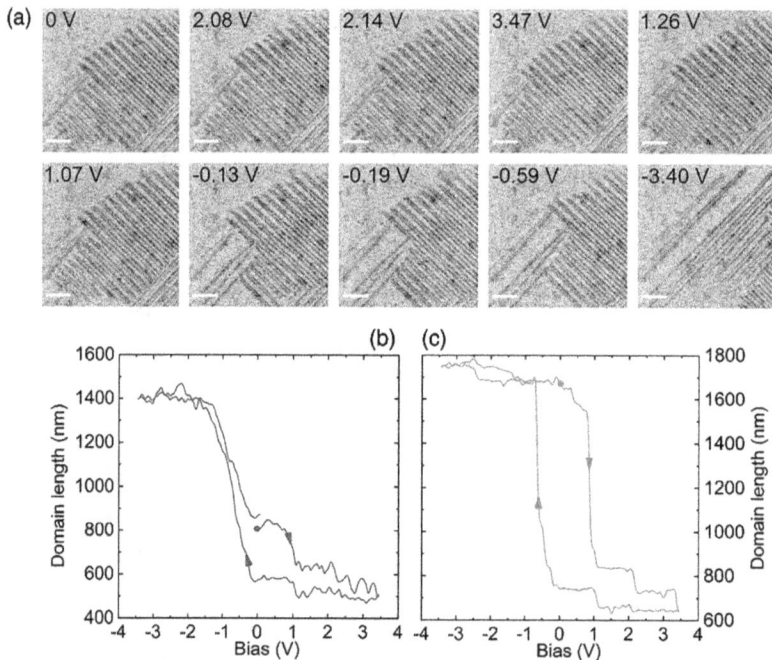

Figure 5.8. Evolution of the $BaTiO_3$ domains as a function of applied voltage imaged by TEM. (Reproduced from [70]. CC BY 4.0.)

5.3.5 Atomic force microscopy

Atomic force microscopy (AFM) provides nanometer-scale spatial resolution, allowing precise imaging of ferroelastic and ferroelectric domain structures. However, as a scanning technique, AFM suffers from limited temporal resolution: switching events are only captured through comparison of successive scans, often separated by seconds or minutes. This makes it challenging to directly resolve the fast dynamics of avalanches in real time.

Nonetheless, this limitation has been ingeniously addressed in recent works. For example, in studies on $Pb(Zr_{0.2}Ti_{0.8})O_3$ thin films, AFM was used to alternate subcoercive electric biasing with piezoresponse imaging, enabling the detection of individual switching events by differencing consecutive scans. The resulting switched areas—interpreted as individual avalanches—were analysed statistically to reveal power-law behavior and distinct regimes depending on the applied field and sample history [72]. A similar protocol has been used to study changes in avalanche dynamics as a function of the concentration of point defects [73].

AFM-based nanoindentation—with nanonewton forces held for hours—reveals jerky 'crackling' events in ferroelectric $PbTiO_3$, behaving analogously to a nanoscale DMA (as shown in figure 5.9), where the squared velocity of indentation fluctuations provides a proxy for avalanche energy and dynamics [71].

5.3.6 Displacement current

In the specific case of ferroelectric/ferroelastic materials, avalanches can be measured with the displacement current, which originates from the changes in polarization of the material when the sample is mounted as a capacitor. Each time a domain changes, a small current is measured as a consequence of the charge compensation in the capacitor electrodes. By knowing the area and integrating the current to compute the charge, the polarization can be calculated. Some studies have used jumps measured in the current to study the avalanche behavior during the ferroelectric/ferroelastic switching [31, 74]. This method could work as well for pure ferroelastic materials, but only in the case where the domains eventually exhibit

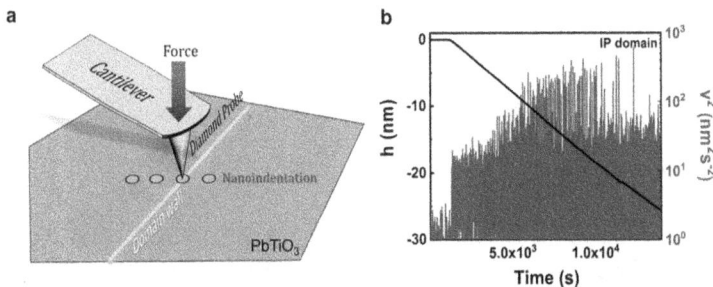

Figure 5.9. (a) Schematic of the AFM nanoindentation method, where a constant nanonewton-scale force is applied via the tip while monitoring surface displacement over time. (b) Representative signal of vertical displacement recorded by the AFM tip, exhibiting discrete jumps indicative of nanoscale avalanche events during local deformation. (Reproduced from [71]. CC BY 4.0.)

large enough dielectric constants. If a domain changes, a significant change in the dielectric constant will occur, which can be measured as a result of the domain switching. In this case, the avalanche will be measured as a change in capacitance.

5.4 Modeling avalanche statistics

Due to the strong role of disorder coupled to the strong non-linearity, deterministic, or even non-linear, time series analyses are unlikely to provide relevantant information on the dynamics of avalanches. We propose stochastic models as a viable alternative to study the bulk statistics of avalanches or characterize avalanche time sequences. The most common approach is to study the statistical properties of avalanches by modeling the sequences as a spatio-temporal point process [75] where events i are punctual in space and time $(t_i, \mathbf{r}_i)$, ordered in time $(t_i < t_{i+1} \ \forall \ i)$, and avalanche profiles are reduced to relevant scalar parameters ($\{X_i\}$) marking the event by its amplitude or energy, extension, and duration. Note that in conceptual models considering uncorrelated heterogeneity, i.e. pure random disorder, and infinite system sizes, avalanche dynamics and statistics map directly to stochastic point processes with an i.i.d. distribution of sizes [36].

In general, we will consider *stochastic marked point processes* (p.p.) with some exogenous factors that introduce heterogeneity in space and time and a non-Markovian dependence in history ($\mathcal{H}_t$), i.e. the whole sequence of previous events up to time t. The point process is perfectly defined by the probability of finding an avalanche in an infinitesimal space–time interval, i.e. an intensity:

$$\mu(\mathbf{r}, t \,|\, \mathcal{H}_t) \, d\mathbf{r} \, dt = \mathrm{E}\{\text{event} \in [(t, t + dt), (\mathbf{r}, \mathbf{r} + d\mathbf{r})]\}. \tag{5.4}$$

Point processes can also uniquely be defined as *duration* and *counting* processes. Duration processes account for the interevent or waiting times between consecutive events in a sequence such as $\delta_i := t_i - t_{i=1}$ for $i > 1$ and $\delta_1 = t_1$. Note that the point process is uniquely recovered as $t_i = \sum_{k \leq i} \delta_k$. Given (5.4) we can provide the probability distribution function of waiting times on a given sequence by defining a binomial process of Bernoulli trials in discretized time intervals ($\delta \to \delta_k = k \, \Delta t$) such that an event occurs with probability $\mu(\mathbf{r}, t + k \, \Delta t \,|\, \mathcal{H}_t) \, \Delta t$ as

$$\rho_\delta(\delta_i = \delta) d\delta = \lim_{\Delta t \to 0} \left\{ \mu_k \, \Delta t \prod_{j=0}^{k} [1 - \mu_j \, \Delta t] \right\} = \mu(\mathbf{r}, t \,|\, \mathcal{H}_t) \, e^{-\int_{t_{i-1}}^{t} \mu(\mathbf{r}, t' \,|\, \mathcal{H}_t) dt'} d\delta. \tag{5.5}$$

Alternatively, counting processes ($N(t)$), register the number of events in time,

$$N(t) = \sum_{i \in \mathbb{N}^+} \mathbb{I}(t_i < t) = \int dt' \sum_i \delta(t' - t_i), \tag{5.6}$$

where $\delta(x)$ is the Dirac delta. Note that, under this definition, and given (5.5), the probability of finding a specific train of events $\{t_i, \mathbf{r}_i\}$ in counting process $N(T)$ reads

$$p_{t_1,\ldots,t_N}(t_1, \ldots, t_N \,|\, \mathcal{H}_T) dt_1 \ldots dt_{N(T)} = \prod_i^{N(T)} \mu(\mathbf{r}, t_i \,|\, \mathcal{H}_t) \, e^{-\int_{t_{i-1}}^{t_i} \mu(\mathbf{r}, t' \,|\, \mathcal{H}_t) dt'} dt_1 \ldots dt_{N(T)}. \tag{5.7}$$

From this expression, we can derive the log-likelihood function from a model defined from an intensity $\mu(\mathbf{r}, t | \mathcal{H}_t, \{\Omega\})$ depending on a parameter set $\{\Omega\}$ as

$$\log \mathcal{L}(T | \mathcal{H}_T) = \sum_{i \in 0, N} \log \mu(\mathbf{r}, t | \mathcal{H}_t) - \int_0^T \mu(\mathbf{r}, t | \mathcal{H}_t) dt. \qquad (5.8)$$

Note that this parameter space can be cumbersome when the intensity is time-modulated, i.e. the parameters have a non-trivial dependence on time $\{\Omega(t)\}$ that only can be properly described through numerical interpolation, multiplying arbitrarily the dimension of the parameter space. This is a typical situation in avalanche processes linked to phase transitions such as ferroelastic switching, where the activity rate changes over time throughout the hysteretic curve. We will introduce here alternative means to characterize the point processes, based on the identification among the most common point process models to describe avalanche dynamics.

5.4.1 Poisson processes

Considering the definition of avalanches as correlated events, the most common assumption is to consider independence between avalanches, i.e. Poisson processes. The intensity reads $\mu(t | \mathcal{H}_t) = \mu(t)$ with, at most, a time-modulation depending on exogenous factors. Because of the independence on the history, the intensity can be evaluated as the mean number of events k found inside the time interval $[t, t + \Delta t)$:

$$\mu(t) = \left[\lim_{\Delta t \to 0} \left\langle \frac{N(t + dt) - N(t)}{\Delta t} \right\rangle_M \right](t). \qquad (5.9)$$

Although the Poisson process hypothesis seems reasonable, one has to account for finite-size effects when massive avalanches occur. Therefore, we expect this to be a reliable null-model to describe subcritical and critical regimes.

If we exclude time-modulation, the Poisson processes is homogeneous with a time-independent intensity $\mu = E(N(t_1) - N(t_2))/(t_2 - t_1)$, and the distribution of waiting times (5.5) is reduced to an exponential decay:

$$P_\delta(\delta) dt = \mu \exp(-\mu\delta) dt. \qquad (5.10)$$

Alternatively one can prove the expression above from a *reductio ad absurdum*. Since a Poisson process is memoryless: $\text{Pr}(\delta_i > \delta_1 + \delta_2) = \text{Pr}(\delta_i > \delta_1) \text{Pr}(\delta_i > \delta_2)$. Only an exponential distribution fulfills this property [76].

5.4.2 Renewal processes

Renewal point processes are those whose $\mu(t, \mathcal{H}_t)$ is reset after each event. Therefore, the occurrence times are Markovian, perfectly determined by the probability distribution of waiting times. The intensity can be factorized for each event i such that

$$\mu_i(t_i = t; \mathcal{H}_t) = \rho_\delta(t - t_{i-1}). \qquad (5.11)$$

Therefore, renewal processes are completely determined by the interevent time distribution (5.5) and the intensity parameter μ can be estimated from any renewal process by means of the elementary renewal theorem (ERT) both as $\mu = E\left(\frac{1}{\delta}\right)$ and $\lim_{T\to\infty} \frac{N}{T} = \lim_{T\to\infty} \frac{E(N)}{T} = \mu$ [77]. As a straightforward example, a periodical process is represented as a delta distribution of waiting times, centered on $\delta = \mu^{-1}$, and homogeneous Poisson processes can also be interpreted as a renewal process with $\{\delta_i\}$ i.i.d. according to (5.10), denoting absolute independence between the events.

Renewal avalanche processes are typically found in some self-organized systems displaying synchronization [78–80] or in supercritical systems such as the characteristic earthquake theory [81]. Although supercriticality in ferroic switching will manifest as a singular massive event, milder forms of quasi-periodicity might appear as consequence of facilitation mechanisms such as inertial effects [82, 83].

Finally, events extracted from the thresholding of a continuous signal $v(t)$ will naturally introduce a dead-time or delay corresponding to the duration of such an excursion. No event can be detected in a shorter time than the duration of the last event. Considering a p.p. of independent events, the null-model will be corrected from a Poisson to a special case of renewal called a *dead-time Poisson* or *delayed Poisson* process [84] introducing the delay in the sampling after each event t_{i-1} defined by the duration mark T_{i-1}. The times expected from this model are i.i.d. from

$$P_{\delta_i}(\delta \mid T_{i-1}, t)dt = \begin{cases} 0 & ; \delta < T_{i-1} \\ \mu(t)\exp(-\mu(t)(\delta - T_{i-1}))dt & ; \delta \geqslant T_{i-1} \end{cases}. \qquad (5.12)$$

Instead of denoting Markovian correlations between the events, this model corrects an artificial correlation that is introduced during recording. Considering the scale-free nature of avalanche durations $\{T\}$ sequences of independent events with dead-times are easily mistaken as Weibull renewal processes. This mistake can be avoided by performing Poisson rejection tests (section 5.4.1) on the interevent times accounting for the duration of the previous event $\delta_i \to \delta_i - T_{i-1}$.

5.4.3 Self-excited processes

More complex avalanche activity might appear to display clustering of signals in short bursts of activity. These can be modeled as a self-excitation model where the intensity $\mu(t, \mathbf{r}; \mathcal{H}_t)$ depends on the previous activity as a linear combination of a Poisson background rate μ_0 and the excitation rate Ψ_i caused by each previous, i.e. parent, event:

$$\mu(t, \mathbf{r}; \mathcal{H}_t) = \mu_0(t, \mathbf{r}) + \sum_{t_i < t} \Psi_i(t, \mathbf{r} \mid t_i, \mathbf{r}_i, \{X_i\}). \qquad (5.13)$$

In terms of avalanche dynamics, this point process models the phenomena of event–event triggering or aftershocks, leaning on seismological terminology [85]. Event-event triggering has been reported in some ferroelastic phenomena. In particular, it has been observed in the martensitic transformation of CuZnAl alloys under thermal driving [15], represented in figure 5.10, yet no explanation nor specific

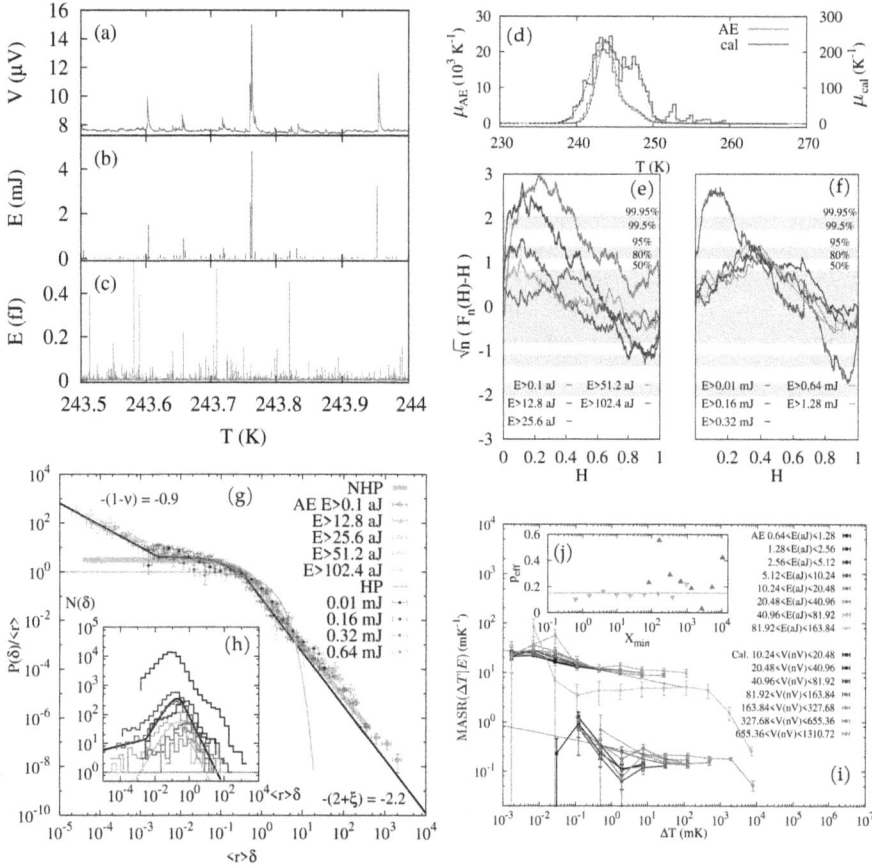

Figure 5.10. Point process characterization of an avalanche catalog during the martensitic transformation of CuZnAl [15]. (a) Raw intermittent cal. and AE signals are discretized in avalanche sequences for (b) cal. (c) and AE. The activity rate is modulated in time during the transformation in both (e) AE and (f) cal. channels. The distribution of interevent times (g) with and (h) without scaling is compatible with a non-trivial double power-law functional form. The MASR exhibits a slight power-law shape, although no productivity dependence on energy is observed. (Reproduced with permission from [15]. Copyright 2014 IOP Publishing.)

characterization of the phenomenon has been provided yet. In contrast, the events during the transformation of NiMnGa appear to be independent [86]. Event-event triggering can be explained as the outcome of interaction mechanisms significantly slower than the fast transformation detected as intermittence in $v(t)$. Therefore, rather than independent avalanches, the $v(t)$ channel is detecting fragments of an avalanche, or sub-avalanches, which can render subcritical sizes at criticality [87, 88].

The epidemic type aftershock sequence model (ETAS) [85], sometimes interpreted as a Hawkes process and sometimes as the outcome of a branching process, is built on the common observations of aftershock statistics as four independent factors:

$$\Psi_i(t, \mathbf{r}, E \,|\, t_i, \mathbf{r}_i, E_i) = \rho(E)\nu(E_i)\Psi_t(t - t_i)\Psi_r(\mathbf{r} - \mathbf{r}_i). \quad (5.14)$$

These are: the scale invariance $\rho(E)$ in energies or event sizes, known as the Gutenberg–Richter law in seismology; a power-law decay of the activity in time since the parent event, known as the Omori law, captured in a temporal kernel $\Psi_t(\tau) = \theta C^\theta (C + \tau)^{-(1+\theta)}$, where the time C imposes a limit on the highest production intensity and θ is a typically positive exponent; an spatial kernel around the parent $\Psi_r(\mathbf{r})$ typically decaying in distance; and the aftershock productivity law $\nu(E_i) = \nu_0(E/E_0)^\alpha$ determining the dependence of the triggering intensity with the energy of the parent event, where ν_0 is the characteristic production for a reference parent energy E_0 and $\alpha > 0$ the so-called productivity exponent.

The ETAS model is a rather sophisticated implementation of the Hawkes self-exciting process which mainly remains as an empirical observational model that fits the data well, rather than a fundamental law derived from first principles. Some aspects, such as the best representation of the spatio-temporal kernels [89], or more complex scaling relations on the production of aftershocks [90, 91] are still under debate. The observation of triggering phenomena in avalanche processes, in particular those phenomenologically far from seismology, such as ferroelasticity, might require alternative triggering kernels. As an example, the ETAS model imposes a power-law relation for short interevent values in the interevent distribution function with exponent $1 + (1 + 1/\theta)^{-1}$ [99] which do not coincide with the fitted exponents in the martensitic transformation of CuZnAl [15] (see figure 5.10).

5.4.4 Statistical characterization of point processes

A first step to study avalanche phenomena as stochastic processes is the identification and characterization of the underlying point process best describing the event statistics and their time-evolution and parameter dependencies. This might be a delicate task due to the limited information and a series of pitfalls if the data are not analysed properly. Nonetheless, misassignation of point processes and their interpretation are among the most common mistakes in studies of avalanche phenomena. We will consider here the study of avalanches obtained through the thresholding of an intermittent signal $v(t)$. Given precise information on the spatial propagation and location will provide more sophisticated methodologies [92], affected by their own pitfalls [93, 94], but these are still in early development in the study of ferric avalanches.

5.4.4.1 Poisson rejection tests

The evolutionary rate measured as $\langle dn/dt \rangle(t)$ in ferroic transitions over cycling, or series of repetitions of the same experiment, can include endogenous intensity terms on top of any temporal evolution of a Poisson rate. One must prove statistical independence between events before stating that the process is local Poisson $\langle dn/dt \rangle(t) = \mu(t)$, and not neglect history dependencies. To that purpose we introduce Poisson rejection tests. Considering independence between events as the null-model for avalanche processes, this is advised to be the first analysis to be performed. Such tests are design to be robust against time-modulated intensities expected in ferroic switching or phase transitions and are highly sensitive to small temporal correlations between events. These methods are based on the analysis of correlations between consecutive events, hence with asymptotically identical $\mu(t)$, interevent times δ_i, δ_{i+1} used to define a new statistical variable consisting in an interevent ratio that is known to be uniform between 0 and 1 if the process is locally Poisson. The R-test [95] defines the variable:

$$R_i = \frac{\delta_{i+1}}{\delta_i + \delta_{i+1}}. \tag{5.15}$$

An statistically significant excess at values $R \to 0$ suggests the existence of clustering, while an excess at $R \to 1$ suggest the presence of longer quiescent times than expected. Both effects typically appear simultaneously in the presence of triggering [19]. The Bi-test [96] uses a slightly different statistical variable (H_i) comparing the interevent time between t_i and the closest event before or after that event ($\delta t_i = \min\{\delta_i, \delta_{i+1}\}$) and $\delta\tau_i$ is the consecutive interevent time in the same temporal direction. It is easy to prove that $p(\delta t|\mu)d\delta t = 2\mu e^{-2\mu\delta t}d\delta t = p(\delta\tau|2\mu)$ (figure 5.11(a)). Therefore, in order to be uniformly distributed, the statistical observable reads

$$H_i = \frac{\delta t_i}{\delta t_i + \frac{1}{2}\delta\tau_i}. \tag{5.16}$$

Note that this object is symmetrical in time and offers an alternative interpretation. Higher regularity between consecutive events, as often expected in renewal processes, appears as an excess around $H = 2/3$, while clustering and quiescence, trademarks of event–event triggering, appear as a simultaneous excess at $H \to 0$ and $H \to 1$. These can be visualized easily using a Kolmogorov–Smirnov (KS) test on $\Delta f(H_i) = F(H_i < H) - H$, with $F(H_i < H)$ the cumulative distribution of H_i values (figure 5.11(b)). Triggering is easily identified in the KS test as a statistically significant rotated 'Z' shape, whilst regularity appears as a rotated 'S' shape (figure 5.11(c)).

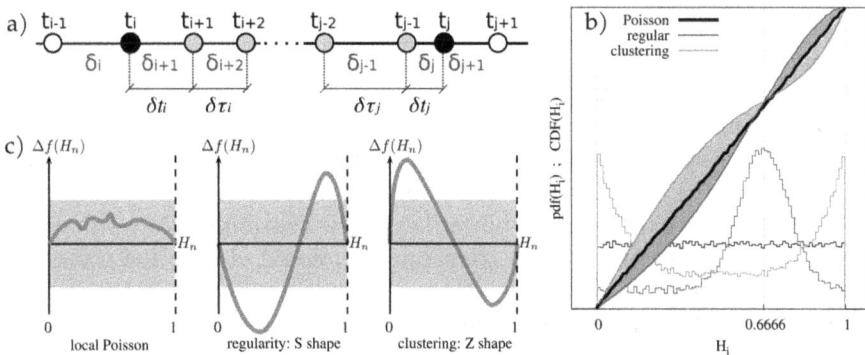

Figure 5.11. (a) Schematic description of δt_i and $\delta\tau_i$ selection in the Bi-test. (b) Expected distributions $pdf(H_i = H)dH$ and $F(H_i < H)$ for Poisson, renewal (regular) and self-excitation processes (clustering). (c) Expected differences $\Delta f(H_i)$ for the three cases.

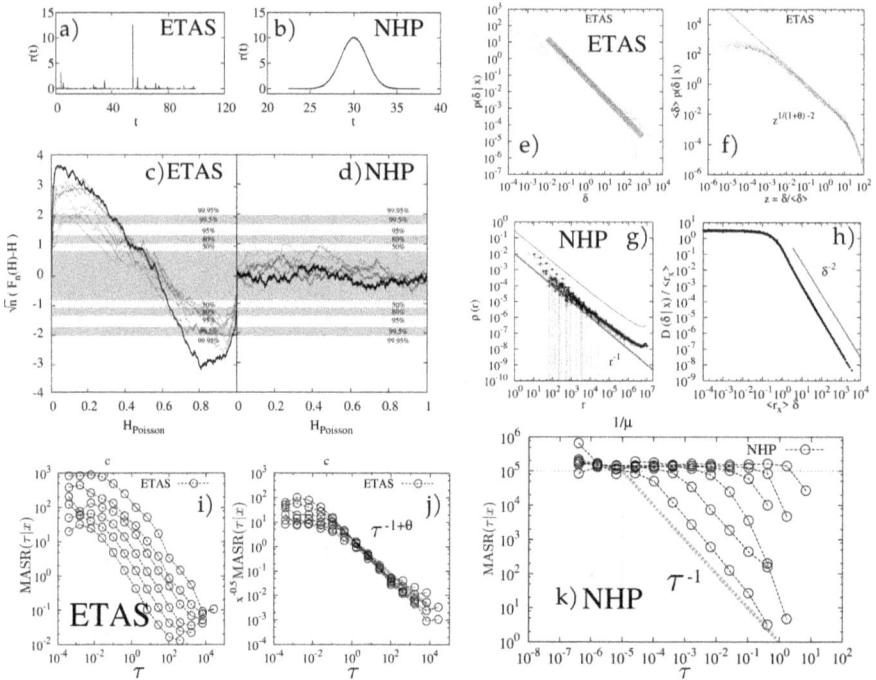

Figure 5.12. Example of the performance of the different methods against an ETAS model and a non-homogeneous Poisson (NHP) with a Gaussian activity profile. (a),(b) Recorded activity rates. (c),(d) KS tests on the Bi-test statistical variable H. (e) Distibution of waiting times in the ETAS model thresholding on mark x (f) rescaled by mean interevent times. Distribution of (g) rates and (h) waiting times in the NHP. (i) MASR in the ETAS model averaging on mark x and (j) rescaled curves according to the imposed productivity law. (k) MASR for the NHP.

In order to avoid resolution problems, it is advised to perform the Poisson rejection tests under thinning through thresholding of the size or magnitude marks x (see figures 5.12(c),(d) and 5.12(e),(f)). This guarantees that the lack of interactions is not an artifact due to data selection, noise or device resolution.

5.4.4.2 Interevent time distributions in modulated intensities

The most common mistake in the temporal analysis of point processes is the misinterpretation of non-exponential distributions of interevent times. These can be either introduced by a renewal processes, or non-Markovian correlations, but also by the modulation of the rate in a non-homogeneous Poisson process. In particular, we highlight that the existence of a power law in the distribution of waiting times is expected in Omori-like event–event triggering, but could also be the trademark of exogenous variations.

Of interest for ferroic transitions are as well the non-homogeneous, or local, Poisson processes, where the intensity $\mu(t)$ evolves over time as expected in the activity during a domain switch or phase transition. Note that the distribution of $\mu(t)\delta$ is an exponential distribution but not the distribution of μ. Since in most empirical data the parameter μ is fitted from $\langle 1/\delta \rangle$, it is difficult to obtain an instantaneous measurement of $\mu(t)$ and instead a long interval mean $\langle\langle\mu\rangle\rangle$ is used. Non-homogeneity in Poisson processes can be identified from the behavior of the distribution $P(\langle\mu\rangle\delta)$ at long δ which can display fat tails when the activity rises smoothly from null to a pick and/or dies out smoothly again, as is usual in ferroic switching. In particular, if we approximate the growth of the Poisson intensity as $\mu(t) \sim t^{1/\xi}$ from $\mu(0) = 0$ to a maximum value μ_{max}, the density of rates will be approximated by $\rho_t(\mu) \propto \mu^{\xi-1}$ for low rates, leading to a distribution of interevent times:

$$\rho(\delta) \propto \int_0^{\mu_{max}} \mu^\xi \rho(\delta|\mu)d\mu = \frac{1}{\delta^{\xi+2}} \int_0^{\delta\mu_{max}} (\delta\mu)^{\xi+2}e^{-\delta\mu}d\delta \propto \begin{cases} \delta^{-\xi-2} & \text{if } \delta \gg \mu_{max}^{-1} \\ e^{-\delta\mu_{max}} & \text{if } \delta \ll \mu_{max}^{-1} \end{cases}. \quad (5.17)$$

This modulation at low intensity will also cause a power law in renewal and self-exciting processes, more remarkably leading to the double power-law shape in the unified scaling law (USL) of earthquake interevent times [18, 97, 98]. This exogenous modulation is typically observed in ferroelastic avalanches either exhibiting temporal correlations or not [15, 86].

Facing any kind of temporal analysis, one has to always consider that only Poisson processes, with or without dead-times, are invariant to random, or conditional, thinning. Other renewal processes will not retain the same properties. However, the interevent times in the ETAS model are a close fit to the USL [97, 98] under thinning through size or energy thresholding. This can be visualized in the collapse of interevent times distribution after scaling with average interevent times $\langle\delta\rangle$ such that $\rho(\delta) = \Phi(\delta/\langle\delta\rangle)/\langle\delta\rangle$ (see figures 5.12(e) and (f)) where $\Phi(x)$ is a universal functional form that typically depends on the triggering kernel and exogenous evolution of background rates.

5.4.4.3 Mean aftershock sequence rates
Given the Omori-like relation in (5.14), one would expect local power-law decays in activity after large events, and scaling relations in a raw analysis of mean aftershock sequence rates. However, exogenous time-modulation can also cause such power-law decays and both phenomena can be easily mistaken for each other.

Note that the actual phenomenological Omori law recorded in catalogs of avalanches is not a direct measure of $\Psi_i(t)$ but needs to account to the contribution in activity rate of all the $g \to \infty$ generations of offspring caused by that same event. The resulting activity, so-called dressed or compound triggering rates [99], results in the convolution of direct triggering rates that can be expressed in terms of Laplace transforms $\mathscr{L}$ and inverse Laplace transforms $\mathscr{L}^{-1}$ such that

$$\Psi_c(\tau|m_i) = \nu(m_i)\sum_{g=1}^{\infty} \mathscr{L}^{-1}(\{n_b\mathscr{L}[\Psi_t(\tau)]\}^g). \quad (5.18)$$

This expression can be solved analytically in simple ETAS models, where a transition is observed between two power-law decays from $1 + \theta$ to $1 - \theta$ for long times [99]. An approximate solution can also be obtained in scale-free aftershock models [91].

Note that the branching description of the ETAS model imposes Markovian correlations in the parent–child relationship. All triggering trees and sub-trees can be considered independently without loss of generality, and are statistical equivalent given the same category of parent, defined by size or energy in the ETAS model. We can measure mean aftershock sequences rates (MASR) [91] as the activity rates conditioned to the presence of a trigger or parent (t_i, x_i), with x_i a mark determining the production term $\nu(x)$ in (5.14), usually the size or energy. The MASR is effectively implemented by performing averages in windows of x. This measure is equivalent to the compound rates plus the contribution of all events without a causal connection to the parent. If the triggering is low enough, we can consider this second contribution as an independent term:

$$\text{MASR}(\tau|x) \approx \Phi_c(\tau|x) + \left\langle \mu_0(t) + \int dt' \mu(t')\phi(t - t'|x) \right\rangle \approx \Phi_c(\tau|x) + \langle \mu(t) \rangle. \quad (5.19)$$

If the background activity is low and the contribution of independent events is small, $\text{MASR}(\tau|x_i) \approx \Phi_c(\tau|x_i)$. This same measure can be used to determine $\nu(x)$ given a generic dependence of production on mark x. This is achieved by simply scaling the MASR curves by a functional form $\nu(x)$ until we obtain the effective temporal kernel with compound rates (see figure 5.12(i),(j) by the ETAS model). This measure will loose reliability the further we get from the parent event, where the contribution of other triggering trees will superimpose to the one under consideration. We use a simple truncation criterion to MASR based on the functional form of terms $\nu(m_i)$ and $\psi_t(t - t_i)$. Conditional rates are evaluated until a larger event than the parent is found. We know that when this condition is found, the activity will be dominated by the new large event. This truncation method by itself introduces other biases in the presence of time-modulation, since low activity rates will typically include longer aftershock sequences. This bias is ease to identify in MASR analysis. Consider a non-homogeneous Poisson process with a Gaussian intensity (figure 5.12(b)). Since $\mu_x(t) = \rho(x)\mu(t)$ depends explicitly on time, we need to integrate over the whole process:

$$\text{MASR}(\tau|x) = \frac{\int_\Delta^\infty d\delta \int_{\mu_x^{\min}}^{\mu_x^{\max}} d\mu_x \frac{dt}{d\mu_x} \mu_x(t)^2 \rho(\delta|\mu_x(t))}{\rho(x)\int_\Delta^\infty d\delta \int_{\mu_x^{\min}}^{\mu_x^{\max}} d\mu_x \frac{dt}{d\mu_x} \mu_x(t)\rho(\delta|\mu_x(t))} \begin{cases} = \frac{\langle \mu_x \rangle}{\rho(x)} = \langle \mu \rangle & \text{for } \tau \ll \frac{1}{\mu_x^{\max}} \\ \propto \tau^{-1} & \text{for } \tau \gg \frac{1}{\mu_x^{\max}} \end{cases} \quad (5.20)$$

The resulting MASR is shown in figure 5.12(k). For low values of τ we obtain the expected plateau, i.e. time independence, expected in a Poisson process, but this measure is biased at longer times for large mark windows x.

Figure 5.10 summarizes the point process analysis of an experimental avalanche process in ferroelasticity. The results correspond to the the martensitic transformation under thermal driving of a CuZnAl alloy studied through two channels, AE and calorimetry (cal), resulting in the sequences represented in figures 5.10(a)–(c). Details can be found in [15]. This is an example where event–event triggering was identified through the use of the Bi-test (figures 5.10(e) and (f)) and characterized through MASR (figure 5.10(i)) without observing significant dependence of the production on event energy E. Interevent time distributions for both channels fit the USL well with a double power-law functional form (figure 5.10(g)) characteristic of triggering phenomena and the underlying non-homogeneous rate of the transition (figure 5.10(d)) beyond artifacts due to low statistics at binning (figure 5.10(h)).

5.4.5 Statistical modeling of scale-free marks

The marginal statistics of marks ($\{X_i\}$) are often represented by a set of statistical parameters $\{\Gamma\}$. In particular, the crackling noise, i.e. scale-free, regime is characterized by multivariate power-law distributions of the marks defining the size of avalanches as defined in (5.2), as can be visualized in bivariate distributions (see figure 5.13). Temporal modulated parameter sets $\{\Gamma\}(t)$ are the trademark of non-linear behaviors, such as the aforementioned criticality and sharp regime transitions in the macroscopic state of the system. Even when these factorize with the intensity, modeling the mark statistics and all the dependencies is a fundamental part in the characterization of the point process. We will focus on the analysis of power-law distributions and relations between marks expected close to criticality. We identify the proximity to a critical transition or critical regime, when the probability distribution function ($\rho(X_i = x)$) of a set of N measurements ($\{X_i\}$) can be expressed as a homogeneous function within a wide domain. With respect to a scale λ: $g(x/\lambda) = \lambda^\kappa g(x)$ within a contiguous range $x \in \Omega$. This is fulfilled by power-law relations of the type $g(x) \sim x^{-\kappa}$. A distribution will be of the type

$$\rho(X_i = x) = \frac{x^{-\kappa}}{\zeta(\kappa, \Omega)}, \qquad (5.21)$$

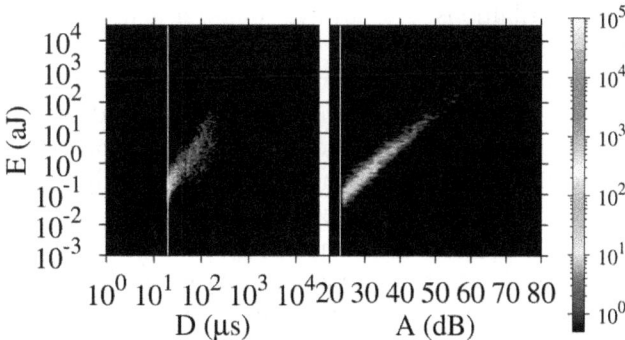

Figure 5.13. Multivariate power-law distribution of amplitude (A), duration (D) and energy (E) of acoustic events recorded during the martensite transition of a CuZnAl alloy after filtering [15], represented as bivariate distributions $\rho(D, E)$ and $\rho(A, E)$.

which can only be normalized by $\zeta(\kappa, \Omega)$ within a finite range (bounded by a maximum $x \leqslant x_{\max}$ or minimum $x \geqslant x_{\min}$ or both), hence indicating an inherent limitation in the domain in which magnitudes can be scale free.

We will focus on scale-free distributions of variables x with exponents $1 < \kappa < 2$ above a lower bound $x_{\min} > 0$. This are the usual exponents found in avalanche magnitudes both in natural processes, experiments and models (e.g. the aforementioned mean-field theory renders a distribution of sizes with $\kappa_S = 1.5$ and durations $\kappa_T = 2$ expected from memoryless branching) although it is also common to find higher exponents in experiments.

5.4.5.1 Estimation of exponents and the domain of scale-free marks

Given the universal nature of critical exponents, their precise estimation provides both an assessment for extreme event statistics and insight into the physical properties of the system. Note that the existence of a finite scale-free domain is not only caused by departure from criticality, but is inherent in experimental and numerical data. Any experimental data are unavoidably constrained to a resolution limit for small events, either natural to the instrument or the digital signal, and instrument saturation for large events. A particular example for ferroelastic experiments, it is difficult to find instruments (amplifiers, voltmeters, etc) providing more than five decades in resolution. Numerical simulations are usually discretized in lattice elements or particles imposing a natural limit to small events. Unavoidable limitations on computational power impose a maximum size limit, introducing finite-size effects to raw statistics that can potentially be addressed by means of finite-size scaling techniques [100].

Considering $\kappa > 1$, the simplest power-law distribution model considers the existence of a lower bound $x_{\min}$ to the scale-free behavior. Considering an open domain ($x_{\max} \to \infty$) the normalization term can be written as

$$\zeta(\kappa, x_{\min}) = \frac{x_{\min}^{1-\tau}}{1 - \kappa}.$$ (5.22)

Such power-law distributions are always well-normalized, but the statistical moments ($E(x^m)$) are only defined for $m < \kappa - 1$. For $1 < \kappa < 2$, no integer moment is well defined. This is a problem when trying to apply traditional parametric statistics. To begin with, the sum of independent measures fails to render normality as expected by the standard central limit theorem, but it complies with the generalized central limit theorem rendering Lévy distributions [101]. On the other hand, graphical methods such as least-square-errors are not adequate and ill-advised to fit statistical parameters. Improvements are achieved by considering cumulative distribution functions or rank-ordering methods [102]. Methods based on direct maximum-likelihood estimation are instead advised to fit both the exponent τ and the scale-free on-set $x_{\min}$. Considering a general scale free distribution of type (5.21), we base the estimation on the optimization of the log-likelihood function ($\ln \mathscr{L}(\{X_i\}; \kappa, \Omega)$) leading to the relation

$$\sum_i \ln X_i = N \frac{\zeta'(\hat{\kappa}; \Omega)}{\zeta(\hat{\kappa}; \Omega)},$$ (5.23)

where $\zeta'(\kappa; \Omega) = \frac{\partial \zeta(\kappa; \Omega)}{\partial \kappa}$. Given the univalued nature of the likelihood function, the uncertainty of the estimation can be obtained from the Fisher information matrix, which returns the standard deviation of the Gaussian approximating $\mathscr{L}(\{X_i\}; \kappa, \Omega)$ such that

$$\sigma_{\hat{\kappa}} = \left(\frac{\partial^2 \ln \mathscr{L}}{\partial \kappa^2}\right)^{-1/2}. \tag{5.24}$$

In the case of a pure power-law distribution represented as (5.22), the estimation of κ has an exact solution,

$$\hat{\kappa}(\{X_i\}; \Omega) = 1 + \frac{N}{\sum_i \ln(X_i/x_{\min})}, \tag{5.25}$$

and the uncertainty is given by $\sigma_{\hat{\kappa}} = N^{-1/2}(\hat{\kappa} - 1)$.

Physical constraints or characteristic scales indicative of a distance to criticality might cause an upper bound ($x_{\max}$) for the scale-free domain in $\rho(x)$ in addition to $x_{\min}$. Therefore the normalization factor reads

$$\zeta(\kappa; \Omega) = \frac{1}{1-\kappa}\left(x_{\max}^{1-\kappa} - x_{\min}^{1-\kappa}\right). \tag{5.26}$$

In this case, the solution to (5.23) must be solved numerically using root-finding techniques. Similarly, $\{X_i\}$ defined in integer or discretized values (such as measurements close to an operational resolution) will render different normalization factors that cannot be solved analytically [103], although models (5.22) and (5.26) are still valid for higher x values. Given the maximum likelihood estimation for $\hat{\kappa}(\{X_i\}; \Omega)$ it is advised to use goodness-of-fit tests to estimate the scale-free domain (Ω) [102, 104, 105]. Although an exact solution cannot be given, in general, the objective is to maximize the contiguous domain (Ω) with a goodness of fit above a reasonably selected p-value (e.g. $p_{\mathrm{th}} = 0.20$) and accompanied by a a consistent $\hat{\kappa}(\omega)$ value for any contiguous subset domain $\omega \subset \Omega$, which is guaranteed for the scale-free nature of the distribution. The advised procedure to compute p-values consists in the resampling of synthetic power-laws with the fitted exponent, and direct comparison with likelihood values. This procedure guarantees that the computation of the p-value is independent of the estimated exponent $\hat{\kappa}$. p-values obtained through classical KS tests considering $\hat{\kappa}$ would typically be positively biased, although the transition from low to high p-values is so sharp that the difference can be neglected in favor of computational costs.

We exemplify the use of maximum estimation methods with the exponents and direct p-value estimation maps [103] in the calorimetric and AE energies of avalanches during the martensitic transformation of CuZnAl [15] (figure 5.14). We observed a power-law tail with an exponent κ_E: $=\varepsilon \approx 2.15$ compatible in the entire resolution range in the calorimetric chanel, and for $E > 10 aJ$ in the acoustic emission channel.

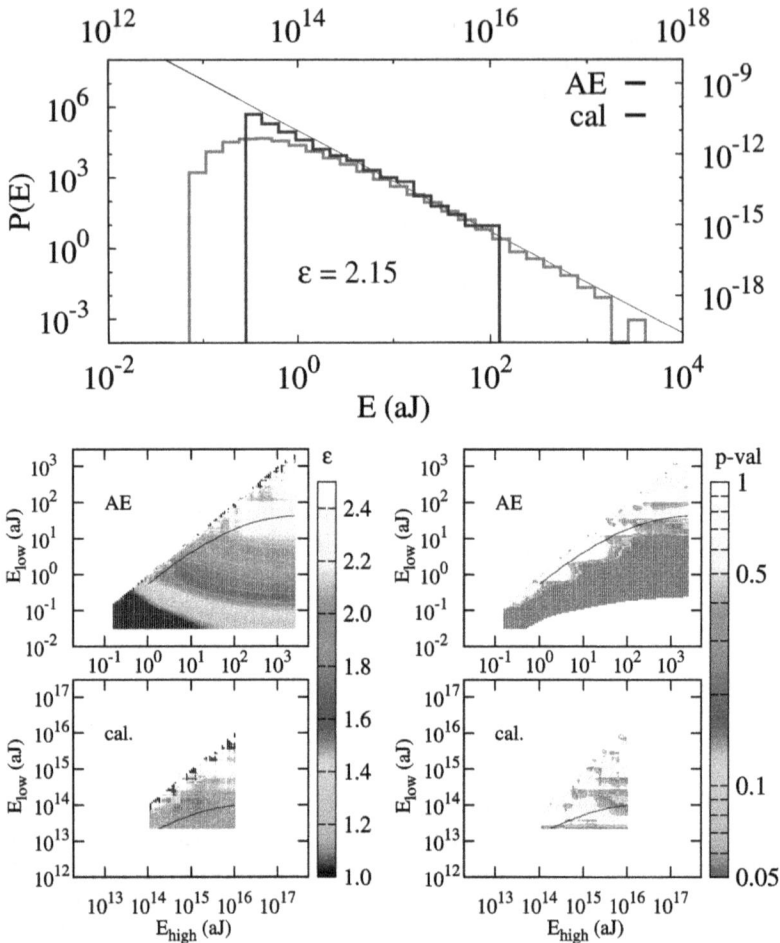

Figure 5.14. Top: Distribution of avalanche energies calculated in the calorimetric and AE channels. Bottom: Map representation of the estimated exponents and *p*-values.

5.5 Summary

Avalanches in ferroelastic materials can be induced by different external stimuli, e.g. temperature, stress, and electric field. They occur either at transitions between different crystallographic phases or when the domain structure reorganizes. They arise as a result of a competition between long-range elastic interactions related to the ferroelastic order and local disorder. Only close to a critical disorder, does avalanche dynamics become relevant in the description of the switching, with a scale-free propagation that can be described by critical exponents. The modeling of the avalanche statistics relies on stochastic models. The most common approach to study the statistical properties of avalanches is to model the sequences as a spatio-temporal point process. Maximum-likelihood estimation of intensities is complex due to the changes in activity rate over time throughout the ferroelastic hysteretic curve. Several alternatives exists: Poisson processes where avalanche events are independent, renewal processes where the intensity of the event is reset after each event, and self-excited processes where avalanches appear as clusters triggered by other switching events. Recording these avalanches remains an experimental challenge: acoustic emission and dynamic mechanical analysis are sensitive to a rather wide range of event sizes and timescales, but do not provide spatial resolution and are inherently indirect; optical microscopy allows individual avalanches to be localized in space but only on a restricted range of event sizes and timescales because of the diffraction-limited resolution and minimum exposure time; and transmission electron microscopy and atomic force microscopy offer exceptional spatial resolution but remain challenging to operate. The recorded signals are often characterized by multivariate power-law distributions, with small deviations inherent to the resolution limit of experimental techniques.

References

[1] Barkhausen H 1919 Zwei mit Hilfe der neuen Verstärker entdeckte Erscheinungen *Phys. Zeit.* **20** 401–3

[2] Babcock K L and Westervelt R M 1990 Avalanches and self-organization in cellular magnetic-domain patterns *Phys. Rev. Lett.* **64** 2168–71

[3] Cote P J and Meisel L V 1991 Self-organized criticality and the Barkhausen effect *Phys. Rev. Lett.* **67** 1334–7

[4] Bak P, Tang C and Wiesenfeld K 1987 Self-organized criticality: an explanation of the $1/f$ noise *Phys. Rev. Lett.* **59** 381

[5] Sethna J P, Dahmen K, Kartha S, Krumhansl J A, Roberts B W and Shore J D 1993 Hysteresis and hierarchies: dynamics of disorder-driven first-order phase transformations *Phys. Rev. Lett.* **70** 3347–50

[6] Sethna J P, Dahmen K A and Myers C R 2001 Crackling noise *Nature* **410** 242–50

[7] Spasojević D, Bukvić S, Miloć S and Stanley H E 1996 Barkhausen noise: elementary signals, power laws, and scaling relations *Phys. Rev.* E **54** 2531–46

[8] Vives E, Ràfols I, Mañosa L, Ortín J and Planes A 1995 Statistics of avalanches in martensitic transformations. I. Acoustic emission experiments *Phys. Rev.* B **52** 12644–50

[9] Zhang L, Salje E K H, Ding X and Sun J 2014 Strain rate dependence of twinning avalanches at high speed impact *Appl. Phys. Lett.* **104** 162906

[10] Salje E K H 2020 Ferroelastic domain walls as templates for multiferroic devices *J. Appl. Phys.* **128** 164104

[11] Salje E K H, Xue D, Ding X, Dahmen K A and Scott J F 2019 Ferroelectric switching and scale invariant avalanches in BaTio$_3$ *Phys. Rev. Mater.* **3** 014415

[12] Shao G, Xu Y, Zhou Y, Ding X, Sun J, Salje E K H, Lookman T and Xue D 2022 Acoustic emission study on avalanche dynamics of ferroelectric switching in lead zirconate titanate ceramics *J. Appl. Phys.* **132** 224102

[13] Rosinberg M-L and Vives E 2011 Metastability, hysteresis, avalanches, and acoustic emission: martensitic transitions in functional materials ed T Kakeshita, T Fukuda, A Saxena and A Planes *Disorder and Strain-Induced Complexity in Functional Materials* (Berlin: Springer) vol 148 pp 249–72

[14] Carrillo L, Mañosa L, Ortín J, Planes A and Vives E 1998 Experimental evidence for universality of acoustic emission avalanche distributions during structural transitions *Phys. Rev. Lett.* **81** 1889–92

[15] Baró J, Martín-Olalla J-M, Romero F J, Gallardo M C, Salje E K H, Vives E and Planes A 2014 Avalanche correlations in the martensitic transition of a Cu–Zn–Al shape memory alloy: analysis of acoustic emission and calorimetry *J. Phys.: Condens. Matter.* **26** 125401

[16] Turcotte D L and Malamud B D 2004 Landslides, forest fires, and earthquakes: examples of self-organized critical behavior *Physica* A **340** 580–9

[17] Lockner D 1993 The role of acoustic emission in the study of rock fracture *Int. J. Rock Mech. Min. Sci. Geomech. Abstr.* **30** 883–99

[18] Baró J, Corral Á, Illa X, Planes A, Salje E K H, Schranz W, Soto-Parra D E and Vives E 2013 Statistical similarity between the compression of a porous material and earthquakes *Phys. Rev. Lett.* **110** 088702

[19] Davidsen J, Goebel T, Kwiatek G, Stanchits S, Baró J and Dresen G 2021 What controls the presence and characteristics of aftershocks in rock fracture in the lab? *J. Geophys. Res. Solid Earth* **126** e2021JB022539

[20] Miguel M C, Vespignani A, Zapperi S, Weiss J and Grasso J-R 2001 Intermittent dislocation flow in viscoplastic deformation *Nature* **410** 667–71

[21] Zaiser M 2006 Scale invariance in plastic flow of crystalline solids *Adv. Phys.* **55** 185–245

[22] Wang G, Chan K C, Xia L, Yu P, Shen J and Wang W H 2009 Self-organized intermittent plastic flow in bulk metallic glasses *Acta Mater.* **57** 6146–55

[23] Schall P, Weitz D A and Spaepen F 2007 Structural rearrangements that govern flow in colloidal glasses *Science* **318** 1895–9

[24] Hu Y, Shu L, Yang Q, Guo W, Liaw P K, Dahmen K A and Zuo J-M 2018 Dislocation avalanche mechanism in slowly compressed high entropy alloy nanopillars *Commun. Phys.* **1** 61

[25] Denisov D V, Lörincz K A, Uhl J T, Dahmen K A and Schall P 2016 Universality of slip avalanches in flowing granular matter *Nat. Commun.* **7** 10641

[26] Berthier E, Kollmer J E, Henkes S E, Liu K, Schwarz J M and Daniels K E 2019 Rigidity percolation control of the brittle–ductile transition in disordered networks *Phys. Rev. Mater.* **3** 075602

[27] Ward L M and Greenwood P E 2007 $1/f$ noise *Scholarpedia* **2** 1537

[28] Dahmen K A 2017 *Mean Field Theory of Slip Statistics* (Cham: Springer International) pp 19–30

[29] Casals B, van Dijken S, Herranz G and Salje E K H 2019 Electric-field-induced avalanches and glassiness of mobile ferroelastic twin domains in cryogenic $SrTiO_3$ *Phys. Rev. Res.* **1** 032025

[30] Puchberger S, Soprunyuk V, Schranz W, Tröster A, Roleder K, Majchrowski A, Carpenter M A and Salje E K H 2017 The noise of many needles: jerky domain wall propagation in $PbZrO_3$ and $LaAlO_3$ *APL Mater.* **5** 046102

[31] Casals B, Nataf G F, Pesquera D and Salje E K H 2020 Avalanches from charged domain wall motion in $BaTiO_3$ during ferroelectric switching *APL Mater.* **8** 011105

[32] Casals B, Nataf G F and Salje E K H 2021 Avalanche criticality during ferroelectric/ferroelastic switching *Nat. Commun.* **12** 345

[33] Harris A B 1974 Effect of random defects on the critical behaviour of Ising models *J. Phys. C: Solid State Phys.* **7** 1671

[34] Vojta T and Hoyos J A 2014 Criticality and quenched disorder: Harris criterion versus rare regions *Phys. Rev. Lett.* **112** 075702

[35] Nattermann T 1998 *Theory of the Random Field Ising Model* (Singapore: World Scientific) pp 277–98

[36] Baró J and Corral Á 2025 A universal route from avalanches in mean-field models with random fields to stochastic Poisson branching events *Chaos* **35** 073107

[37] Harris T E 1951 First passage and recurrence distributions *Technical Report* Rand Corp, Santa Monica, CA

[38] Harris T E 1963 *The Theory of Branching Processes* (Berlin: Springer)

[39] Alstrøm P 1988 Mean-field exponents for self-organized critical phenomena *Phys. Rev.* A **38** 4905–6

[40] Grassberger P 2022 Revisiting a low-dimensional model with short-range interactions and mean-field critical behavior *Europhys. Lett.* **136** 26002

[41] Beggs J M and Plenz D 2003 Neuronal avalanches in neocortical circuits *J. Neurosci.* **23** 11167–77

[42] Baró J, Pouragha M, Wan R and Davidsen J 2021 Quasistatic kinetic avalanches and self-organized criticality in deviatorically loaded granular media *Phys. Rev.* E **104** 024901

[43] van Kessenich L M, Luković M, de Arcangelis L and Herrmann H L 2018 Critical neural networks with short- and long-term plasticity *Phys. Rev.* E **97** 032312

[44] Paczuski M, Maslov S and Bak P 1996 Avalanche dynamics in evolution, growth, and depinning models *Phys. Rev.* E **53** 414–43

[45] Alessandro B, Beatrice C, Bertotti G and Montorsi A 1990 Domain-wall dynamics and Barkhausen effect in metallic ferromagnetic materials. I. Theory *J. Appl. Phys.* **68** 2901–7

[46] Hébraud P and Lequeux F 1998 Mode-coupling theory for the pasty rheology of soft glassy materials *Phys. Rev. Lett.* **81** 2934–7

[47] Karmakar S, Lerner E and Procaccia I 2010 Statistical physics of the yielding transition in amorphous solids *Phys. Rev.* E **82** 055103

[48] Lu G, Cordero F, Hideo K, Ding X, Xu Z, Chu R, Howard C J, Carpenter M A and Salje E K H 2024 Elastic precursor softening in proper ferroelastic materials: a molecular dynamics study *Phys. Rev. Res.* **6** 013232

[49] Salje E K H, Ding X, Zhao Z, Lookman T and Saxena A 2011 Thermally activated avalanches: jamming and the progression of needle domains *Phys. Rev.* B **83** 104109

[50] He X, Ding X, Sun J and Salje E K H 2016 Parabolic temporal profiles of non-spanning avalanches and their importance for ferroic switching *Appl. Phys. Lett.* **108** 072904

[51] Zhao Z, Ding X, Sun J and Salje E K H 2014 Thermal and athermal crackling noise in ferroelastic nanostructures *J. Phys.: Condens. Matter.* **26** 142201

[52] He X, Li S, Ding X, Sun J, Selbach S M and Salje E K H 2019 The interaction between vacancies and twin walls, junctions, and kinks, and their mechanical properties in ferroelastic materials *Acta Mater.* **178** 26–35

[53] He X, Salje E K H, Ding X and Sun J 2018 Immobile defects in ferroelastic walls: wall nucleation at defect sites *Appl. Phys. Lett.* **112** 092904

[54] Lu G and Salje E K H 2024 Ferroelastic twin walls for neuromorphic device applications *Front. Mater.* **11** 2024

[55] Lu G, Hideo K, Ding X, Xu Z, Chu R, Nataf G F and Salje E K H 2023 Influence of kinks on the interaction energy between ferroelastic domain walls in membranes and thin films *Microstructures* **3** 2023033

[56] Salje E K H, Wang X, Ding X and Scott J F 2017 Ultrafast switching in avalanche-driven ferroelectrics by supersonic kink movements *Adv. Funct. Mater.* **27** 1700367

[57] Salje E K H, Wang X, Ding X and Sun J 2014 Simulating acoustic emission: the noise of collapsing domains *Phys. Rev.* B **90** 064103

[58] Ji H and Robbins M O 1992 Percolative, self-affine, and faceted domain growth in random three-dimensional magnets *Phys. Rev.* B **46** 14519

[59] Sethna J P, Dahmen K A and Myers C R 2001 Crackling noise *Nature* **410** 242

[60] Sornette D and Sammis C G 1995 Complex critical exponents from renormalization group theory of earthquakes: implications for earthquake predictions *J. Physique* **5** 607–19

[61] Casals B, Dahmen K A, Gou B, Rooke S and Salje E K H 2021 The duration-energy-size enigma for acoustic emission *Sci. Rep.* **11** 5590

[62] Vu C-C and Weiss J 2020 Asymmetric damage avalanche shape in quasibrittle materials and subavalanche (aftershock) clusters *Phys. Rev. Lett.* **125** 105502

[63] Bronstein E, Faran E, Talmon R and Shilo D 2024 Uncovering avalanche sources via acceleration measurements *Nat. Commun.* **15** 747

[64] Salje E K H, Dul'kin E and Roth M 2015 Acoustic emission during the ferroelectric transition $Pm3m$ to $P4mm$ in $BaTiO_3$ and the ferroelastic transition $R3m$-$C2/c$ in $Pb_3(PO_4)_2$ *Appl. Phys. Lett.* **106** 152903

[65] Xu Y, Xue D, Zhou Y, Su T, Ding X, Sun J and Salje E K H 2019 Avalanche dynamics of ferroelectric phase transitions in $BaTiO_3$ and $0.7Pb(Mg_{2/3}Nb_{1/3}))O_3$-$0.3PbTiO_3$ single crystals *Appl. Phys. Lett.* **115** 022901

[66] Planes A, Mañosa L and Vives E 2013 Acoustic emission in martensitic transformations *J. Alloys Compd.* **577** S699–704

[67] Zammit-Mangion L J and Saunders G A 1984 Acoustic emission and domain wall dynamics in ferroelectric-ferroelastic gadolinium and terbium molybdate *J. Phys. C: Solid State Phys.* **17** 2825

[68] Harrison R J and Salje E K H 2010 The noise of the needle: avalanches of a single progressing needle domain in $LaAlO_3$ *Appl. Phys. Lett.* **97** 021907

[69] Scott J J R, Casals B, Luo K-F, Haq A, Mariotti D, Salje E K H and Arredondo M 2022 Avalanche criticality in $LaAlO_3$ and the effect of aspect ratio *Sci. Rep.* **12** 14818

[70] Ignatans R, Damjanovic D and Tileli V 2021 Individual Barkhausen pulses of ferroelastic nanodomains *Phys. Rev. Lett.* **127** 167601

[71] Nguyen C-P T, Schoenherr P, Salje E K H and Seidel J 2023 Crackling noise microscopy *Nat. Commun.* **14** 4963

[72] Tückmantel P, Gaponenko I, Caballero N, Agar J C, Martin L W, Giamarchi T and Paruch P 2021 Local probe comparison of ferroelectric switching event statistics in the creep and depinning regimes in $Pb(Zr_{0.2}Ti_{0.8})O_3$ thin films *Phys. Rev. Lett.* **126** 117601

[73] Bulanadi R, Cordero-Edwards K, Tückmantel P, Saremi S, Morpurgo G, Zhang Q, Martin L W, Nagarajan V and Paruch P 2024 Interplay between point and extended defects and their effects on jerky domain-wall motion in ferroelectric thin films *Phys. Rev. Lett.* **133** 106801

[74] Tan C D, Flannigan C, Gardner J, Morrison F D, Salje E K H and Scott J F 2019 Electrical studies of Barkhausen switching noise in ferroelectric PZT: critical exponents and temperature dependence *Phys. Rev. Mater.* **3** 034402

[75] Daley D J and Vere-Jones D 2007 *An Introduction to the Theory of Point Processes: General Theory and Structure* **vol 2** (Berlin: Springer Science)

[76] Saichev A and Sornette D 2007 Theory of earthquake recurrence times *J. Geophys. Res. Solid Earth* **112** B04313

[77] Smith W L 2018 Renewal theory and its ramifications *J. R. Stat. Soc.* B **20** 243–84

[78] Mirollo R E and Strogatz S H 1990 Synchronization of pulse-coupled biological oscillators *SIAM J. Appl. Math.* **50** 1645–62

[79] Barbieri R and Brown E N 2008 Application of dynamic point process models to cardiovascular control *Biosystems* **93** 120–5

[80] Buck J B 1938 Synchronous rhythmic flashing of fireflies *Q. Rev. Biol.* **13** 301–14

[81] Kagan Y Y 1993 Statistics of characteristic earthquakes *Bull. Seismol. Soc. Am.* **83** 7–24

[82] di Santo S, Burioni R, Vezzani A and Muñoz M A 2016 Self-organized bistability associated with first-order phase transitions *Phys. Rev. Lett.* **116** 240601

[83] Dahmen K A, Ben-Zion Y and Uhl J T 2009 Micromechanical model for deformation in solids with universal predictions for stress–strain curves and slip avalanches *Phys. Rev. Lett.* **102** 175501

[84] Müller J W 1974 Some formulae for a dead-time-distorted Poisson process: to André Allisy on the completion of his first half century *Nucl. Instrum. Methods* **117** 401–4

[85] Ogata Y 1988 Statistical models for earthquake occurrences and residual analysis for point processes *J. Am. Stat. Assoc.* **83** 9–27

[86] Baró J, Dixon S, Edwards R S, Fan Y, Keeble D S, Mañosa L, Planes A and Vives E 2013 Simultaneous detection of acoustic emission and Barkhausen noise during the martensitic transition of a Ni-Mn-Ga magnetic shape-memory alloy *Phys. Rev.* B **88** 174108

[87] Baró J and Davidsen J 2018 Universal avalanche statistics and triggering close to failure in a mean-field model of rheological fracture *Phys. Rev.* E **97** 033002

[88] Baró J, Dahmen K A, Davidsen J, Planes A, Castillo P O, Nataf G F, Salje E K H and Vives E 2018 Experimental evidence of accelerated seismic release without critical failure in acoustic emissions of compressed nanoporous materials *Phys. Rev. Lett.* **120** 245501

[89] Ogata Y and Zhuang J 2006 Space–time ETA models and an improved extension *Tectonophysics* **413** 13–23

[90] Davidsen J and Baiesi M 2016 Self-similar aftershock rates *Phys. Rev.* E **94** 022314

[91] Baró J and Davidsen J 2017 Are triggering rates of labquakes universal? Inferring triggering rates from incomplete information *Eur. Phys. J. Spec. Top.* **226** 3211–25

[92] Baiesi M and Paczuski M 2004 Scale-free networks of earthquakes and aftershocks *Phys. Rev.* E **69** 066106

[93] Baró J 2020 Topological properties of epidemic aftershock processes *J. Geophys. Res.: Solid Earth* **125** e2019JB018530

[94] Puy A, Baró J, Davidsen J and Pastor-Satorras R 2025 Leveraging spurious Omori–Utsu relation in the nearest-neighbor declustering method *Phys. Rev.* E **111** 024307

[95] van der Elst N J and Brodsky E E 2010 Connecting near-field and far-field earthquake triggering to dynamic strain *J. Geophys. Res.: Solid Earth* **115** B07311

[96] Bi H, Börner G and Chu Y 1989 Correlations in the absorption lines of the quasar Q0420-388 *Astron. Astrophys.* **218** 19–23

[97] Bak P, Christensen K, Danon L and Scanlon T 2002 Unified scaling law for earthquakes *Phys. Rev. Lett.* **88** 178501

[98] Corral Á 2004 Long-term clustering, scaling, and universality in the temporal occurrence of earthquakes *Phys. Rev. Lett.* **92** 108501

[99]   Helmstetter A and Sornette D 2002 Subcritical and supercritical regimes in epidemic models of earthquake aftershocks *J. Geophys. Res.: Solid Earth* **107** ESE 10–1–ESE 10-21

[100] Pérez-Reche F J and Vives E 2003 Finite-size scaling analysis of the avalanches in the three-dimensional Gaussian random-field Ising model with metastable dynamics *Phys. Rev.* B **67** 134421

[101] Brookes B C 2016 Théorie de l'Addition de Variables Aléatoires. By Paul Lévy Pp. xx 385. Second edition 1954. 1200f. (Gauthier-Villars, Paris) *Math. Gazette* **39** 344

[102] Corral Á, Serra I and Ferrer-i-Cancho R 2020 Distinct flavors of Zipf's law and its maximum likelihood fitting: rank-size and size-distribution representations *Phys. Rev.* E **102** 052113

[103] Baró J and Vives E 2012 Analysis of power-law exponents by maximum-likelihood maps *Phys. Rev.* E **85** 066121

[104] Clauset A, Shalizi C R and Newman M E J 2009 Power-law distributions in empirical data *SIAM Rev.* **51** 661–703

[105] Deluca A and Corral Á 2013 Fitting and goodness-of-fit test of non-truncated and truncated power-law distributions *Acta Geophys.* **61** 1351–94

IOP Publishing

Ferroelastic Materials

Guillaume F Nataf, Blai Casals and Ekhard K H Salje

Chapter 6

Functional properties of ferroelastic domain walls

Guillaume F Nataf, Kumara Cordero-Edwards, Guangming Lu and Ekhard K H Salje

6.1 Introduction

Ferroelastic domains are dominated by elastic interactions which are long-ranging and lead to universal domain structures. For this reason, ferroelastic domain patterns are remarkably similar with few variations between different crystal structures. The domains form characteristic patterns such as needles and right-angle configurations [1]. In figure 6.1(a), the typical domain structure with simple parallel domain walls is shown. Figure 6.1(b) shows the right-angle domains with local wall bending and widening details in figure 6.1(c).

Once a domain is generated inside a large crystal, below the phase transition, its lifetime is virtually unlimited. There is no mechanism in a stress-free crystal by which the crystal could get rid of the domain wall. The only possible decay channel is the lateral movement of the wall until it disappears through the surface, but this movement is prevented by lattice pinning. Several domain walls can then combine to form different hierarchical structures (figure 6.2): needle domains (figure 6.1), comb and tartan patterns (figure 6.2(a)), and more complex patterns such as tweed [3–5], which is a dense interwoven array of domain walls (figure 6.2(c)).

In this chapter, we describe some of the fundamental characteristics of ferroelastic domain walls, such as their thicknesses and distinct mechanical properties. We highlight emerging properties at domain walls, including (super-)conductivity, polarity, and magnetic properties. We then show that domain walls can be used as defects to reversibly control thermal conductivity.

doi:10.1088/978-0-7503-6089-0ch6
6-1

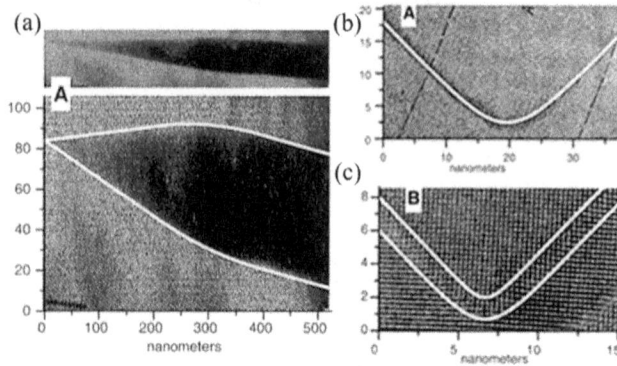

Figure 6.1. (a) Needle and (b),(c) right-angle domains in ferroelastic samples. The top image in (a) gives the real aspect ratio of a needle in $PbZrO_3$. The bottom image in (a) enhances the vertical length scale to show the functional form of the needle tip. (b) Right-angle domains in $Gd_2(MoO_4)_3$ samples showing the bending of the domain wall together with (c) the widening of the wall in $YBa_2Cu_3O_{7-\delta}$. (Reproduced with permission from [2].)

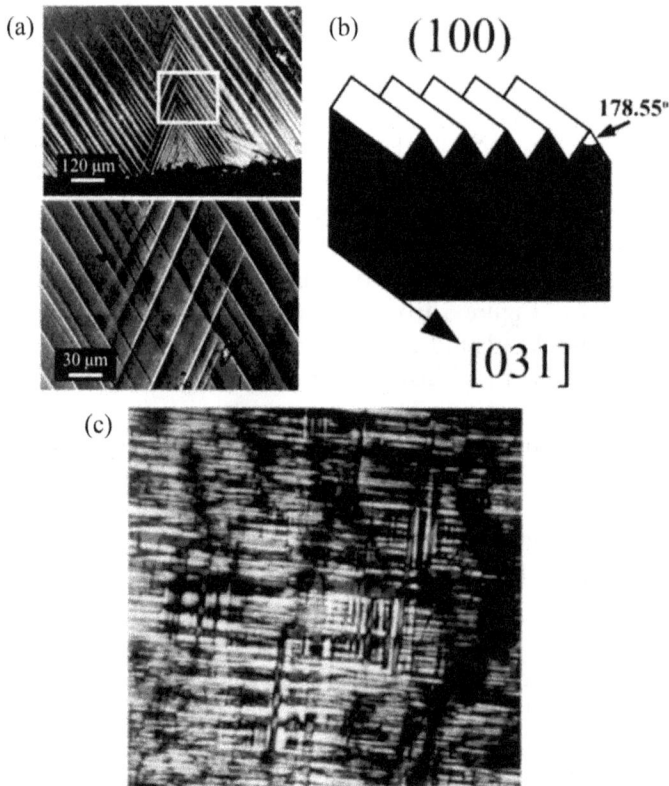

Figure 6.2. (a) Ferroelastic domains in $Pb_3(PO_4)_2$ under an optical microscope with crossed polarizers. (b) Schematics of ferroelastic domains in $Pb_3(PO_4)_2$. The domains are oriented parallel to [031]. A (100) cleavage surface forms a zig-zag profile exhibiting ridges and valleys with a typical angle of $178.55°$. (c) Tweed microstructure in microcline ($KAlSi_3O_8$). Interwoven local domain-like structures are common in ferroelastics. ((a) and (b) Reproduced with permission from [6]. Copyright 1997 IOP Publishing. (c) Reproduced with permission from [7]. Copyright 1994 Taylor and Francis.)

6.2 Domain wall thickness

As discussed in detail in chapter 3, symmetry aspects can be used to predict the domain structure. However, they do not predict the thickness of the wall. Theoretical estimates of domain wall thicknesses can be obtained using *ab initio* calculations [8] and phenomenological treatments [9–11]. For BaTiO$_3$, using the Ginzburg–Landau–Devonshire phenomenological model, it was found that all ferroelastic walls have a thickness below 4 nm [10].

Experimentally, the only technique with a sufficient spatial resolution to directly measure the thickness of domain walls is probably transmission electron microscopy (TEM) [12–14]. However, it requires invasive sample preparation to obtain a thin sample where domain walls may exhibit different thicknesses and properties than the bulk. As an alternative, diffuse x-ray scattering has been used successfully. X-ray diffraction (XRD) is a powerful analytical technique used to study the atomic structure of crystalline materials. It works by measuring the way x-rays are diffracted when they interact with the periodic lattice of a crystalline substance. This diffraction pattern provides information about the crystal structure, phase, and other properties. In the case of ferroelastic materials, this technique exploits differences in the lattice strain between domains, which leads to diffraction peaks that can reveal the distribution of domain orientations. It has also been implemented to investigate domain wall thicknesses by studying the diffuse scattering from domain walls [15–19].

In general, to study these domain structures it is necessary to have an array of parallel, non-periodically, spaced walls. A monochromatic, focused x-ray beam (Cu Kα_1) is diffracted by the sample before being registered by a (2θ) position sensitive detector or an area detector. The specimen position with respect to the beam is changed in three different angular directions, as well as in two translational directions. The experiment consists of collecting diffracted intensities around Bragg reflections by varying one of the angles (here: ω) and constructing rocking curves from the $\omega - 2\theta$ data by integration, e.g. along the 2θ direction. The two-dimensional intensity profiles are transferred into reciprocal space and fitted with two Gaussian functions, to obtain the relevant parameters. Figure 6.3 shows the typical dog-bone structure obtained from the diffraction signal, where the intensity in the handle of the dog bone is directly related to the thickness and internal structure of the domain wall.

Within a theoretical framework, we can calibrate the wall thickness. The total Gibbs free energy in the vicinity of a phase transition is given by the Landau energy and the elastic energy contribution [20]. To describe situations where the order parameter Q varies nonuniformly in space, an additional term $\frac{1}{2}g(\nabla Q)^2$ must be introduced to the potential. This Ginzburg term is the energy related to the spatial fluctuations in Q, and g is often called the Ginzburg parameter. Equilibrium solutions of the total energy describe the variation of the order parameter across a domain boundary and can be shown to be of the form

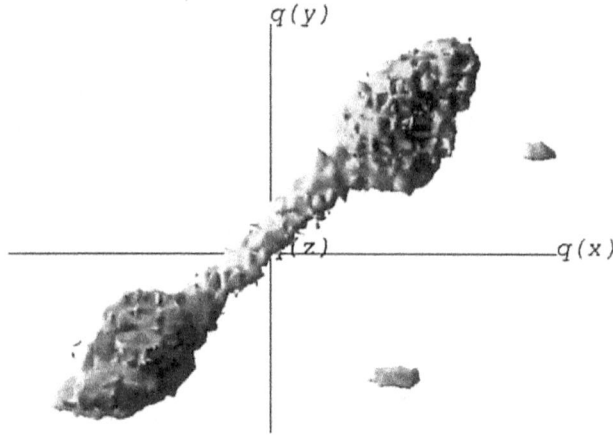

Figure 6.3. Isointensity surface plot of the (400)/(040) Bragg peak in a crystal of WO_3 at room temperature. The Bragg peak is 'inside' the isosurface; the dog-bone structure stems from the scattering from domain walls. (Reproduced with permission from [15]. Copyright 1998 Taylor and Francis.)

Table 6.1. Domain wall widths in units of the crystallographic repetition length normal to the twin plane.

Material	Method	$2W$ (nm)	Reference
WO_3	X-ray diffraction	1.2	[15]
$Gd(MoO_4)_3$	TEM	<1	[12]
$Pb_3(PO_4)_2$	TEM	0.8	[21]
	X-ray diffraction	<1.3	[22]
Sm_2O_3	TEM	0.3	[13]
KH_2PO_4	X-ray diffraction	0.4	[23]
$PbTiO_3$	TEM	0.4–1	[14]
$YBa_2Cu_3O_{7-x}$	X-ray diffraction	0.2	[18]
$(Na,K)AlSi_3O_8$	X-ray diffraction	0.3	[17]

$$Q = Q_0 \tanh\left(\frac{x}{W}\right), \tag{6.1}$$

where Q_0 is the value of the order parameter in the bulk, x is the spatial variable orthogonal to the wall, and W is the wall width. Table 6.1 shows that, independent of the material, the typical wall thickness in units of the crystallographic repetition length normal to the domain plane varies between 0.2 and 2 nm at $T \ll T_c$ (T_c = temperature of the transition to the paraelastic phase).

In addition, this Landau theory predicts the wall thickness dependence with temperature, such that W becomes larger with increasing temperature and diverging at T_c. Figure 6.4 shows the temperature dependence of the wall thickness measured

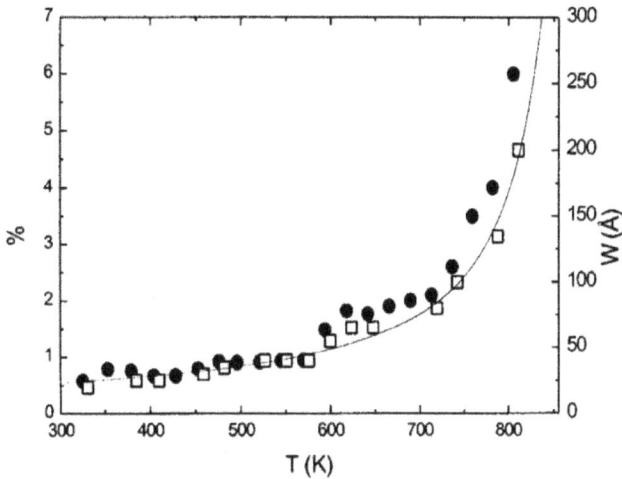

Figure 6.4. Temperature evolution of wall thickness in LaAlO$_3$. (Reproduced with permission from [16]. Copyright 1999 AIP Publishing.)

in ferroelastic lanthanum aluminate (LaAlO$_3$), with $T_c = 875$ K, confirming the prediction of this theory. It would be interesting to repeat such measurement with TEM to confirm the evolution of the thickness of the wall with temperature through a direct observation of the atoms.

Recent TEM experiments have also revealed that domain walls are not smooth, as is usually assumed in theoretical calculations. Instead, they exhibit meanders and atomic steps called kinks [24–27], which can have an influence on the interactions between domain walls.

6.3 Wall–wall interaction

Probing the interactions between ferroelastic domain walls is experimentally challenging. Still, with *in situ* TEM, it was shown in BaTiO$_3$ that the movement of 90° domain walls under an electric field is different when the domains are close to each other [28]. Parallel needle domains move relatively smoothly (figure 6.5). Instead, when domain walls approach orthogonal domain walls, without touching them, they are first strongly pinned, before exhibiting a large jump associated with the annihilation of the periodic orthogonal domains (figure 6.5). The depolarizing and strain fields at the needle tip are supposed to play a key role in this mechanism [28].

The interaction between domain walls has also been investigated by simulations. What matter greatly in such simulations are the boundary conditions: clamped ferroelastics develop different domain structures from those where the surfaces can relax to a local energy minimum. The conditions under which computer simulations are performed are hence crucial. Periodic boundary conditions and open boundary conditions are most widely used, sometimes with one surface clamped to mimic thin films on substrates. In addition, local defect fields, such as holes or rough surfaces,

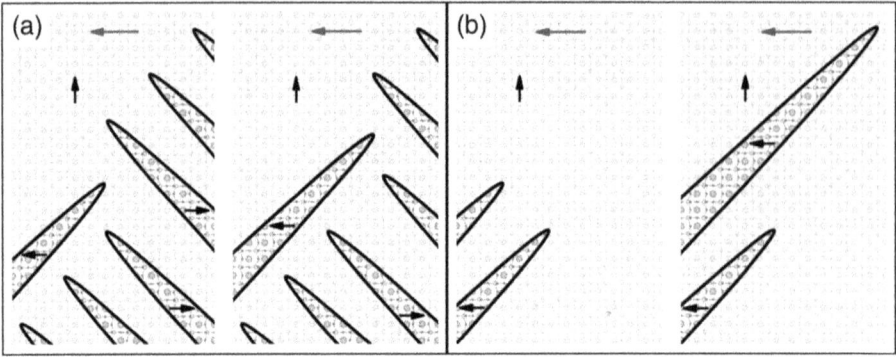

Figure 6.5. (a) Needle domain configuration in a metastable electromechanical equilibrium before a large jump of the domain wall and new equilibrium after the application of external electric field in the polar direction of the needle domains. (b) Configuration of two needle domains located far from orthogonal domain walls. The electric field direction is depicted with blue arrows and the polarization directions in all domains are shown with black arrows. (Reproduced from [28]. CC BY 4.0.)

were introduced to stabilize specific domain structures [29]. For the temporal evolution of domain structures, the key issue is that domains often interact strongly and move collectively, forming avalanches. Simulations are best performed using molecular dynamics (MD), which focuses on the long-range interactions rather than the short-range as in density functional theory (DFT) [30, 31]. MD simulations also have the advantage that fast-moving domain walls are detected easily, which would be more difficult in force field calculations [32–34]. The control parameters involve the elastic properties of the sample, the boundary conditions, the thickness of the domain walls and their interactions [35].

The MD model used here was introduced in [36]. The atomic arrangement is shown in figure 6.6. The potential energy $U(r)$ contains four terms: the harmonic first nearest atomic interactions $U(r) = 0.1(r - 1)^2$ (black springs); the anharmonic second-nearest Landau-type double-well interactions $U(r) = -0.05(r - \sqrt{2})^2 + 40(r - \sqrt{2})^4$ (red springs) along the diagonals in the lattice unit; the fourth-order third-nearest interactions $U(r) = 0.04(r - 2)^4$ (blue arrow); and anharmonic fourth-nearest Landau-type double-well interactions $U(r) = -0.05(r - \sqrt{5})^2 + 25.5(r - \sqrt{5})^4$ (green arrow), where r is the distance between atoms. The first- and third-nearest interactions are related to the elasticity and constitute the elastic background. An equilibrium shear angle of $2°$ is maintained by the second-nearest interactions while the additional fourth-nearest Landau-type interactions help to obtain a reasonable domain wall thickness and stability [16]. The model parameters were inspired by the well-known second-order phase transition of $SrTiO_3$ with a typical ferroelastic shear angle of $2°$ [37]. Similar potential forms have been applied successfully to investigate the internal frictions accompanying the dynamic motions of ferroelastic domains [38, 39], interactions of fine microstructures inside the domain walls [40, 41], piezoelectricity [42, 43], ferroelectricity [29] and magnetism [44] emerging from the static and dynamic polar ferroelastic domain walls [45, 46].

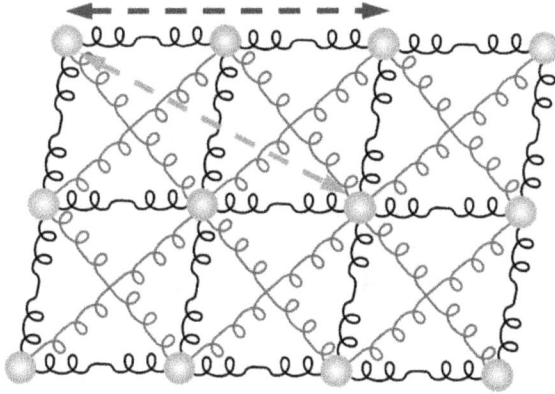

Figure 6.6. Interatomic potential for a generic ferroelastic model. The model consists of nearest-neighbours (black springs), next-nearest-neighbours (red springs), third-nearest-neighbours (blue arrow) and fourth-nearest-neighbours (green arrow). The red springs are Landau springs with a double-well potential so that the energy has a minimum value when the lattice is sheared with respect to the cubic unit cell. (Reproduced from [47]. Copyright 2013 AIP Publishing.)

The ferroelastic domain pattern stems then from stretching one diagonal spring and contracting the other. In the high symmetry paraelastic state the spring model has symmetry which conforms to plane group *p4mm*. The distorted structure shown in figure 6.6 has the symmetry of one of the subgroups of this, namely *c2mm*. The three possible components in the strain tensor that describe possible lattice distortions of the *p4mm* parent structure are e_1, e_2 and e_3. In symmetry-adapted form they are $e_1 + e_2$ (Γ_1), $e_1 - e_2$ (Γ_2) and e_3 (Γ_3). $e_1 + e_2$ correspond to a volume strain (or an area strain in two dimensions) while $e_1 - e_2$ and e_3 are shear strains. Likewise, the parent structure has three non-zero elastic constants (symmetry-adapted)

$$\begin{bmatrix} C_{11} + C_{12} & 0 & 0 \\ 0 & C_{11} - C_{12} & 0 \\ 0 & 0 & C_{33} \end{bmatrix}. \tag{6.2}$$

$(C_{11} + C_{12})$ is the 'bulk' modulus in two dimensions with a symmetry of Γ_1 (identity), while $(C_{11} - C_{12})$ and C_{33} are shear constants with symmetries of Γ_2 and Γ_3, respectively. The symmetry-breaking shear strain is e_3 and the elastic constant expected to go to zero at the critical temperature, T_c, is C_{33}. Taking e_3 as the order parameter, coupling with the other two strains would have the form $\lambda_1(e_1 + e_2)e_3^2$ and $\lambda_2(e_1 - e_2)^2 e_3^2$. In the *c2mm* structure, the expected strain relationships are then $(e_1 + e_2) \propto e_3^2$ and $(e_1 - e_2) = 0$. Investigation of the spring model has shown that there is a second elastic instability involving weak softening of $(C_{11} - C_{12})$. This would give a transition *p4mm* → *p2mm*, with $(e_1 - e_2)$ as the symmetry-breaking shear strain (Γ_2 active).

Ferroelastic domain walls often contain atomic kinks (see figure 6.7). A typical configuration contains one horizontal domain wall and a kink. The kink inside the domain wall locally bends the lattice planes and curves lattices on the top and bottom of the domain wall [40]. We now consider two interacting kinks inside the same domain wall (figure 6.8). The interaction is attractive for kink–antikink motions (with

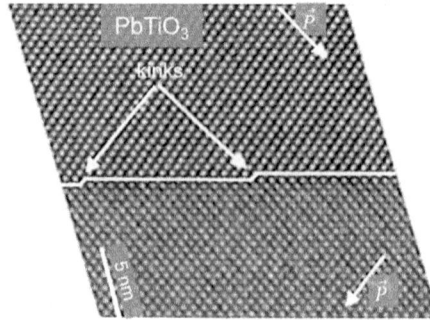

Figure 6.7. Kinks in a 90° domain wall in lead titanate observed by high-resolution transmission electron microscopy. (Reproduced with permission from [40]. Copyright 2022 the American Physical Society.)

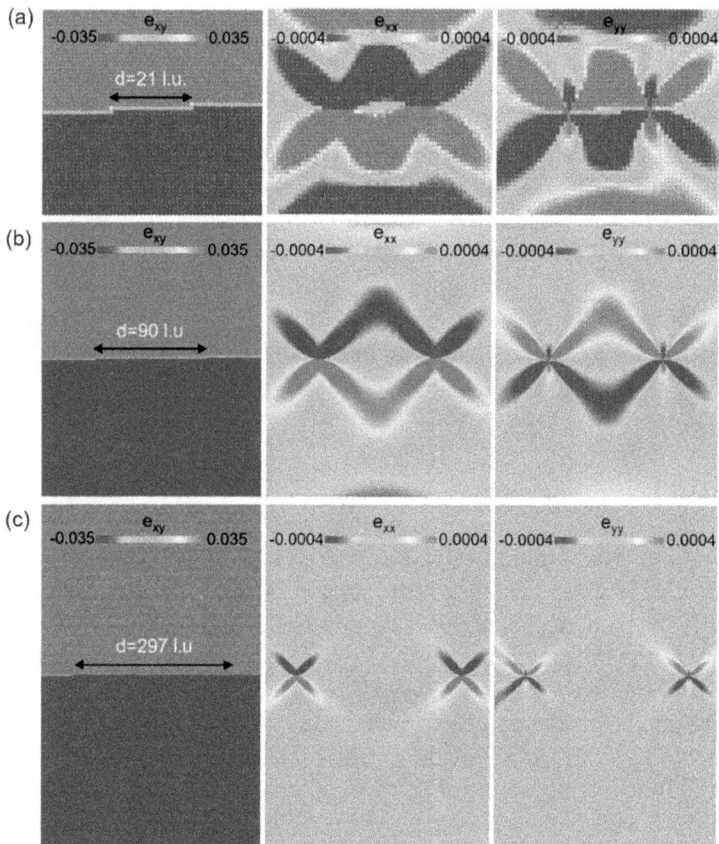

Figure 6.8. Strain maps of kink–kink configurations inside a horizontal wall with equilibrium separation distances resulting from repulsive interactions between kinks. The system sizes in the y-direction are (a) 51, (b) 201 and (c) 601 lattice units. Strain colour maps are coded by atomic-level normal strains (e_{xx} and e_{yy}) and shear strain (e_{xy}). (Reproduced with permission from [40]. Copyright 2022 the American Physical Society.)

kink amplitudes in opposite directions) and repulsive for kink–kink configurations (with kink amplitudes in the same direction). When a kink–kink configuration is initialized inside the wall, the repulsive interactions push the two kinks apart from each other to reduce the elastic energy.

When the sample thickness in y-direction extends to over 200 lattice units, the potential energy minimum is smooth enough to evaluate the interaction energy that follows a typical dipole–dipole interaction $\Delta E \sim d^{-2}$ for distances smaller than the equilibrium distance. This dipolar decay is the same in the case of two kinks, and kinks and antikinks in two different parallel domain walls, instead of a single wall. This effect is extremely size dependent (figure 6.9). Thick samples are dominated by 'monopolar' interactions and scale as $1/d$ rather than $1/d^2$ as in very thin samples. The crossover between these two regimes is shown in figure 6.9. The thick samples are in this simulation characterised by sample sizes > 1400 lattice units, which can be estimated in real materials as around 200 nm. Such sizes are common in free-standing thin samples. They are thicker than typical samples in TEM which

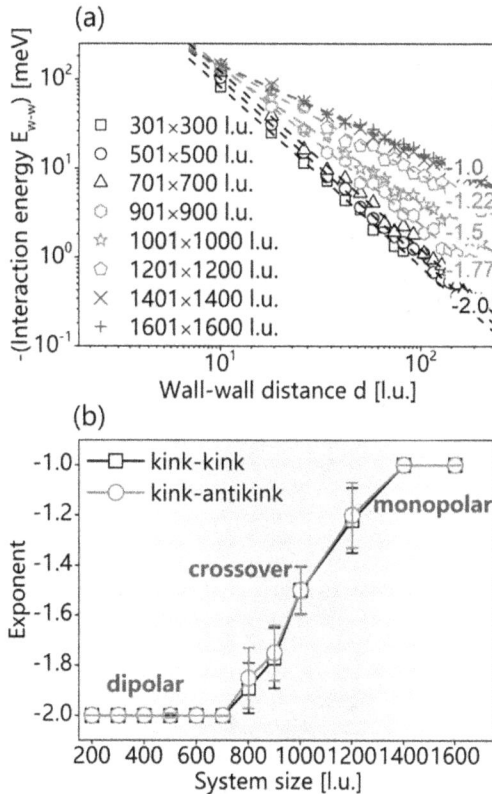

Figure 6.9. Interaction energies of kink–kink configurations residing inside two parallel walls as a function of the wall–wall distance d. (a) Interaction energy on logarithmic scales with the fitted scaling exponents. (b) Scaling exponents as a function of sample sizes. The thickness scaling changes from d^{-2} for thin samples to d^{-1} for thick samples. (Reproduced from [41]. CC BY 4.0.)

indicates that extreme caution must be taken when images are analysed to obtain interaction energies, quantify domain patterns, and extrapolate between such samples and bulk crystals. Moreover, when bending effects are disallowed by clamping, the effect is that of 'thick' samples because the surface bending represents the image force which allows the sample to relax and display dipolar interactions.

6.4 Mechanical properties

By definition, along at least one direction, the spontaneous strain on both sides of a ferroelastic domain wall, i.e. in two adjacent ferroelastic domains (twins), is different. Thus, the mechanical properties, such as the Young's modulus, must change between adjacent domains, as experimentally reported in $BaTiO_3$ [48, 49] and $PbTiO_3$ [50]. In these works, it was found that domains with in-plane ferroelectric polarization exhibit a higher elastic modulus compared to domains with out-of-plane polarization [49, 50].

In addition, a distinct mechanical contrast has been observed at domain walls in several ferroelectric and ferroelastic systems, whose origin remains debated [50–54]. It was consistently found that 180° domain walls in $LiNbO_3$, $BaTiO_3$, $PbTiO_3$ and $Pb(Zr,Ti)O_3$ are mechanically softer than the domains (figure 6.10(a), (b)) [51–53]. It was also found in $PbTiO_3$ and $(K_{0.5}Na_{0.5})NbO_3$ single crystals, that 90° ferroelastic domain walls exhibit different mechanical properties compared to the surrounding domains (figure 6.10(c), (d)) [50, 54]. In tetragonal $PbTiO_3$, ferroelastic ridges have a smaller elastic modulus than domains, while ferroelastic valleys show the highest value [50]. In orthorhombic $(K_{0.5}Na_{0.5})NbO_3$, 90° domain walls also exhibit either a higher or a lower Young's modulus depending on the side of the domain [54]. 60° domain walls do not show a contrast. All the results are summarized in table 6.2.

Different techniques have been used to measure the mechanical properties. In [48, 50–53], researchers rely on the influence of tip–sample interaction forces

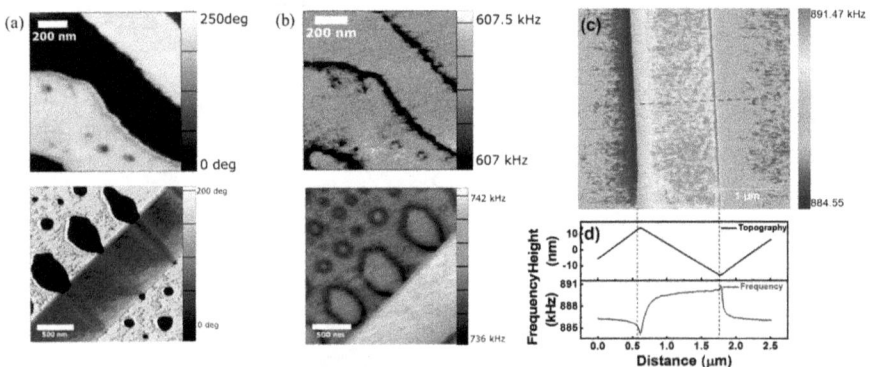

Figure 6.10. (a) Piezoresponse force microscopy images show the polarization of 90° and 180° domain walls, and (b) contact resonance frequency images demonstrate changes in the frequency between domains and domain walls in $BaTiO_3$. (c) Contact-resonant force microscopy map of 90° domain walls in $PbTiO_3$ and (d) horizontal line cut along the red line in (c) additional to the topography line cut at the same location. ((a)–(b) Reproduced from [53]. CC BY 4.0. (c)–(d) Reproduced from [50]. CC BY 4.0.)

Table 6.2. Elastic modulus of domains and domain walls for different ferroelectric and/or ferroelastic materials.

| Material | Structure | Elastic modulus (GPa) | | | |
		Contact resonance frequency	Force–distance	Nano indentation	Reference
$BaTiO_3$	a	275.1			[48]
	c-down	172.9			[48]
	c-up	158.7			[48]
	180° domain wall	153.8			[48]
$BaTiO_3$	a			147.6	[49]
	c-down			135.3	[49]
	c-up			135.3	[49]
$BaTiO_3$	c	63.6			[53]
	180° domain wall	51.2			[53]
$PbTiO_3$	a	58.6	63.0	71.9	[50]
	c	48.3	50.3	54.2	[50]
	90° domain wall ridge	43.5	44.3	71.7	[50]
	90° domain wall valley	60.7	67.5	73.3	[50]
$(K_{0.5}Na_{0.5})NbO_3$	Domain		120		[54]
	90° domain wall (1)		100		[54]
	90° domain wall (2)		140		[54]

on the acoustic vibration modes of the cantilever of an atomic force microscope. The technique is called contact resonance frequency microscopy mode or atomic force acoustic microscopy [55, 56]. In this technique, the tip is in contact with the sample and excites it mechanically through a piezo element placed at the base of the cantilever. The resonance frequency of the system depends on both the characteristics of the tip and the tip–sample mechanical contact. By keeping the force between the tip and the sample constant during the measurement, it is possible to access the mechanical properties of the material. Higher resonance frequencies correlate with stiffer contact areas and, conversely, lower resonance frequencies indicate that the material is softer [55]. In [50] and [54], again with an atomic force microscope, force–distance curves were recorded as a complementary method to measure mechanical properties. The PeakForce Quantitative Nano-Mechanics (PeakForce QNM) technique provides both topographical imaging and quantitative mechanical property measurements at the nanoscale. It consists of two main components. In the PeakForce Tapping mode, the maximum force applied to the sample is used as a feedback signal to generate high-resolution surface images. Simultaneously, the QNM functionality utilizes the force–distance curves generated during PeakForce Tapping to extract quantitative material properties. The measured Young's modulus depends on the cantilever's spring constant and deflection sensitivity. To ensure accuracy, a calibration sample with a known Young's modulus, preferably close to that of the unknown material, is required. The reduced elastic modulus is determined by fitting part of the unloading curve of the force–distance curves with the Hertzian model [50, 54, 57]. If the Poisson's ratio is known, the Young's modulus of the unknown sample can be accurately determined. In addition, AFM-based nano-indentation on domains and on domain walls was performed [49, 50].

The mechanical contrast at domain walls could originate from point defects. It is indeed known that domain walls tend to attract defects [58–60], such as oxygen vacancies [59, 60]. These defects would induce a local disorder, which would reduce the stiffness at the domain wall [61]. However, in $LiNbO_3$, $BaTiO_3$ and $PbTiO_3$, it was found that the mechanical contrast of the 180° walls exists only in domains where the polarization lies out of the plane with respect to the surface, while it disappears at 180° domain walls separating domains with in-plane polarization with respect to the surface [53]. Since the number of defects should be similar in both domain walls, one would instead expect the same mechanical response for both domain walls, if defects were the dominating mechanism. Defects seems thus unlikely to be at the origin of the mechanical contrast at 180° domain walls. It is known that defects can behave differently at ferroelastic 90° domain walls, compared to 180° domains, due to the difference in elastic energy [59, 60]. For example, for 90° domain walls in $PbTiO_3$, it was found by first-principles calculations that oxygen vacancies are more stable on the tail side of the domain rather than the head side [60]. One could thus still assume that defects play a role in the mechanical contrast observed at ferroelastic domain walls. However, experiments also show that there is no contrast at 90° domain walls when the polarization is parallel to the surface (a_1/a_2 domain structure) [50], ruling out a strong influence of defects [50].

The mechanical contrast could also arise from the motion of the domain wall, due to the inhomogeneous pressure applied by the tip for the mechanical measurement. It is known that moving domain walls soften materials [62]. Finite element simulations confirm that the local stress induced by the tip can lead to a reversible shift of domain walls, resulting in a 'strain dip' that contributes to the softening of the material [53]. In this scenario, both ridges and valleys would give the same contrast, since it would be associated with the movement of the wall. This is not what is observed experimentally at ferroelastic domain walls in $PbTiO_3$ and $(K_{0.5}Na_{0.5})NbO_3$ where the contrast at the wall depends on the side of the domain [50, 54].

Molecular dynamics simulations predict a different Young's modulus at the intersection between ferroelastic domain walls and the surface (figure 6.11) [63]. The moduli decay from the middle of the domain wall over 15 nm on each side of the wall [63]. The simulated change in the Young's modulus, compared to domains, is around 0.7%: ridges exhibit a higher Young's modulus, while valleys exhibit a lower

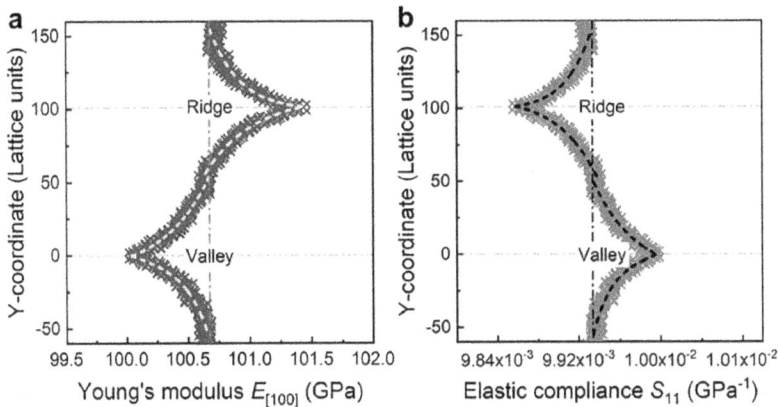

Figure 6.11. The profiles of the (a) elastic modulus and (b) elastic compliance of the surface layer. Softening of moduli at the domain wall valley and hardening at the domain wall ridge are observed. (Reproduced from [63]. CC BY 4.0.)

value. It is, however, contradicting experimental results, where it was found that the ridge domain wall has the lowest elastic modulus while the valley wall has the highest value [50].

All the samples investigated so far are not only ferroelastics but also ferroelectrics, and hence piezoelectrics. When the tip applies a pressure, it induces, by the piezoelectric effect, a change in polarization and the appearance of bound charges. These charges create a depolarizing field, which implies an electrostatic energy cost and appears as a stiffening of the mechanical response [53]. This response will depend on the effective piezoelectric coefficient of the area below the tip, and hence on the domain and domain wall configuration. A detailed analysis in [53] shows that this scenario could explain the apparent 180°-domain wall softening, including the different behaviour between out-of-plane and in-plane ferroelectric domains. However, it could not describe the asymmetric response between ridges and valleys at 90° domain walls [50].

None of the explanations above can explain all the experimental results, and not all measurements report the same trends and absolute elastic moduli values (table 6.2). One must consider possible measurement artefacts as well. Atomic force microscopy measurements are prone to crosstalk responses from the topography, which could be particularly critical when investigating ferroelastic surfaces with valleys and ridges, where the contact area between the tip and the sample changes. Indentation-based measurements may lead to a large shift, or even the suppression, of the domain wall.

One way to clarify the mechanisms at play would be to investigate non-ferroelectric ferroelastic materials, to rule out the influence of changes in polarization and screening charges. Among them, materials where domain walls are strongly pinned, and exhibit low mobility, would be ideal to avoid mechanical softening due to the movement of the wall under the tip. Measurements with different techniques, on the same domain walls, would also bring more confidence to the results.

6.5 Enhanced electrical conductivity in insulating materials

Enhanced electrical conductivity at domain walls has been reported in several ferroelectric and/or ferroelastic materials [64, 65]. It can originate from three mechanisms, acting individually or together [64]. First, as mentioned in the previous section, domain walls attract defects that can create intra-bandgap electronic states. Second, the different symmetry at domain walls can lead to an intrinsic reduction of the bandgap compared to the bulk. Third, at head-to-head or tail-to-tail configurations of the electric polarization, the local electric fields associated with the polarization discontinuities can shift the energies of the electronic band structure, allowing conduction bands to dip below the Fermi energy, or valence bands to peak above it. In addition, the interaction of domain walls with the surface can further enhance the electrical conductivity, as discussed in chapter 7.

Most studies have been performed on non-ferroelastic 180° domain walls [66–70], in particular in hexagonal manganites [71–77] and in uniaxial ferroelectrics such as lithium niobate [78–81]. The first spatially resolved measurement revealing an enhanced electrical conductivity at ferroelectric domain walls was performed by conductive atomic force microscopy on ferroelastic 109° domain walls in rhombohedral bismuth ferrite [66], followed by several measurements on 71° domain walls [82–84]. The mechanism at the origin of this different conductivity is still debated. An intrinsic bandgap lowering at the wall was proposed based on density functional theory calculations [85] and supported by scanning tunnelling spectroscopy measurements of the local density of states [84], but further first-principles calculations suggested that the bandgap closing is too small to result in enhanced conductivity [86]. The role of defects was also investigated. Oxygen vacancies, attracted by strain gradients at ferroelastic domain walls, could reduce the Schottky barrier around the walls, leading to a higher measured current using conductive atomic force microscopy [82, 83, 87]. In addition, bismuth vacancies have been shown to accumulate at the wall, and to change the oxidation state of iron, offering possibilities for conduction through electron hopping [58, 88]. This idea is supported by the observation of persistent conductive path signals when conducting domain walls were moved by applied electric fields [88]. However, in another experiment, the negligible time required for the increase in current argued against ionic motion [89, 90].

In lead zirconate titanate, it was shown that the presence of kinks, which lead to localized polarization charge, are essential to explain the enhanced conductivity at 90° domain walls, as schematized in figure 6.12 [91, 92]. In the probed thin films, the ferroelastic domain's width shrinks gradually from the film surface to the interface with the substrate. As a result, the walls bend through the formation of several kinks, as evidence by scanning transmission electron microscopy images [92]. Although, on average the polarization keeps the neutral head-to-tail configuration across the walls, the presence of kinks breaks the charge neutrality locally. These bound charges at the wall induce a local electronic band bending that can either enhance or suppress the conduction, depending on the polarization orientation and the electrode work-function.

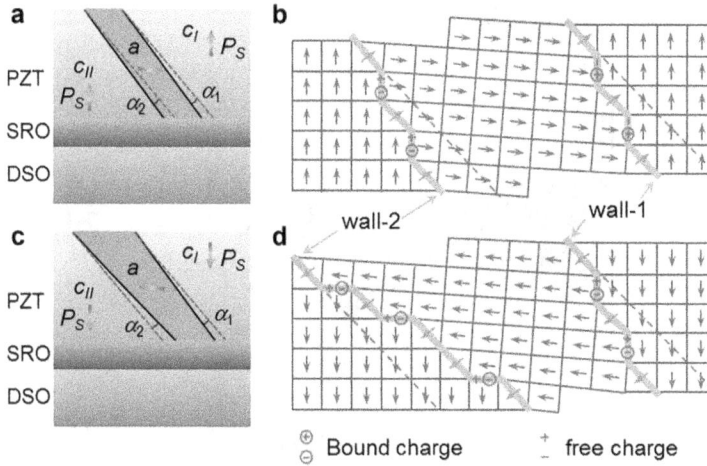

Figure 6.12. Schematic domain wall bending and charging in PZT films. (a) The morphology and (b) unit-cell model for the in-phase bent domain walls in the thinner film. (c) The morphology and (d) unit-cell model for the out-of-phase bent domain walls in the thicker film. The blue and pink dashed lines and the black and green solid lines indicate the positions of the neutral and real domain walls, respectively. The blue arrows in (b) and (d) denote the directions of P_S in the ferroelastic domains. (Reproduced with permission from [92]. Copyright 2017 the American Chemical Society.)

Historically, one should note that enhanced electrical conductivity at domain walls had already been reported in 1998 in doped tungsten trioxide [93]. Sodium and oxygen vacancies were successfully introduced in the material by vapour phase reaction with sodium, leading to a change in colour of the sample, from yellow-green initially to bluish-black. Optical images in reflection showed that the regions around the domain walls exhibited a stronger colour change, indicating an accumulation of oxygen vacancies, and thus a stronger reduction at the wall. Conductivity measurements as a function of temperature revealed superconductivity (electrical resistance going to zero) with a transition temperature near 3 K (figure 6.13). Measurements under a magnetic field showed that the critical field was above 15 T. Since no magnetization was found, even at low temperature, it was assumed that the superconductivity does not arise from the bulk but is confined to the domain walls.

This result was partially confirmed later by local conductive atomic force microscopy measurements that revealed a steep increase of the current across the walls at room temperature [94]. The investigated sample had been heated for 90 min at 750 °C in a high vacuum (1.5×10^{-6} mbar) to release oxygen from the sample and reduce the sample surface. After several weeks under ambient conditions, the sample showed a decrease of the current at the wall attributed to a reoxygenation of the domain wall [94].

The facility with which oxygen is released under reducing conditions in tungsten trioxide is less related to the chemical bonding of oxygen but rather to the low energy required to transfer the valence state of localized surplus electrons on the W^{6+} sites to W^{5+}. This tendency to form W^{5+} states near surfaces was directly confirmed by x-ray photoelectron spectroscopy experiments [95] and indirectly by scanning

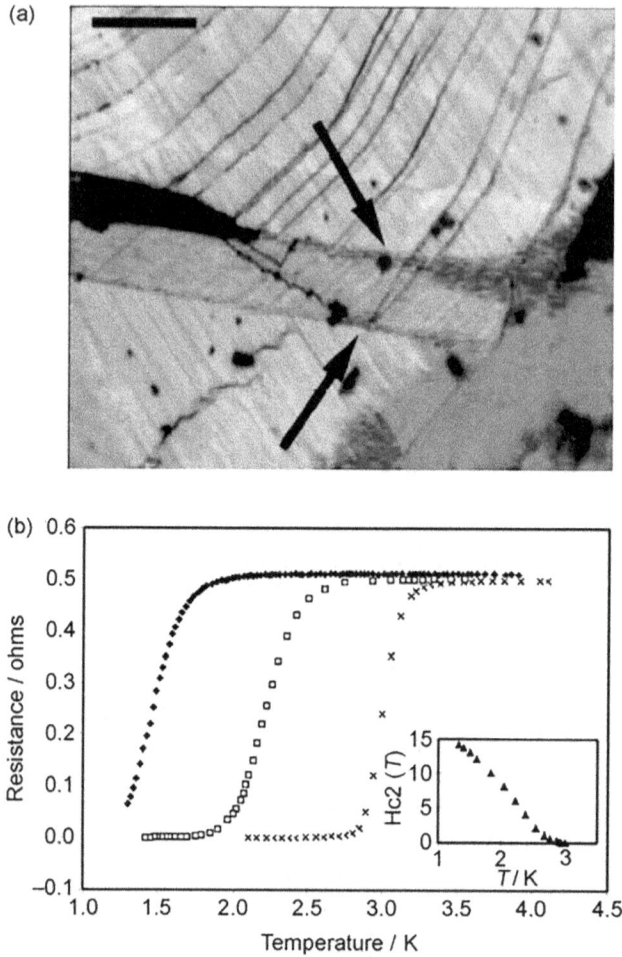

Figure 6.13. (a) Superconducting domain walls (arrows) in WO_3 close to the crystal surface. The scale bar in the top left corner is 50 μm. (b) Resistance of the superconducting twin wall in WO_3. The onset of superconductivity is 3 K, the critical field H_{c2} increases to 15 T at low temperature. (Reproduced with permission from [93]. Copyright 1998 IOP Publishing.)

tunnelling microscopy imaging [96]. These W^{5+} states are not localized, however, and form bipolarons in the low-temperature phase [97–100]. These polarons may in fact be a key ingredient for enhanced conductivity, and even superconductivity, at domain walls. In bismuth ferrite, it has been argued, based on first-principles calculations, that ferroelastic domain walls localize free excess electrons and create polaronic trapped states in the bandgap [101]. Screening charges for charged sections of domain walls, as described in bismuth ferrite, could also take the form of localized small polarons. Overall it would be interesting to investigate further the role of polarons and bipolarons in changes in conductivity at domain walls, e.g. in lithium niobate, where they are known to exist in the bulk already [102].

6.6 Polarity in non-polar materials

The polar character of ferroelastic domain walls is a general result derived from symmetry and group theory [103–108]. By describing the symmetries of domain walls with a layer group, it is possible to show that all strain-compatible ferroelastic domain walls are non-centrosymmetric, and to predict the orientation of the polarization [108]. Quantitative values of the polarization can then be calculated by considering the flexoelectric effect and the strain gradient that always exist at ferroelastic domain walls [46, 109, 110]. In addition, for materials with multiple order parameters, a biquadratic coupling between the primary order parameter and polarization, as a second order parameter, is always allowed and yields two energetically equivalent signs for the wall polarity [110, 111]. Whether or not a domain wall polarization can be switched depends on the respective strengths of these two mechanisms: symmetry/flexoelectricity will lead to a polarization defined by the strain gradient at the wall that can be switched only by switching adjacent ferroelastic domains (which is energetically costly), while the competition between order parameters at the wall (a symmetry lowering) will lead to a polarization that is switchable without modifying the domain structure. It is clear that for most applications, such as memory devices [112, 113], the second mechanism is required.

The first experimental works to demonstrate the existence of a polarization at ferroelastic domain walls were performed on the perovskites $SrTiO_3$ and $CaTiO_3$, where it has been proposed that one of the oxygens' octahedra tilts goes to zero at the wall, allowing for the emergence of Ti off-centering and hence polarization, seen as a competing secondary order parameter. Both are incipient ferroelectrics [114, 115] with a latent polar instability and are therefore indeed prone to exhibit a switchable domain wall polarization.

Aberration-corrected TEM imaging [117] has been implemented in combination with exit wave reconstruction [118] to study domain walls in $CaTiO_3$ [116]. The exit wave reconstruction process is performed using TrueImage software. Following reconstruction, residual aberrations are corrected using the standard techniques available in TrueImage. The amplitude and phase of the reconstructed exit wave are presented in figures 6.14(a) and 6.14(b), respectively, with a resolution of 0.8 Å. While the combination of exit wave reconstruction and aberration-corrected TEM significantly enhances visual interpretability, it does not provide precise quantitative measurements for atomic column positions. To achieve this level of precision, statistical parameter estimation techniques must be applied [119–121]. As can be seen in figure 6.15 these methods enable the visualization of atomic displacements directly inside domain walls, as well as measurements of atomic column positions with an accuracy of a few picometers. The distances orthogonal to the domain wall (see figure 6.15(b)) show deviations from a constant lattice parameter for the Ti–Ti distances in the vicinity of the wall, corresponding to an off-centre displacement of Ti by 6 pm, whereas no conclusions can be drawn for the Ca–Ca distances. In the case of distances parallel to the domain wall, as shown in figure 6.15(c), no significant deviations from a constant for both types of atomic columns are observed.

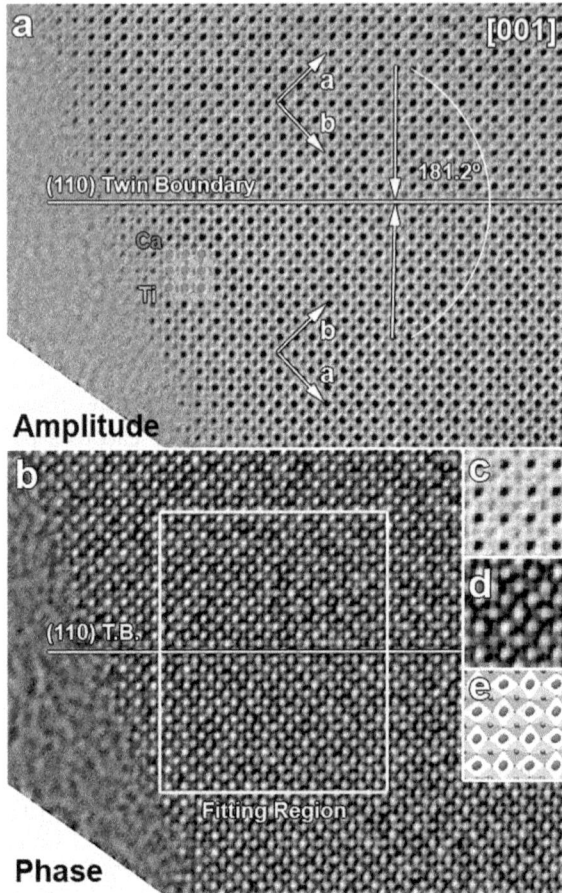

Figure 6.14. (a) Amplitude of the reconstructed exit wave. The CaTiO$_3$ crystal is imaged along the [001] zone axis orientation, the (110) twin boundary is indicated by the horizontal white line. The Ca and Ti column positions are marked by red and green dots, respectively. The angle of 181.2° reveals the twin relation over the interface. (b) The phase of the reconstructed exit wave; the region used for the statistical parameter estimation is indicated by the white rectangle. (c) 4 × 4 unit cell magnified from the amplitude image. (d) 4 × 4 unit cell magnified from the phase image. (e) 4 × 4 unit-cell structure model (the same colour scheme as in (a), with tilted oxygen octahedra depicted in shades of white). (Reproduced with permission from [116]. Copyright 2012 John Wiley and Sons.)

The loss of inversion symmetry at domain walls in CaTiO$_3$ was later confirmed by second harmonic generation (SHG). SHG is a nonlinear process in which two photons at the same frequency ω interact with the material to generate a new photon at twice the frequency of the initial photons [122]. The relation between the electric field E^{ω} of fundamental waves and the induced nonlinear polarization $P^{2\omega}$ is described by the following expression [123, 124]:

$$P_i^{2\omega} = \varepsilon_0 d_{ijk} E_j^{\omega} E_k^{\omega}, \tag{6.3}$$

(a)

(b) mean intercolumn distance ⊥(Å)

(c) mean intercolumn distance ⊢(Å)

twin wall

— Ca-Ca
— Ti-Ti

— Ca-Ca
— Ti-Ti

(d)

(e) mean displacement along x (pm)

(f) mean displacement along y (pm)

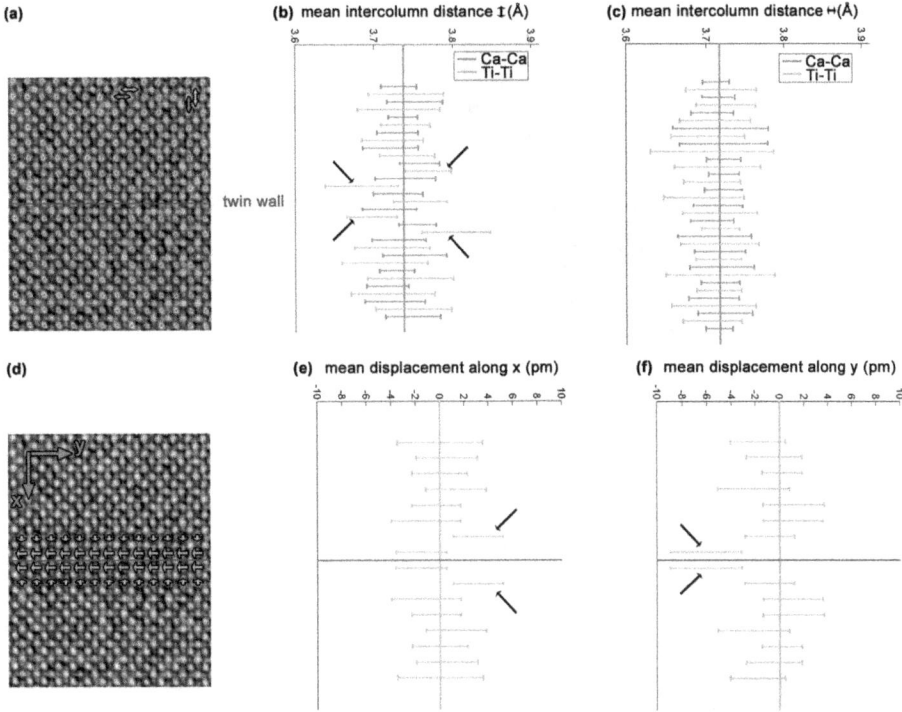

Figure 6.15. (a) Reproduction of the boxed region in figure 6.14(b). The estimated Ca and Ti atomic column positions are indicated in red and green, respectively. The blue horizontal line shows the position of the twin wall. (b) Mean interatomic Ca–Ca and Ti–Ti column distances perpendicular to the twin wall, averaged in the direction parallel with the twin wall, together with their 90% confidence intervals. The vertical red line corresponds to the overall mean Ca–Ca interatomic distance. Note the deviations from this line for particular Ti–Ti atomic column distances close to the twin wall (arrows). (c) Mean interatomic Ca–Ca and Ti–Ti column distances parallel with the twin wall, averaged in the direction parallel with the twin wall, together with their 90% confidence intervals. The vertical red line corresponds to the overall mean Ca–Ca interatomic distance. No deviations from this line are observed for the Ca–Ca or the Ti–Ti atomic column distances. (d) Mean displacements of the Ti atomic columns from the centre of the four neighbouring Ca atomic columns and indicated by green arrows. (e), (f) Displacements of Ti atomic columns in the x- and y-directions averaged along and in mirror operation with respect to the twin wall together with their 90% confidence intervals. (Reproduced with permission from [116]. Copyright 2012 John Wiley and Sons.)

where d_{ijk} ($i, j, k = 1$–3) denotes the second-order nonlinear SHG tensor and ε_0 is the dielectric constant of vacuum. The SHG intensity $I^{2\omega}$ is given by following equation:

$$I^{2\omega} \propto |P^{2\omega}|^2. \qquad (6.4)$$

Since the second-order nonlinear SHG tensor is a third-rank tensor, and considering equations (6.3) and (6.4), a centrosymmetric material where there is no break of space inversion will inherently be non-SHG active. Non-ferroelectric ferroelastic materials, such as $CaTiO_3$, should thus exhibit no SHG signal, except at domain walls where a polarization emerges. This is what is found in $CaTiO_3$, both for

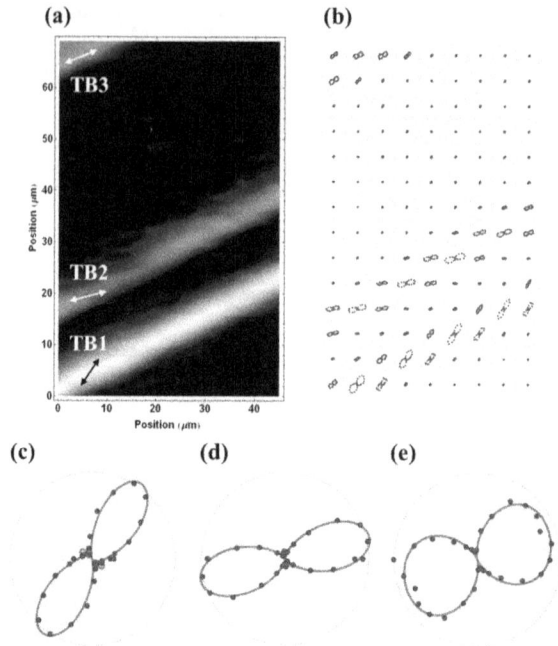

Figure 6.16. (a) SHG image of an area that contains three SHG active domain walls (TB) of type A. The corresponding polar diagram mapping is shown in (b). (c)–(e) Fitting results for TB1, TB2, and TB3, respectively. (Reproduced with permission from [125]. Copyright 2014 the American Physical Society.)

domain walls in as-grown samples, and for domain walls induced by an applied stress (figure 6.16) [125, 126]. SHG can also be used to determine the direction of the polarization vector of the domain walls, through polar diagrams, following equations (6.3) and (6.4) (figures 6.16(b)–(e)). The SHG intensity is recorded for different directions of the polarization of the incident light, while keeping the polarization directions of the fundamental and the second harmonic waves parallel. For $CaTiO_3$, the symmetry is found to be m [125, 126]. It has also been reported that the polarization can lie out of the domain wall plane when there is another domain wall in the vicinity, suggesting an interaction between domain walls, as discussed in section 6.3.

Macroscopic measurements using resonant piezoelectric spectroscopy (RPS) also provide evidence for polar domain walls in $CaTiO_3$ [127]. Figure 6.17(a) shows a schematic diagram of the RPS set-up consisting of a sample held between two piezoelectric transducers, without bonding agents. Silver electrodes, on the sides of the sample, are used to apply a small ac-voltage (1–20 V) across the sample at frequencies from 0.05 to 2 MHz. One of the transducers records the mechanical oscillations of the sample. Peaks in the spectrum occur at the frequencies of each natural resonance of the sample. Since the sample is excited electrically, and not mechanically, they are linked to a piezoelectric or an electrostrictive response. The electrostrictive response can be excluded given that the response to the electric field is linear and not quadratic. The origin of the piezoelectric response could be three-fold:

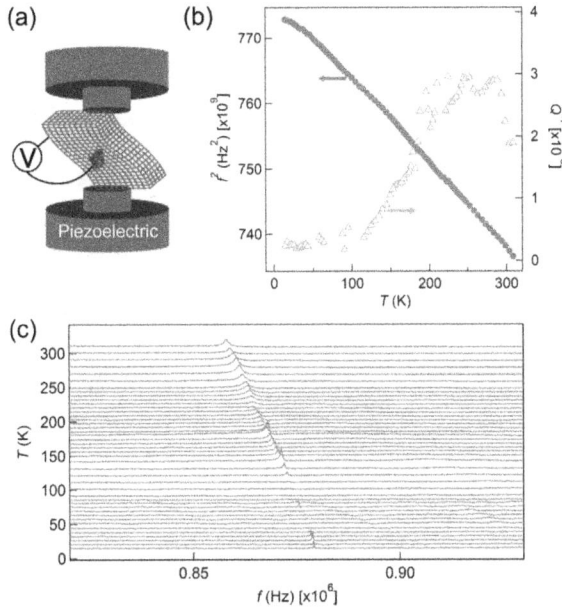

Figure 6.17. Mechanical resonances induced by the piezoelectric effect. (a) Experimental set-up for resonant piezoelectric spectroscopy. (b) Temperature dependence of a resonance near 870 kHz. Squared frequency is indicated by filled circles and the inverse mechanical factor by empty triangles. (c) Resonance amplitude as a function of frequency for temperatures between 10 K and 310 K. (Reproduced with permission from [127]. Copyright 2017 the American Physical Society.)

- Polar defects, such as in the incipient ferroelectric $KTaO_3$, where switchable defect dipoles freeze with decreasing temperature and induce macroscopic polarity [128].
- Precursor polar nanoregions, as in the cubic phase of $BaTiO_3$ and $Pb(Sc,Ta)O_3$, where the applied voltage in RPS interacts with piezoelectric regions inside a precursor structure and generates mechanical resonances [129, 130].
- Polar domain walls, when the ac-voltage leads to the oscillation of the wall through the piezoelectric effect and creates a strain field around them. The resulting elastic wave becomes resonant at a natural frequency for the sample, enhancing considerably the amplitude of the elastic wave.

Precursor polar structures and polar defects have not been reported before in $CaTiO_3$. The resonances observed in RPS have thus been attributed to polar domain walls. They shift to lower frequencies with increasing temperature (figure 6.17(b, c)). In fact, the squared frequency decreases almost linearly with increasing temperature. The elastic losses are low between 10 K and 100 K but show a rapid increase above 100 K (figure 6.17(b)). This is consistent with the behaviour observed previously by resonant ultrasound spectroscopy in $CaTiO_3$ [131] and pulse-echo ultrasonic technique [132], confirming that the resonances seen in RPS correspond to the natural mechanical resonance frequencies of the sample (and are not induced by a surface effect).

It is interesting to note that the macroscopic piezoelectric response, and hence the RPS signal, should vanish if the polar domain walls were statistically perfectly distributed. However this compensation is known to usually be imperfect in piezoelectric materials, and it is very likely that domain walls are distributed unequally, in particular considering the comparatively large domain sizes, thereby explaining the existence of the RPS signal.

The same technique has been used to investigate polar domain walls in $SrTiO_3$. There, the possibility to measure as a function of temperature is crucial since $SrTiO_3$ is ferroelastic only below 105 K. In the ferroelastic phase the material forms a dense network of domains [133]. When the sample is excited mechanically, by resonant ultrasound spectroscopy, resonances are observed in the whole investigated temperature range (120 K–15 K) [134]. Instead, in RPS, resonances appear only below 80 K [134]. They are attributed to polar domain walls. The onset of polarity in the domain walls is supposed to be gradual, as well as the mobility, which may explain why the piezoelectric signal is detected at temperatures slightly below the transition temperature.

Below ~40 K, there is a strong elastic softening and an increase in damping associated with an enhanced mobility of the domain walls. Interestingly, isothermal measurements reveal that this softening is not stable with time, below 50 K. Instead, a systematic stiffening of the structure as a function of time is observed [135]. It is described well by a stretched exponential, typical for glassy relaxations representing pinning or depinning events [136, 137]. Attenuation is also observed to change during the relaxation, with an exponential decrease with time. All these observations are indicative of dynamic interactions within a dense network of domain walls and slowly evolving coupling of ferroelastic and ferroelectric order parameters.

As a complementary local measurement, scanning stress microscopy has been used. This technique maps the electrical response to local stress as a function of lateral position [138]. This microscope used the van der Pauw configuration and the tip of the scanning SQUID chip, which is a non-conducting silicon tip that is rastered over a conducting interface. With piezoelectric elements the tip is pushed into contact with the sample and applies local stress with a contact area of 0.1–1 μm^2. This area is calculated using the Hertzian contact formula:

$$a^3 = \frac{3F}{8} \frac{\frac{(1-v_1)}{E_1} + \frac{(1-v_2)}{E_2}}{1/d_1},$$ (6.5)

where F is the applied force, and v, E, and d are the Poisson ratio, Young's modulus and radius of curvature, respectively. The subscript '1' refers to the silicon tip [139] and '2' to the $SrTiO_3$ substrate [140, 141].

Based on the estimated contact area, the force range applied was kept within the elastic regime, up to 2 μN, exerting stress gradients that decay as the square root of the distance from the contact point. The macroscopic four-probe voltage of the device was hence mapped as a function of the location of the tip for a specific stress at each point (figure 6.18). When the tip pressed on the sample at specific locations the transport value of the whole square changed significantly, thus obtaining a map

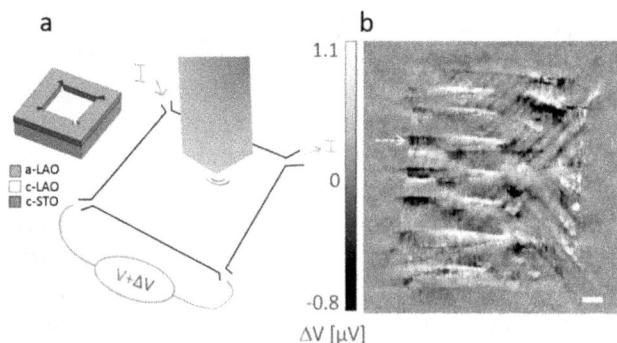

Figure 6.18. (a) Illustration of the device and the experiment. The prefix 'a' stands for amorphous, and 'c' for crystalline. A non-conducting silicon tip is brought in contact with the sample. The scanning and vertical stress application were performed using piezo elements. Voltage was measured in opposing leads (bottom pair) to current injection (top pair) detecting the voltage change ΔV. (b) ΔV as a function of location of the contact point reveals strong responses on domain walls. In this case the tip stress was 0.4 μN uniformly over the image. The background voltage V is 925 μV at 4.2 K. The white arrow points to a needle [100] domain on which ΔV changes sign. The scale bar is 20 μm. (Reproduced with permission from [138]. Copyright 2017 Springer Nature.)

of the change in the global device resistance versus the location of the applied stress (figure 6.18(b)). When the tip scans above the sample without contact, or the stress is applied outside the square, the resistance of the sample did not change, as expected.

Using this technique, Frenkel *et al* provided direct evidence of polarity at the walls by imaging wall polarity below 40 K, demonstrating that the walls are polar. They suggest that this polarity drives the previously reported modulated current flow at the LaAlO$_3$/SrTiO$_3$ interface [142]. This hypothesis is supported by the emergence of spatial modulations in the current flow exclusively below 40 K, coinciding with the onset of strong wall polarity [134].

All these results have been obtained in incipient ferroelectrics. Further works have been dedicated to demonstrating the existence of polar domain walls in non-ferro-electric ferroelastic materials without latent polar instability [143–145]. In Pb$_3$(PO$_4$)$_2$, WO$_3$ and BiVO$_4$, SHG suggests polar domain walls with the symmetry m [143, 144, 146] while the symmetry $3m$ is found for LaAlO$_3$ [145]. An attempt to gain further control over the polarization in the domain walls was performed by doping Pb$_3$(PO$_4$)$_2$ with divalent-metal ions [147]. Ca-doped and Mg-doped Pb$_3$(PO$_4$)$_2$ show an enhanced SHG activity at the domain walls. Since SHG is not element specific, it is difficult to determine which atom plays a major role in enhancing the SHG signal. A possibility is that Ca and Mg ions in domain walls exhibit larger displacements than Pb, due to their smaller ionic radii, leading to a higher polarization [147].

While the presence of polar domain walls in non-polar ferroelastic materials is now well established, the experimental demonstration of the bi-stable polarity of domain walls and its switching has not yet been realized. Using perovskite calcium titanate as a paradigm, molecular dynamics calculations indicate that high electric fields broaden the ferroelastic domain walls, thereby reducing flexoelectricity (as the domain wall strain gradient is inversely proportional to the wall width), eventually enabling switching [148]. The structural response (twinning angle and polarization) of a $CaTiO_3$ supercell with two domain walls is used in [148]. Before applying an electric field, the flexoelectric polarization is positive for one wall, negative for the other, and zero on average, so there is no net remnant polarization. Applying an electric field gradually polarizes the crystal structures of the domains and the domain walls. Concomitant with this increase in dielectric polarization, there is a decrease in spontaneous strain, and a gradual broadening of the domain walls. Both these events decrease flexoelectricity. At some critical field, the weak flexoelectric polarization is overcome by the biquadratic coupling between the primary order parameter and polarization, and the polarization jumps abruptly. Upon retrieving the electric field, the polar phase remains trapped in a metastable state, and eventually reverts to their initial state as the field is further lowered. Applying the electric field along the opposite direction, the domain wall dipoles behave analogously, thus generating an antiferroelectric-like double-hysteresis loop.

These calculations suggest that the polarization of ferroelastic domain walls is switchable. In addition, the switching is not ferroelectric-like but antiferroelectric-like.

Surface states are also generally polar, and the polarity in the domain wall strongly interacts with surface polarity near the intersection between the domain wall and the surface, as discussed in detail in chapter 7.

6.7 Magnetism in non-magnetic materials

It is possible to induce a magnetization in a non-magnetic material with a time-dependent electric polarization [149], following the classical induction of a magnetic field by a circulating current. The ionic magnetic moment m of a unit cell is indeed given by

$$m = \sum_i \gamma_i \mathbf{L}_i,$$ (6.6)

where $\mathbf{L}_i$ is the angular momentum arising from the motion of an ion and γ_i is the gyromagnetic ratio tensor of the ion, both summed over all ions in the unit cell [150]. A direct consequence of this relation is that a moving ferroelectric domain wall can result in a dynamic magnetic field [150]. If we consider the example of a ferroelastic ferroelectric 90° domain wall, the angular momentum when the wall moves emerges from the rearrangement of the atomic positions of the ions in each domain, as shown in figure 6.19(a). In the chosen case of a Néel-type domain wall, where the ferroelectric polarization rotates within the surface plane, the movement of the domain wall is described by two dimensionless amplitudes Q_x and Q_y (figure 6.19 (b)), along two orthogonal directions. Their evolution with time leads to the emergence of a net magnetic moment, which can reach up to 24.7×10^{-6} μB per formula unit (fu^{-1}) for the fastest domain walls.

Beyond domain walls, molecular dynamics modelling reveals that the movement of a single kink in a non-ferroelectric ferroelastic domain wall can induce a magnetic moment [44, 151]. Because of the movement of the kink, the cations and anions near the domain wall are displaced (figure 6.20). As in the previous case of ferroelectric

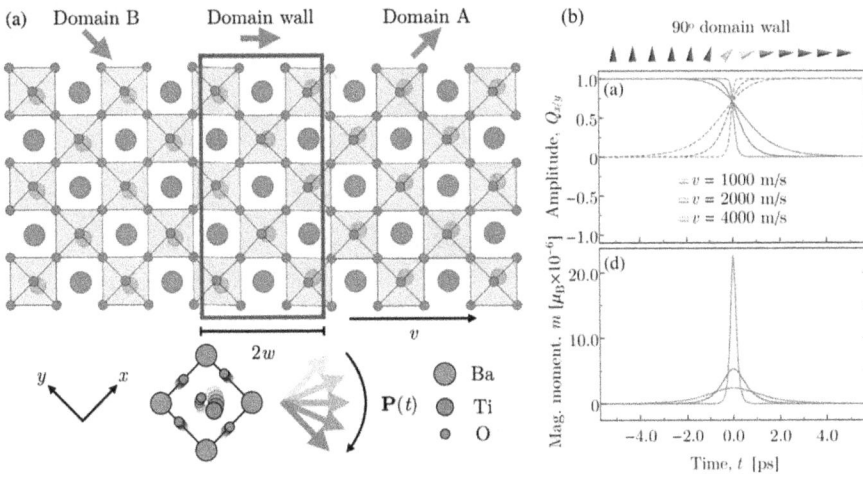

Figure 6.19. (a) Schematic of a 90° domain wall in the x–y plane of $BaTiO_3$ with a thickness $2w$ and velocity v separating the two domains A and B. At the bottom, the rearrangement of the titanium and oxygen ions in the moving domain wall is depicted, as well as the change of ferroelectric polarization in time (thick arrows). The displacements of the oxygen and titanium ions are exaggerated for illustration purposes. Note that atomic radii, not ionic, are shown. (b) Time evolution of a 90° domain wall. Shown are the dimensionless amplitudes Q_x (solid lines) and Q_y (dashed lines). (Reproduced with permission from [150]. Copyright 2019 the American Physical Society.)

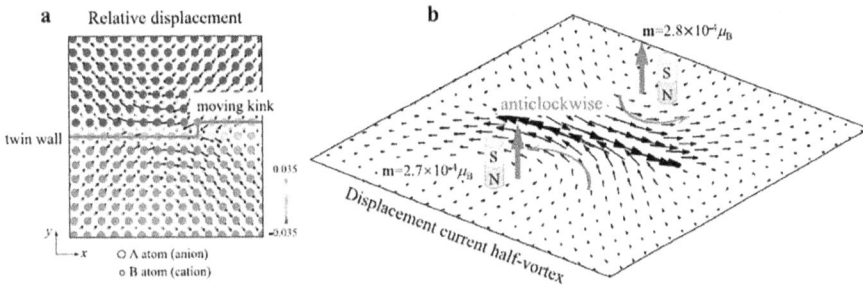

Figure 6.20. (a) Relative displacements of cations and anions near the moving kink during a time interval of 0.5 ps. The colours are coded by the atomic-level shear strain. Ionic displacements are amplified by a factor of 50 for clarity. (b) Two half-vortices of the displacement currents rotate in the same direction in both domains. The current density was calculated based on the relative displacements of anions and cations. The local displacement current is $\sim 10^{-19}$ A. The magnetic field is $\sim 2.7 \times 10^{-4}$ μB for each half-vortex. The current vectors are amplified by a factor of 2×10^{17} for clarity. (Reproduced from [44]. CC BY 4.0.)

domain walls, this relative displacement results in turns into displacement currents that form vortices on either side of the domain wall and thus a magnetic moment. Contrary to the case of domain walls, these moments are not confined to the kink but extend in the surrounding bulk. Importantly, the two displacement currents rotate in the same direction, enhancing the total value of the generated magnetic moment.

To obtain a significant magnetic field in the material, many kinks, in the same area, are required. This can be achieved for example at the tip of a ferroelastic needle domain. The magnetic moment depends then on the number of kinks and the distance between them: in particular, it increases with increasing kink–kink distance. The magnetic moment can be even further enhanced by considering a comb pattern, consisting of several adjacent needle domains. For such a pattern, the simulations revealed a moment density of 3.9×10^{-5} μB fu^{-1} [44], very small compared to magnetic materials with typical moment densities of 2 μB fu^{-1}, but in principle experimentally measurable.

In general, structural defects carrying strong strain gradients, such as moving kinks, needles, domain walls and their junctions, should induce displacement currents and hence magnetic moments. Confirming this experimentally would open the door to a whole new field for multiferroic functional materials. However, it requires combining both high sensitivity to magnetic fields and high spatial resolution, which is far from trivial. It has been proposed that high-resolution SQUID pick-up loops could be used [44], by analogy with existing measurements performed on SrTiO$_3$-based heterointerfaces discussing the role of ferroelastic domain walls [152]. Alternatively, one could work with a nitrogen-vacancy micro-scope to measure the magnetic stray field [150]. A time-dependent electric field or stress field, used to move a single domain wall across the tip of the microscope, would produce a magnetic stray field with changing sign depending on the direction of the domain wall motion. The resulting oscillatory magnetic signal at the probe tip, tuned to the resonance frequency of the nitrogen-vacancy centre, would induce Rabi oscillations that should be above the sensitivity of existing equipment [150].

6.8 Control of thermal conductivity

Several simulations predict that ferroelastic domain walls should have an influence on the thermal conductivity of a material [153–156]. Indeed, domain walls, with different structures, symmetries and mechanical properties, can be considered as planar defects, and as such act as scattering centres for phonons conducting heat. The general picture is that increasing the number of domain walls increases the number of interactions between phonons and domain walls, and decreases the thermal conductivity [153]. The nature of the interaction itself remains complex. In some cases, high frequencies phonons (18–27 THz) are reported to be more affected by domain walls than low frequency phonons [153]. In other cases, it is found that optical modes of low to medium frequencies (3–20 THz) undergo a stronger scattering than high frequency phonons [153]. It has also been proposed that longitudinal phonons could be more affected than transverse phonons [153], opening the possibility for using domain walls as phonon polarizers [157, 158].

Seminal experimental works on bulk single crystals mention the influence of domain walls on the thermal conductivity at low temperature, but without knowledge of the precise orientation and density of domain walls [159–161]. Instead, in LaAlO$_3$, a systematic study of the influence of domain walls on thermal conductivity has been performed [156]. Different thermal treatments were applied to single crystals to tune

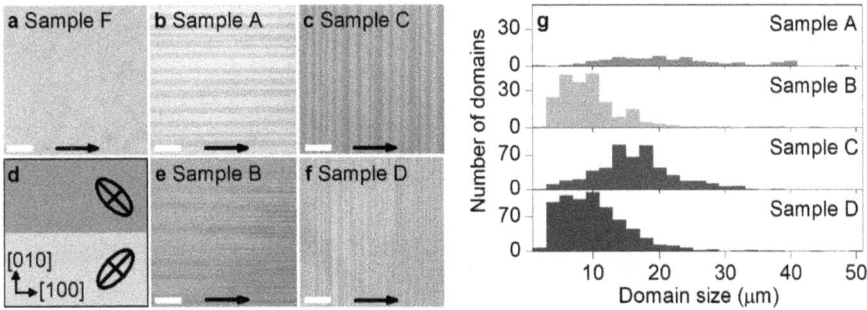

Figure 6.21. Optical images in reflection of selected single crystals of LaAlO$_3$ (a) after annealing, (b), (c) as-received, and (e), (f) after quenching. The panels are oriented such that the heat flow in thermal conductivity measurements goes from the left to the right, as indicated by the black arrows. (d) Schematics of the domain structure showing two domains, where ellipses indicate the orientation of their axes of compression (minor axis) and extension (major axis), projected into the plane of the schematic. Scale bars correspond to 100 μm. (g) Histogram of the number of domains with given sizes, measured from optical images. (Reproduced with permission from [156]. Copyright 2023 the American Physical Society.)

the domain structure. Figure 6.21 shows optical images in reflection of the five single crystals studied. Sample F (figure 6.21(a)) has been annealed in air at 1200 K for 6 h to obtain a monodomain state, i.e. free of domain walls, achieved thanks to a redistribution of defects at high temperature, where defects act as pinning centres for domain walls [162]. Figure 6.21(b) (sample A) and 6.21(c) (sample C) are as-received commercial single crystals with a regular domain pattern. Domain walls are strictly orthogonal to the surface. Domain structures observed in figure 6.21(e) (sample B) and figure 6.21(f) (sample D) are obtained after quenching the as-received single crystals from 680 K to room temperature in a silicon-oil/water mixture. They exhibit an increased density of domain walls, which remain mostly orthogonal to the surface.

The thermal conductivity of these samples—from room temperature down to 2 K —is shown in figure 6.22(a). In the low-temperature boundary scattering regime, a decrease of the thermal conductivity is observed in samples with domain walls parallel to the heat flow (A and B). This decrease is stronger when domain walls are orthogonal to the flow (C and D). To further analyse the influence of domain walls, specific heat measurements (figure 6.22(c)) are used to calculate the temperature dependence of the phonon mean free path L (figure 6.22(d)). At low temperatures, the mean free path saturates differently depending on the orientation and density of domain walls. L is found to be more constrained when domain walls are orthogonal to the heat flow and their density is high, as in sample D compared to sample C, rather than in the case of domain walls parallel to the heat flow as in samples A and B. L is also always lower in the monodomain sample F. An interesting result is thus that domain walls have a different influence on the thermal conductivity depending on their orientation with respect to the heat flow. Domain walls orthogonal to the heat flow reduce the thermal conductivity drastically, while domain walls parallel to the heat flow do it to a lesser extent. From a modelling point of view, it is equivalent to say that the distribution of domain walls implies a distribution of thermal

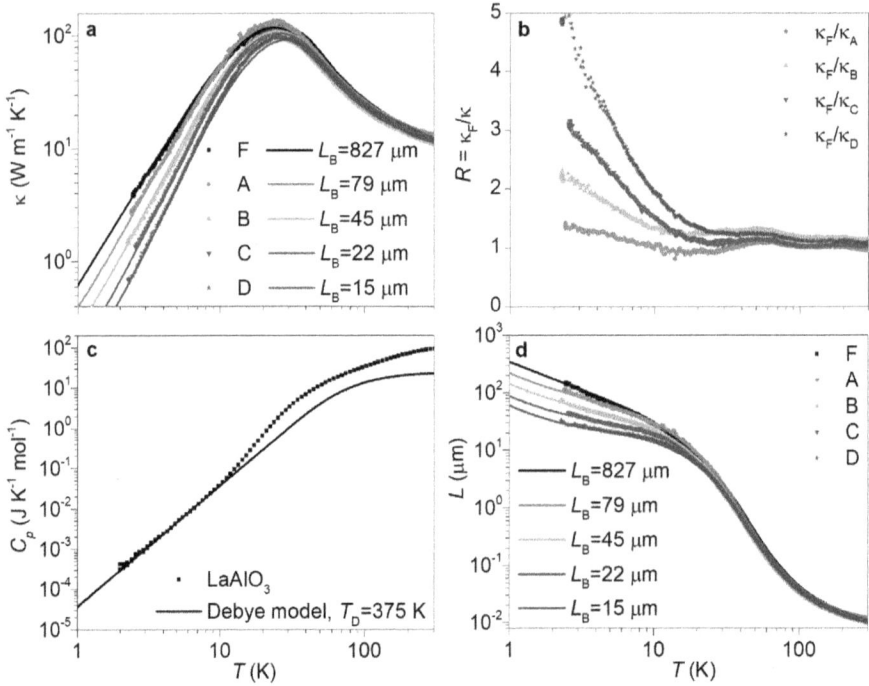

Figure 6.22. Thermal conductivity as a function of temperature of LaAlO$_3$ samples displaying distinct domain wall densities and orientations. (a) Sample F is monodomain whereas in samples A and B domain walls are parallel to the heat flow, and orthogonal to it in C and D. (b) The switching ratio R illustrates the enhancement of the monodomain thermal conductivity compared to the others. (c) Specific heat as a function of temperature measured in a single crystal of LaAlO$_3$. The line corresponds to the Debye model which accounts for the acoustic phonons' contribution. (d) Temperature dependence of the mean free path L deduced from $\kappa = \frac{1}{3} C_{vol} v L$ with the inferred sound velocity $v = 4773$ ms^{-1} and $C_{vol} = C_p \left(\frac{M}{V \, M_{mol}} \right)$. (Reproduced with permission from [156]. Copyright 2023 by the American Physical Society.)

conductivities considered either in series or in parallel depending on the orientation of the walls with respect to the heat flow.

The influence of domain walls on thermal conductivity in LaAlO$_3$ is dominant only at low temperature, where the mean free path of the phonons is large, and comparable to the size of the domains. At higher temperature, phonon–phonon interactions govern the thermal conductivity. To see a strong effect of domain walls on thermal conductivity at room temperature, it is necessary to decrease the domain size. One strategy consists in working with materials where domains are small, even in the bulk [163–165]. For instance, in the relaxor Pb(Mg$_{1/3}$Nb$_{2/3}$)O$_3$−xPbTiO$_3$, thermal conductivity has been measured in samples with domain sizes in the submicrometer range [164]. Using alternating current poling, different densities of domain walls were stabilized leading to different thermal conductivities. As in previous studies, the lowest thermal conductivity was observed for the highest density of domain walls [164].

Another strategy takes advantage of the decrease in domain size with decreasing thickness in ferroelectric materials (Kittel's law [166]). Thus, in thin films, domain sizes can be reduced to 10–100 nm. For example, in $PbZr_{0.3}Ti_{0.7}O_3$ films, tens of nanometer-spaced ferroelastic domain walls were obtained [167]. The density of domain walls was then modified with an applied electric field, leading to a controllable thermal conductivity. In addition, in epitaxial thin films, it is possible to control the density and type of domain walls with the strain originating from the substrate [168, 169]. This has been successfully achieved in $PbTiO_3$, where tensile and compressive strains were used to vary the size of the domains from 20 to 60 nm (figure 6.23). It was found that the larger the domain wall density, the lower the thermal conductivity. In addition, the reduction in thermal conductivity was stronger for 90° domain walls than for 180° domain walls [168]. The thermal resistance from domain walls was reported to be comparable to the thermal resistance of grain boundaries in oxides, and even higher in epitaxial $BiFeO_3$ films [169].

Overall, all measurements demonstrate that (ferroelastic) domain walls reduce thermal conductivity. However, the exact mechanism by which domain walls interact

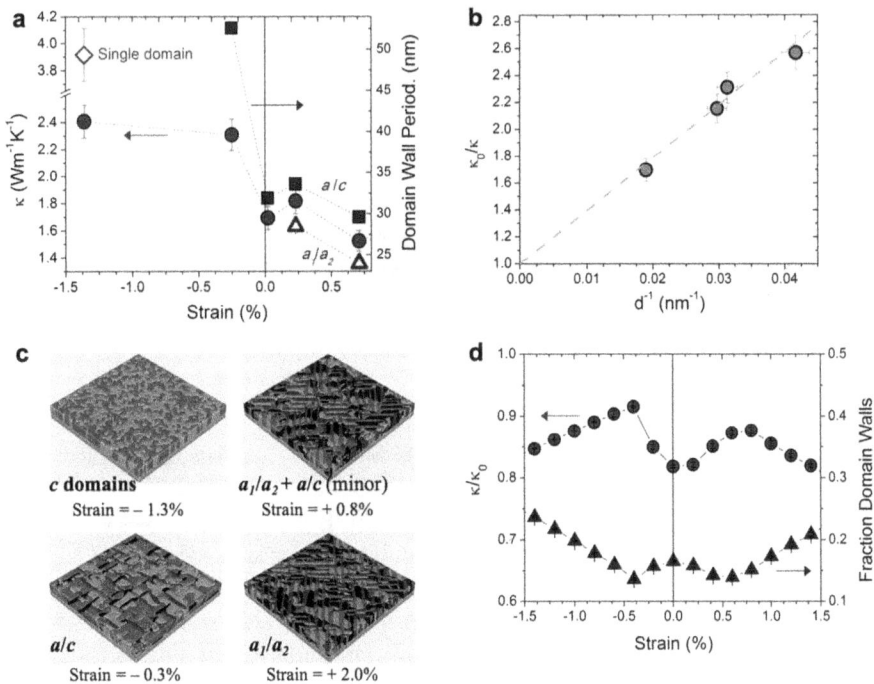

Figure 6.23. The effect of ferroelectric domain walls on the thermal conductivity of $PbTiO_3$ at room temperature. (a) Room-temperature thermal conductivity and domain wall periodicity (obtained from piezoresponce force microscopy (PFM) analysis) of $PbTiO_3$ as a function of epitaxial strain. (b) Fitting of the thermal conductivity to the Kapitza resistance model. κ_0 stands for the thermal conductivity of the single domain. (c) Strain-dependent domain patterns obtained from phase-field simulations. The fraction of a-domains increases progressively with respect to c-domains, as tensile strain increases. (d) Strain-dependent thermal conductivity (normalized by κ_0) and the fraction of domain walls with regard to domains of $PbTiO_3$ predicted by phase-field simulations. (Reproduced with permission from [168]. Copyright 2019 American Chemical Society.)

with phonons remains debated. This is partly because most measurements are performed on samples with complex domain structures or on thin films where the influence of residual strain, substrate and defects can be substantial. Furthermore, the thermal conductivity is often assessed only in one direction, missing out anisotropies induced by different orientations of domains and domain walls. This is particularly critical given that in ferroelectric materials, the thermal conductivity depends on the direction of propagation of heat with respect to the electric polarization direction [170]. The main application envisioned in these fundamental studies remains a thermal conductivity switch, where different states of thermal conductivity can be obtained under the application of an external field. How far could one go with the ability to control heat with ferroelectrics/ferroelastics? One long-term idea is to develop a new type of computing paradigm based on thermal currents rather than electric currents [171]. Heat computing could solve many bottlenecks faced by electronic computing, such as high energy consumption, overheating and lack of robustness in harsh environments. Indeed, heat is a pervasive form of energy and is one of the main actors in a plethora of scenarios, ranging from exothermic chemical reactions to mechanical friction. It could thus be harvested from the environment and there would be no need for energy-consuming sources, as in electronic architectures.

6.9 Summary

Ferroelastic domain walls are narrow interfaces, with widths at the unit-cell level, and unique mechanical properties. They usually exhibit emergent functionalities, such as enhanced electrical conductivity, local polarization that may be switched, or magnetization. They can combine to form different hierarchical structures where wall–wall interactions are important, and which can be controlled by an applied external field. Their ability to scatter phonons can be used to control reversibly thermal conductivity.

Acknowledgements

Guangming Lu is grateful for the financial support by the National Natural Science Foundation of China (Grant No. 12304130), Shandong Provincial Natural Science Foundation (ZR2024QA146) and the Doctoral Starting Fund of Yantai University (Grant No. 1115-2222006). EKHS is grateful to EPSRC (EP/P024904/1) and the European Union's Horizon 2020 programme under the Marie Sklodowska–Curie Grant (861153). Kumara Cordero-Edwards acknowledges funding by grant No. 964931 (TSAR, https://doi.org/10.3030/964931) from the European Union's Horizon 2020 research and innovation program and grant CEX2021-001214-S from the Severo Ochoa program.

Bibliography

[1] Salje E K H and Ishibashi Y 1996 Mesoscopic structures in ferroelastic crystals: needle twins and right-angled domains *J. Phys. Condens. Matter* **8** 8477

[2] Salje E K H, Buckley A, van Tendeloo G, Ishibashi Y and Nord G L 1998 Needle twins and right-angled twins in minerals; comparison between experiment and theory *Am. Mineral.* **83** 811

[3] Bratkovsky A M, Marais S C, Heine V and Salje E K H 1994 The theory of fluctuations and texture embryos in structural phase transitions mediated by strain *J. Phys. Condens. Matter* **6** 3679

[4] Bratkovsky A M, Salje E K H and Heine V 1994 Overview of the origin of tweed texture *Phase Transit.* **52** 77

[5] Marais S, Heine V, Nex C and Salje E 1991 Phenomena due to strain coupling in phase transitions *Phys. Rev. Lett.* **66** 2480

[6] Bosbach D, Putnis A, Bismayer U and Güttler B 1997 An AFM study on ferroelastic domains in lead phosphate, $Pb_3(PO_4)_2$ *J. Phys. Condens. Matter* **9** 8397

[7] Putnis A and Salje E 1994 Tweed microstructures: experimental observations and some theoretical models *Phase Transit.* **48** 85

[8] Meyer B and Vanderbilt D 2002 *Ab initio* study of ferroelectric domain walls in $PbTiO_3$ *Phys. Rev.* B **65** 104111

[9] Hlinka J and Márton P 2006 Phenomenological model of a $90°$ domain wall in $BaTiO_3$-type ferroelectrics *Phys. Rev.* B **74** 104104

[10] Marton P, Rychetsky I and Hlinka J 2010 Domain walls of ferroelectric $BaTiO_3$ within the Ginzburg–Landau–Devonshire phenomenological model *Phys. Rev.* B **81** 144125

[11] Hlinka J and Marton P 2008 Ferroelastic domain walls in barium titanate—quantitative phenomenological model *Integr. Ferroelectr.* **101** 50

[12] Yamamoto N, Yagi K and Honjo G 1977 Electron microscopic studies of ferroelectric and ferroelastic $Gd_2(MoO_4)_3$. I. General features of ferroelectric domain wall, antiphase boundary, and crystal defects *Phys. Status Solidi* **41** 523

[13] Boulesteix C, Yangui B, Nihoul G and Bourret A 1983 High resolution and conventional electron microscopy studies of repeated wedge microtwins in monoclinic rare earth sesquioxides *J. Microsc.* **129** 315

[14] Stemmer S, Streiffer S K, Ernst F and Rüuhle M 1995 Atomistic structure of $90°$ domain walls in ferroelectric $PbTiO_3$ thin films *Philos. Mag.* A **71** 713

[15] Locherer K R, Chrosch J and Salje E K H 1998 Diffuse x-ray scattering in WO_3 *Phase Transit.* **67** 51

[16] Chrosch J and Salje E K H 1999 Temperature dependence of the domain wall width in $LaAlO_3$ *J. Appl. Phys.* **85** 722

[17] Hayward S A, Chrosch J, Salje E K H and Carpenter M A 1997 Thickness of pericline twin walls in anorthoclase: an x-ray diffraction study *Eur. J. Mineral.* **8** 1301

[18] Chrosch J and Salje E K H 1994 Thin domain walls in $YBa_2Cu_3O_{7-\delta}$ and their rocking curves an x-ray diffraction study *Physica* C **225** 111

[19] Vlokh R, Uesu Y, Yamada Y, Skab I and Vlokh O 1998 Appearance of paraelastic phase as ferroelastic domain walls in $K_2Cd_2(SO_4)_3$ single crystals *J. Phys. Soc. Japan* **67** 3335

[20] Salje E K 1991 *Phase Transitions in Ferroelastic and Co-Elastic Crystals* (Cambridge: Cambridge University Press)

[21] Torrès J, Roucau C and Ayroles R 1982 Investigation of the interactions between ferroelastic domain walls and the structural transition in lead phosphate observed by electron microscopy: I. Experimental results *Phys. Status Solidi* **70** 659

[22] Wruck B, Salje E K H, Zhang M, Abraham T and Bismayer U 1994 On the thickness of ferroelastic twin walls in lead phosphate $Pb_3(PO_4)_2$ an x-ray diffraction study *Phase Transit.* **48** 135

[23] Andrews S R and Cowley R A 1986 X-ray scattering from critical fluctuations and domain walls in KDP and DKDP *J. Phys. C: Solid State Phys.* **19** 615

[24] Gonnissen J, Batuk D, Nataf G F, Jones L, Abakumov A M, Van Aert S, Schryvers D and Salje E K H 2016 Direct observation of ferroelectric domain walls in $LiNbO_3$: wall-meanders, kinks, and local electric charges *Adv. Funct. Mater.* **26** 7599

[25] Shur V Y, Pelegova E V, Turygin A P, Kosobokov M S and Alikin Y M 2021 Forward growth of ferroelectric domains with charged domain walls. Local switching on non-polar cuts *J. Appl. Phys.* **129** 044103

[26] Condurache O, Dražić G, Sakamoto N, Rojac T and Benčan A 2021 Atomically resolved structure of step-like uncharged and charged domain walls in polycrystalline $BiFeO_3$ *J. Appl. Phys.* **129** 054102

[27] Jia C-L, Mi S-B, Urban K, Vrejoiu I, Alexe M and Hesse D 2008 Atomic-scale study of electric dipoles near charged and uncharged domain walls in ferroelectric films *Nat. Mater.* **7** 57

[28] Ignatans R, Damjanovic D and Tileli V 2021 Individual Barkhausen pulses of ferroelastic nanodomains *Phys. Rev. Lett.* **127** 167601

[29] Lu G, Li S, Ding X, Sun J and Salje E K H 2019 Ferroelectric switching in ferroelastic materials with rough surfaces *Sci. Rep.* **9** 15834

[30] Prosandeev S, Wang D, Ren W, Íñiguez J and Bellaiche L 2013 Novel nanoscale twinned phases in perovskite oxides *Adv. Funct. Mater.* **23** 234

[31] Warwick A R, Íñiguez J, Haynes P D and Bristowe N C 2019 First-principles study of ferroelastic twins in halide perovskites *J. Phys. Chem. Lett.* **10** 1416

[32] Arul Kumar M, Beyerlein I J and Tomé C N 2016 Effect of local stress fields on twin characteristics in HCP metals *Acta Mater.* **116** 143

[33] Ardeljan M, McCabe R J, Beyerlein I J and Knezevic M 2015 Explicit incorporation of deformation twins into crystal plasticity finite element models *Comput. Methods Appl. Mech. Eng.* **295** 396

[34] Lubk A, Rossell M D, Seidel J, Chu Y H, Ramesh R, Hÿtch M J and Snoeck E 2013 Electromechanical coupling among edge dislocations, domain walls, and nanodomains in $BiFeO_3$ revealed by unit-cell-wise strain and polarization maps *Nano Lett.* **13** 1410

[35] Emelyanov A Y, Pertsev N A and Salje E K H 2001 Effect of finite domain-wall width on the domain structures of epitaxial ferroelectric and ferroelastic thin films *J. Appl. Phys.* **89** 1355

[36] Salje E K H, Ding X, Zhao Z, Lookman T and Saxena A 2011 Thermally activated avalanches: jamming and the progression of needle domains *Phys. Rev. B* **83** 104109

[37] Hayward S A and Salje E K H 1999 Cubic-tetragonal phase transition in $SrTiO_3$ revisited: Landau theory and transition mechanism *Phase Transit.* **68** 501

[38] He X, Li S, Ding X, Sun J, Kustov S and Salje E K H 2022 Internal friction in complex ferroelastic twin patterns *Acta Mater.* **228** 117787

[39] Zhao Z, Ding X, Lookman T, Sun J and Salje E K H 2013 Mechanical loss in multiferroic materials at high frequencies: friction and the evolution of ferroelastic microstructures *Adv. Mater.* **25** 3244

[40] Lu G, Ding X, Sun J and Salje E K H 2022 Wall–wall and kink–kink interactions in ferroelastic materials *Phys. Rev. B* **106** 144105

[41] Lu G, Hideo K, Ding X, Xu Z, Chu R, Nataf G F and Salje E K H 2023 Influence of kinks on the interaction energy between ferroelastic domain walls in membranes and thin films *Microstructures* **3** 2023033

[42] Lu G, Li S, Ding X and Salje E K H 2019 Piezoelectricity and electrostriction in ferroelastic materials with polar twin boundaries and domain junctions *Appl. Phys. Lett.* **114** 202901

[43] Lu G, Li S, Ding X, Sun J and Salje E K H 2020 Enhanced piezoelectricity in twinned ferroelastics with nanocavities *Phys. Rev. Mater.* **4** 074410

[44] Lu G, Li S, Ding X, Sun J and Salje E K H 2020 Current vortices and magnetic fields driven by moving polar twin boundaries in ferroelastic materials *NPJ Comput. Mater.* **6** 145

[45] Lu G, Li S, Ding X, Sun J and Salje E K H 2019 Electrically driven ferroelastic domain walls, domain wall interactions, and moving needle domains *Phys. Rev. Mater.* **3** 114405

[46] Salje E K H, Li S, Stengel M, Gumbsch P and Ding X 2016 Flexoelectricity and the polarity of complex ferroelastic twin patterns *Phys. Rev. B* **94** 024114

[47] Salje E K H, Ding X and Zhao Z 2013 Noise and finite size effects in multiferroics with strong elastic interactions *Appl. Phys. Lett.* **102** 152909

[48] Checa M, Stefani C, Kelley K, Balke N, Collins L, Catalan G, Jesse S and Domingo N 2025 Nanoscale polarization-dependent Young's modulus of ferroelectric $BaTiO_3$ single crystals *ACS Nano* **19** 9835

[49] Park Y-B, Dicken M J, Xu Z-H and Li X 2007 Nanoindentation of the *a* and *c* domains in a tetragonal $BaTiO_3$ single crystal *J. Appl. Phys.* **102** 083507

[50] Nguyen C T, Schoenherr P and Seidel J 2023 Intrinsic mechanical compliance of 90° domain walls in $PbTiO_3$ *Adv. Funct. Mater.* **33** 2211906

[51] Tsuji T, Saito S, Fukuda K, Yamanaka K, Ogiso H, Akedo J and Kawakami Y 2005 Significant stiffness reduction at ferroelectric domain boundary evaluated by ultrasonic atomic force microscopy *Appl. Phys. Lett.* **87** 071909

[52] Tsuji T, Ogiso H, Akedo J, Saito S, Fukuda K and Yamanaka K 2004 Evaluation of domain boundary of piezo/ferroelectric material by ultrasonic atomicforce microscopy *Jpn. J. Appl. Phys.* **43** 2907

[53] Stefani C, Ponet L, Shapovalov K, Chen P, Langenberg E, Schlom D G, Artyukhin S, Stengel M, Domingo N and Catalan G 2020 Mechanical softness of ferroelectric 180° domain walls *Phys. Rev. X* **10** 041001

[54] Žiberna K, Koblar M, Bradeško A, Bah M, Levassort F, Dražić G, Uršič H and Benčan A 2025 Elastic properties of domains and domain walls in $(K_{0.5}Na_{0.5})NbO_3$ single crystal *J. Eur. Ceram. Soc.* **45** 117566

[55] Rabe U, Kopycinska M, Hirsekorn S, oz Salda a J M, Schneider G A and Arnold W 2002 High-resolution characterization of piezoelectric ceramics by ultrasonic scanning force microscopy techniques *J. Phys. D: Appl. Phys.* **35** 2621

[56] Rabe U, Amelio S, Hirsekorn S and Arnold W 2002 Imaging of ferroelectric domains by atomic force acoustic microscopy *Acoustic Imaging* **vol 25** (Boston, MA: Springer) pp 253–60

[57] Kwaśniewska A, Świetlicki M, Prószyński A and Gładyszewski G 2021 The quantitative nanomechanical mapping of starch/kaolin film surfaces by peak force AFM *Polymers* **13** 244

[58] Rojac T *et al* 2017 Domain-wall conduction in ferroelectric $BiFeO_3$ controlled by accumulation of charged defects *Nat. Mater.* **16** 322

[59] Chandrasekaran A, Damjanovic D, Setter N and Marzari N 2013 Defect ordering and defect–domain-wall interactions in $PbTiO_3$: a first-principles study *Phys. Rev. B* **88** 214116

[60] Chandrasekaran A, Wei X-K, Feigl L, Damjanovic D, Setter N and Marzari N 2016 Asymmetric structure of 90° domain walls and interactions with defects in $PbTiO_3$ *Phys. Rev. B* **93** 144102

[61] He X, Li S, Ding X, Sun J, Selbach S M and Salje E K H 2019 The interaction between vacancies and twin walls, junctions, and kinks, and their mechanical properties in ferroelastic materials *Acta Mater.* **178** 26

[62] Schranz W and Kityk A V 2008 Domain wall dynamics in ferroelastic crystals *Ferroelectrics* **375** 178

[63] He X, Ding X, Sun J, Nataf G F and Salje E K H 2023 Elastic softening and hardening at intersections between twin walls and surfaces in ferroelastic materials *APL Mater.* **11** 071114

[64] Nataf G F, Guennou M, Gregg J M, Meier D, Hlinka J, Salje E K H and Kreisel J 2020 Domain-wall engineering and topological defects in ferroelectric and ferroelastic materials *Nat. Rev. Phys.* **2** 634

[65] Catalan G, Seidel J, Ramesh R and Scott J F 2012 Domain wall nanoelectronics *Rev. Mod. Phys.* **84** 119

[66] Seidel J *et al* 2009 Conduction at domain walls in oxide multiferroics *Nat. Mater.* **8** 229

[67] Sluka T, Tagantsev A K, Bednyakov P and Setter N 2013 Free-electron gas at charged domain walls in insulating $BaTiO_3$ *Nat. Commun.* **4** 1808

[68] Guyonnet J, Gaponenko I, Gariglio S and Paruch P 2011 Conduction at domain walls in insulating $Pb(Zr_{0.2}Ti_{0.8})O_3$ thin films *Adv. Mater.* **23** 5377

[69] Gaponenko I, Tückmantel P, Karthik J, Martin L W and Paruch P 2015 Towards reversible control of domain wall conduction in $Pb(Zr_{0.2}Ti_{0.8})O_3$ thin films *Appl. Phys. Lett.* **106** 162902

[70] Tselev A, Yu P, Cao Y, Dedon L R, Martin L W, Kalinin S V and Maksymovych P 2016 Microwave AC conductivity of domain walls in ferroelectric thin films *Nat. Commun.* **7** 11630

[71] Choi T, Horibe Y, Yi H T, Choi Y J, Wu W and Cheong S-W 2010 Insulating interlocked ferroelectric and structural antiphase domain walls in multiferroic $YMnO_3$ *Nat. Mater.* **9** 253

[72] Meier D, Seidel J, Cano A, Delaney K, Kumagai Y, Mostovoy M, Spaldin N A, Ramesh R and Fiebig M 2012 Anisotropic conductance at improper ferroelectric domain walls *Nat. Mater.* **11** 284

[73] Wu W, Horibe Y, Lee N, Cheong S-W and Guest J R 2012 Conduction of topologically protected charged ferroelectric domain walls *Phys. Rev. Lett.* **108** 077203

[74] Småbråten D R, Meier Q N, Skjærvø S H, Inzani K, Meier D and Selbach S M 2018 Charged domain walls in improper ferroelectric hexagonal manganites and gallates *Phys. Rev. Mater.* **2** 114405

[75] Schoenherr P, Shapovalov K, Schaab J, Yan Z, Bourret E D, Hentschel M, Stengel M, Fiebig M, Cano A and Meier D 2019 Observation of uncompensated bound charges at improper ferroelectric domain walls *Nano Lett.* **19** 1659

[76] Turner P W, McConville J P V, McCartan S J, Campbell M H, Schaab J, McQuaid R G P, Kumar A and Gregg J M 2018 Large carrier mobilities in $ErMnO_3$ conducting domain walls revealed by quantitative Hall-effect measurements *Nano Lett.* **18** 6381

[77] Du Y, Wang X L, Chen D P, Dou S X, Cheng Z X, Higgins M, Wallace G and Wang J Y 2011 Domain wall conductivity in oxygen deficient multiferroic $YMnO_3$ single crystals *Appl. Phys. Lett.* **99** 252107

[78] Schröder M, Haußmann A, Thiessen A, Soergel E, Woike T and Eng L M 2012 Conducting domain walls in lithium niobate single crystals *Adv. Funct. Mater.* **22** 3936

[79] Godau C, Kämpfe T, Thiessen A, Eng L M and Haußmann A 2017 Enhancing the domain wall conductivity in lithium niobate single crystals *ACS Nano* **11** 4816

[80] Schröder M, Chen X, Haußmann A, Thiessen A, Poppe J, Bonnell D A and Eng L M 2014 Nanoscale and macroscopic electrical AC transport along conductive domain walls in lithium niobate single crystals *Mater. Res. Express* **1** 035012

[81] Werner C S, Herr S J, Buse K, Sturman B, Soergel E, Razzaghi C and Breunig I 2017 Large and accessible conductivity of charged domain walls in lithium niobate *Sci. Rep.* **7** 9862

[82] Farokhipoor S and Noheda B 2011 Conduction through 71° domain walls in $BiFeO_3$ thin films *Phys. Rev. Lett.* **107** 127601

[83] Farokhipoor S and Noheda B 2012 Local conductivity and the role of vacancies around twin walls of (001)–$BiFeO_3$ thin films *J. Appl. Phys.* **112** 052003

[84] Chiu Y-P *et al* 2011 Atomic-scale evolution of local electronic structure across multiferroic domain walls *Adv. Mater.* **23** 1530

[85] Lubk A, Gemming S and Spaldin N A 2009 First-principles study of ferroelectric domain walls in multiferroic bismuth ferrite *Phys. Rev.* B **80** 104110

[86] Diéguez O, Aguado-Puente P, Junquera J and Íñiguez J 2013 Domain walls in a perovskite oxide with two primary structural order parameters: first-principles study of $BiFeO_3$ *Phys. Rev.* B **87** 024102

[87] Seidel J *et al* 2010 Domain wall conductivity in La-doped $BiFeO_3$ *Phys. Rev. Lett.* **105** 197603

[88] Stolichnov I, Iwanowska M, Colla E, Ziegler B, Gaponenko I, Paruch P, Huijben M, Rijnders G and Setter N 2014 Persistent conductive footprints of 109° domain walls in bismuth ferrite films *Appl. Phys. Lett.* **104** 132902

[89] Maksymovych P, Seidel J, Chu Y H, Wu P, Baddorf A P, Chen L-Q, Kalinin S V and Ramesh R 2011 Dynamic conductivity of ferroelectric domain walls in $BiFeO_3$ *Nano Lett.* **11** 1906

[90] Yang C-H *et al* 2009 Electric modulation of conduction in multiferroic Ca-doped $BiFeO_3$ films *Nat. Mater.* **8** 485

[91] Stolichnov I *et al* 2015 Bent ferroelectric domain walls as reconfigurable metallic-like channels *Nano Lett.* **15** 8049

[92] Wei X-K, Sluka T, Fraygola B, Feigl L, Du H, Jin L, Jia C-L and Setter N 2017 Controlled charging of ferroelastic domain walls in oxide ferroelectrics *ACS Appl. Mater. Interfaces* **9** 6539

[93] Aird A and Salje E K H 1998 Sheet superconductivity in twin walls: experimental evidence of WO_{3-x} *J. Phys. Condens. Matter* **10** L377

[94] Kim Y, Alexe M and Salje E K H 2010 Nanoscale properties of thin twin walls and surface layers in piezoelectric WO_{3-x} *Appl. Phys. Lett.* **96** 032904

[95] Leftheriotis G, Papaefthimiou S, Yianoulis P and Siokou A 2001 Effect of the tungsten oxidation states in the thermal coloration and bleaching of amorphous WO_3 films *Thin Solid Films* **384** 298

[96] Jones F, Rawlings K, Foord J, Egdell R, Pethica J, Wanklyn B M, Parker S and Oliver P 1996 An STM study of surface structures on WO_3(001) *Surf. Sci.* **359** 107

[97] Schirmer O F and Salje E 1980 The W^{5+} polaron in crystalline low temperature WO_3 ESR and optical absorption *Solid State Commun.* **33** 333

[98] Schirmer O F and Salje E 1980 Conduction bipolarons in low-temperature crystalline WO_{3-x} *J. Phys. C: Solid State Phys.* **13** L1067

[99] Salje E and Güttler B 1984 Anderson transition and intermediate polaron formation in WO_{3-x} transport properties and optical absorption *Philos. Mag.* B **50** 607

[100] Salje E K H 2020 Polaronic states and superconductivity in WO_{3-x} *Condens. Matter* **5** 32

[101] Körbel S, Hlinka J and Sanvito S 2018 Electron trapping by neutral pristine ferroelectric domain walls in $BiFeO_3$ *Phys. Rev.* B **98** 100104

[102] Imlau M, Badorreck H and Merschjann C 2015 Optical nonlinearities of small polarons in lithium niobate *Appl. Phys. Rev.* **2** 040606

[103] Janovec V 1976 A symmetry approach to domain structures *Ferroelectrics* **12** 43

[104] Janovec V and Kopský V 1997 Layer groups, scanning tables and the structure of domain walls *Ferroelectrics* **191** 23

[105] Janovec V and Přívratská J 2006 Domain structures *International Tables for Crystallography* (Chester: International Union of Crystallography) pp 449–505

[106] Kopský V 2008 The scanning for layer groups and positional dependence of domain wall energy and structure *Ferroelectrics* **376** 168

[107] Janovec V and Litvin D B 2011 Symmetry-allowed atomic displacements in a ferroelastic domain wall of rhombohedral $BaTiO_3$ *Phase Transit.* **84** 760

[108] Janovec V, Richterová L and Přívratská J 1999 Polar properties of compatible ferroelastic domain walls *Ferroelectrics* **222** 73

[109] Gu Y, Li M, Morozovska A N, Wang Y, Eliseev E A, Gopalan V and Chen L-Q 2014 Flexoelectricity and ferroelectric domain wall structures: phase-field modeling and DFT calculations *Phys. Rev.* B **89** 174111

[110] Goncalves-Ferreira L, Redfern S A T, Artacho E and Salje E K H 2008 Ferrielectric twin walls in $CaTiO_3$ *Phys. Rev. Lett.* **101** 097602

[111] Conti S, Müller S, Poliakovsky A and Salje E K H 2011 Coupling of order parameters, chirality, and interfacial structures in multiferroic materials *J. Phys. Condens. Matter* **23** 142203

[112] Salje E K H and Scott J F 2014 Ferroelectric Bloch-line switching: a paradigm for memory devices? *Appl. Phys. Lett.* **105** 252904

[113] Salje E and Zhang H 2009 Domain boundary engineering *Phase Transit.* **82** 452

[114] Lemanov V V, Sotnikov A V, Smirnova E P, Weihnacht M and Kunze R 1999 Perovskite $CaTiO_3$ as an incipient ferroelectric *Solid State Commun.* **110** 611

[115] Kvyatkovskii O E 2001 Quantum effects in incipient and low-temperature ferroelectrics (a review) *Phys. Solid State* **43** 1401

[116] Van Aert S, Turner S, Delville R, Schryvers D, Van Tendeloo G and Salje E K H 2012 Direct observation of ferrielectricity at ferroelastic domain boundaries in $CaTiO_3$ by electron microscopy *Adv. Mater.* **24** 523

[117] Haider M, Uhlemann S, Schwan E, Rose H, Kabius B and Urban K 1998 Electron microscopy image enhanced *Nature* **392** 768

[118] Coene W M J, Thust A, Op de Beeck M and Van Dyck D 1996 Maximum-likelihood method for focus-variation image reconstruction in high resolution transmission electron microscopy *Ultramicroscopy* **64** 109

[119] den Dekker A J, Van Aert S, van den Bos A and Van Dyck D 2005 Maximum likelihood estimation of structure parameters from high resolution electron microscopy images. Part I: a theoretical framework *Ultramicroscopy* **104** 83

[120] Bals S, Van Aert S, Van Tendeloo G and Ávila-Brande D 2006 Statistical estimation of atomic positions from exit wave reconstruction with a precision in the picometer range *Phys. Rev. Lett.* **96** 096106

[121] Van Aert S, Batenburg K J, Rossell M D, Erni R and Van Tendeloo G 2011 Three-dimensional atomic imaging of crystalline nanoparticles *Nature* **470** 374

[122] Boyd R 2008 *Nonlinear Optics* (Amsterdam: Elsevier)

[123] Kaneshiro J, Kawado S, Yokota H, Uesu Y and Fukui T 2008 Three-dimensional observations of polar domain structures using a confocal second-harmonic generation interference microscope *J. Appl. Phys.* **104** 054112

[124] Yokota H and Uesu Y 2021 Optical second-harmonic generation microscopy as a tool for ferroelastic domain wall exploration *J. Appl. Phys.* **129** 014101

[125] Yokota H, Usami H, Haumont R, Hicher P, Kaneshiro J, Salje E K H and Uesu Y 2014 Direct evidence of polar nature of ferroelastic twin boundaries in $CaTiO_3$ obtained by second harmonic generation microscope *Phys. Rev.* B **89** 144109

[126] Yokota H, Niki S, Haumont R, Hicher P and Uesu Y 2017 Polar nature of stress-induced twin walls in ferroelastic $CaTiO_3$ *AIP Adv.* **7** 085315

[127] Nataf G F *et al* 2017 Control of surface potential at polar domain walls in a nonpolar oxide *Phys. Rev. Mater.* **1** 074410

[128] Aktas O, Crossley S, Carpenter M A and Salje E K H 2014 Polar correlations and defect-induced ferroelectricity in cryogenic $KTaO_3$ *Phys. Rev.* B **90** 165309

[129] Aktas O, Carpenter M A and Salje E K H 2013 Polar precursor ordering in $BaTiO_3$ detected by resonant piezoelectric spectroscopy *Appl. Phys. Lett.* **103** 2011

[130] Aktas O, Salje E K H, Crossley S, Lampronti G I, Whatmore R W, Mathur N D and Carpenter M A 2013 Ferroelectric precursor behavior in $PbSc_{0.5}Ta_{0.5}O_3$ detected by field-induced resonant piezoelectric spectroscopy *Phys. Rev.* B **88** 174112

[131] Perks N J, Zhang Z, Harrison R J and Carpenter M A 2014 Strain relaxation mechanisms of elastic softening and twin wall freezing associated with structural phase transitions in $(Ca,Sr)TiO_3$ perovskites *J. Phys. Condens. Matter* **26** 505402

[132] Placeres-Jiménez R, Gonçalves L G V, Rino J P, Fraygola B, Nascimento W J and Eiras J A 2012 Low-temperature elastic anomalies in $CaTiO_3$: dynamical characterization *J. Phys. Condens. Matter* **24** 475401

[133] Stirling W G, Cowley R A and Bruce A D 1974 Lattice dynamics and phase transition of strontium titanate *Ferroelectrics* **7** 401

[134] Salje E K H, Aktas O, Carpenter M A, Laguta V V and Scott J F 2013 Domains within domains and walls within walls: evidence for polar domains in cryogenic $SrTiO_3$ *Phys. Rev. Lett.* **111** 247603

[135] Pesquera D, Carpenter M A and Salje E K H 2018 Glasslike dynamics of polar domain walls in cryogenic $SrTiO_3$ *Phys. Rev. Lett.* **121** 235701

[136] Chauve P, Giamarchi T and Le Doussal P 2000 Creep and depinning in disordered media *Phys. Rev.* B **62** 6241

[137] Díaz-Sánchez A, Pérez-Garrido A, Urbina A and Catalá J D 2002 Stretched exponential relaxation for growing interfaces in quenched disordered media *Phys. Rev.* E **66** 031403

[138] Frenkel Y, Haham N, Shperber Y, Bell C, Xie Y, Chen Z, Hikita Y, Hwang H Y, Salje E K H and Kalisky B 2017 Imaging and tuning polarity at $SrTiO_3$ domain walls *Nat. Mater.* **16** 1203

[139] Wortman J J and Evans R A 1965 Young's modulus, shear modulus, and Poisson's ratio in silicon and germanium *J. Appl. Phys.* **36** 153

[140] Biegalski M D *et al* 2008 Critical thickness of high structural quality $SrTiO_3$ films grown on orthorhombic (101) $DyScO_3$ *J. Appl. Phys.* **104** 114109

[141] Scott J F and Ledbetter H 1997 Interpretation of elastic anomalies in $SrTiO_3$ at 37 K *Z. Phys.* **B 104** 635

[142] Kalisky B *et al* 2013 Locally enhanced conductivity due to the tetragonal domain structure in $LaAlO_3/SrTiO_3$ heterointerfaces *Nat. Mater.* **12** 1091

[143] Yokota H, Matsumoto S, Salje E K H and Uesu Y 2019 Polar nature of domain boundaries in purely ferroelastic $Pb_3(PO_4)_2$ investigated by second harmonic generation microscopy *Phys. Rev.* **B 100** 024101

[144] Eckstein J T, Yokota H, Domingo N, Catalan G, Aktas O, Carpenter M A and Salje E K H 2024 Domain wall dynamics in tungsten trioxide: evidence for polar domain walls *Phys. Rev.* **B 110** 094107

[145] Yokota H, Matsumoto S, Salje E K H and Uesu Y 2018 Symmetry and three-dimensional anisotropy of polar domain boundaries observed in ferroelastic $LaAlO_3$ in the complete absence of ferroelectric instability *Phys. Rev.* **B 98** 024101

[146] Yokota H, Hasegawa N, Glazer M, Salje E K H and Uesu Y 2020 Direct evidence of polar ferroelastic domain boundaries in semiconductor $BiVO_4$ *Appl. Phys. Lett.* **116** 3

[147] Yokota H, Matsumoto S, Hasegawa N, Salje E and Uesu Y 2020 Enhancement of polar nature of domain boundaries in ferroelastic $Pb_3(PO_4)_2$ by doping divalent-metal ions *J. Phys. Condens. Matter* **32** 345401

[148] Lu G, Catalan G and Salje E K H 2025 Antiferroelectric switching inside ferroelastic domain walls *Phys. Rev. Lett.* **135** 206103

[149] Juraschek D M, Fechner M, Balatsky A V and Spaldin N A 2017 Dynamical multi-ferroicity *Phys. Rev. Mater.* **1** 014401

[150] Juraschek D M, Meier Q N, Trassin M, Trolier-McKinstry S E, Degen C L and Spaldin N A 2019 Dynamical magnetic field accompanying the motion of ferroelectric domain walls *Phys. Rev. Lett.* **123** 127601

[151] Lu G, Li S, Ding X, Sun J and Salje E K H 2021 Tip-induced flexoelectricity, polar vortices, and magnetic moments in ferroelastic materials *J. Appl. Phys.* **129** 084104

[152] Christensen D V *et al* 2019 Strain-tunable magnetism at oxide domain walls *Nat. Phys.* **15** 269

[153] Li S, Ding X, Ren J, Moya X, Li J, Sun J and Salje E K H 2014 Strain-controlled thermal conductivity in ferroic twinned films *Sci. Rep.* **4** 6375

[154] Ding X and Salje E K H 2015 Heat transport by phonons and the generation of heat by fast phonon processes in ferroelastic materials *AIP Adv.* **5** 053604

[155] Liu C, Chen Y and Dames C 2019 Electric-field-controlled thermal switch in ferroelectric materials using first-principles calculations and domain-wall engineering *Phys. Rev. Appl.* **11** 044002

[156] Limelette P, El Kamily M, Aramberri H, Giovannelli F, Royo M, Rurali R, Monot-Laffez I, Íñiguez J and Nataf G F 2023 Influence of ferroelastic domain walls on thermal conductivity *Phys. Rev.* **B 108** 144104

[157] Royo M, Escorihuela-Sayalero C, Íñiguez J and Rurali R 2017 Ferroelectric domain wall phonon polarizer *Phys. Rev. Mater.* **1** 051402

[158] Seijas-Bellido J A, Escorihuela-Sayalero C, Royo M, Ljungberg M P, Wojdeł J C, Íñiguez J and Rurali R 2017 A phononic switch based on ferroelectric domain walls *Phys. Rev.* **B 96** 140101

[159] Mante A J H and Volger J 1971 Phonon transport in barium titanate *Physica* **52** 577

[160] Schnelle W, Fischer R and Gmelin E 2001 Specific heat capacity and thermal conductivity of $NdGaO_3$ and $LaAlO_3$ single crystals at low temperatures *J. Phys. D: Appl. Phys.* **34** 846

[161] Weilert M A, Msall M E, Wolfe J P and Anderson A C 1993 Mode dependent scattering of phonons by domain walls in ferroelectric KDP *Z. Phys.* B **91** 179

[162] Wang X, Helmersson U, Birch J and Ni W-X 1997 High resolution x-ray diffraction mapping studies on the domain structure of $LaAlO_3$ single crystal substrates and its influence on $SrTiO_3$ film growth *J. Cryst. Growth* **171** 401

[163] Pang Y, Li Y, Gao Z, Qian X, Wang X, Hong J and Jiang P 2023 Thermal transport manipulated by vortex domain walls in bulk H-$ErMnO_3$ *Mater. Today Phys.* **307** 100972

[164] Negi A, Kim H P, Hua Z, Timofeeva A, Zhang X, Zhu Y, Peters K, Kumah D, Jiang X and Liu J 2023 Ferroelectric domain wall engineering enables thermal modulation in PMN–PT single crystals *Adv. Mater.* **35** 2211286

[165] Belrhiti-Nejjar R *et al* 2025 Domain-wall driven suppression of thermal conductivity in a ferroelectric polycrystal *Adv. Sci.* **06931** 1–7

[166] Catalan G, Lukyanchuk I, Schilling A, Gregg J M and Scott J F 2009 Effect of wall thickness on the ferroelastic domain size of $BaTiO_3$ *J. Mater. Sci.* **44** 5307

[167] Ihlefeld J F, Foley B M, Scrymgeour D A, Michael J R, McKenzie B B, Medlin D L, Wallace M, Trolier-McKinstry S and Hopkins P E 2015 Room-temperature voltage tunable phonon thermal conductivity via reconfigurable interfaces in ferroelectric thin films *Nano Lett.* **15** 1791

[168] Langenberg E *et al* 2019 Ferroelectric domain walls in $PbTiO_3$ are effective regulators of heat flow at room temperature *Nano Lett.* **19** 7901

[169] Hopkins P E, Adamo C, Ye L, Huey B D, Lee S R, Schlom D G and Ihlefeld J F 2013 Effects of coherent ferroelastic domain walls on the thermal conductivity and Kapitza conductance in bismuth ferrite *Appl. Phys. Lett.* **102** 121903

[170] Féger L *et al* 2024 Lead-free room-temperature ferroelectric thermal conductivity switch using anisotropies in thermal conductivities *Phys. Rev. Mater.* **8** 094403

[171] Nataf G F, Volz S, Ordonez-Miranda J, Íñiguez-González J, Rurali R and Dkhil B 2024 Using oxides to compute with heat *Nat. Rev. Mater.* **9** 530

IOP Publishing

Ferroelastic Materials

Guillaume F Nataf, Blai Casals and Ekhard K H Salje

Chapter 7

Ferroelastic domain walls at surfaces

Anna Morozovska, Eugène Eliseev and Nick Barrett

7.1 Introduction

Elastic strain engineering of perovskite thin film ferroics is conditioned by the epitaxial mismatch between the film and the substrate [1]. Domain wall engineering, on the other hand, is based on topological defects in ferroics, ferroelectrics and ferroelastics, and has attracted sustained attention [2]. Intrinsic topological defects in ferroelectric and ferroelastic are domain walls (also called twin boundaries), and for the purposes of this chapter, domain wall–surface junctions.

Ferroelectric and ferroelastic domains minimize electrostatic and elastic contributions to the free energy and their appearance and organization depend on boundary conditions and chemistry. They break translational symmetry, leading to distinct physical properties and functionality. In ferroelastic materials, the domain walls are known as twin boundaries from the discontinuity and symmetry relationship of the strain tensor from one domain to another. The classical ferroelastic study is of the martensitic transition (chapter 8), they are also of crucial importance in the emergence of order parameters in oxides, notably in perovskite oxides exhibiting ferroelectricity and/or ferroelasticity. At twin boundaries, the spontaneous strain changes sign over only a few nanometers [3] giving rise to strong gradients which can generate new properties quite distinct from those of the adjacent domains. Superconductivity [4], polarity [5, 6], and chirality [6] have all been reported in twin walls. Such emerging functionalities are absent in the bulk [7, 8] and provide a new perspective of 'the material is the machine' [9]. In addition, the nanometric domain wall width makes them potentially quasi-2D functional objects, with a high potential for industrial innovation [10]. Domains and domain walls are therefore potentially elemental functional bricks in ferroelectrics/ferroelastics. In addition, being intrinsic to the ordered phases, they suppress all risk usually associated with, for example, the functionalization of nanoparticles for specific tasks.

The understanding and characterization of the surface discontinuity of ferroelastic domain walls is motivated by the confluence of several fundamental or

doi:10.1088/978-0-7503-6089-0ch7

7-1

potentially technological questions. The loss of inversion symmetry in the twin walls can give rise to unique physical properties quite distinct from those of the bulk material, in themselves fascinating examples of natural engineering. As indicated above, conductivity and even superconductivity can emerge in distinctive structure of the twin walls. This is particularly interesting in the case of the antidistortive ferroelastic $CaTiO_3$ which is non-polar and centrosymmetric in the bulk. Thus, nanometric thick twin walls could provide a natural source of a 2D semiconducting or even metallic-like sheets in a high-quality dielectric. However, in the case of real devices, practical and economic questions of volume, footprint and integration potential make it essential to consider thin films on the scale of device sizes, rather than bulk crystals. The interface between the functional material and, typically, a metallic contact or electrode is especially important. Not only then do the specific twin wall properties need to be understood but also how the surface discontinuity can modulate these same properties with the possibility that only at the surface can certain twin wall properties emerge and also how the interface with a metallic contact may in turn modify or transmit these characteristic surface properties.

In this chapter, we analyse elastic fields caused by the twin wall–surface junction in ferroelastics (such as model $CaTiO_3$ and $SrTiO_3$) in the framework of the Landau–Ginzburg–Devonshire (LGD) approach, and consider the improper ferroelectricity, carrier accumulation and vacancy segregation appearing at the junctions. We then focus on experimental demonstrations of polarity in ferroelastic twin walls at the surface of bulk single crystal $CaTiO_3$ (the experimental work on bulk twin walls is reported in chapter 6). The insights gained from the study of the single crystal surfaces provide the foundation on which engineering twin polarity in nanometric thick films may be envisaged and also how the interface with a metallic contact may in turn modify or transmit these characteristic surface properties.

7.2 Elastic fields at the twin wall–surface junction

The flexoelectric effect describes the coupling of the ferroelectric polarization with the strain gradient (direct flexoelectric effect) and the coupling of the polarization gradient with the strain (inverse flexoelectric effect) [11]. Theoretical studies of flexoelectricity in various ferroelectrics [12–14] and ferroelastics [15] performed in the framework of LGD approach agree with experimental trends [16]. Yudin *et al* [17] and Eliseev *et al* [18] showed that the flexoelectric coupling in a rhombohedral $BaTiO_3$ can induce the strongly anisotropic Bloch-type and Neel-type polarization components with chiral and/or achiral structures, which are qualitatively different from the classical Bloch-type wall structure (see figure 7.1).

As a next step, it was shown [19, 20] that the flexoelectric coupling and rotostriction can lead to the appearance of inhomogeneous electric fields (called the roto-flexo field), which are proportional to the structural antiferrodistortive order parameter (oxygen octahedral tilts) gradient in perovskite ferroelastics. The roto-flexo field can exist in a wide class of materials exhibiting oxygen octahedra rotations [21], such as $SrTiO_3$ and $CaTiO_3$. The field may contribute significantly to

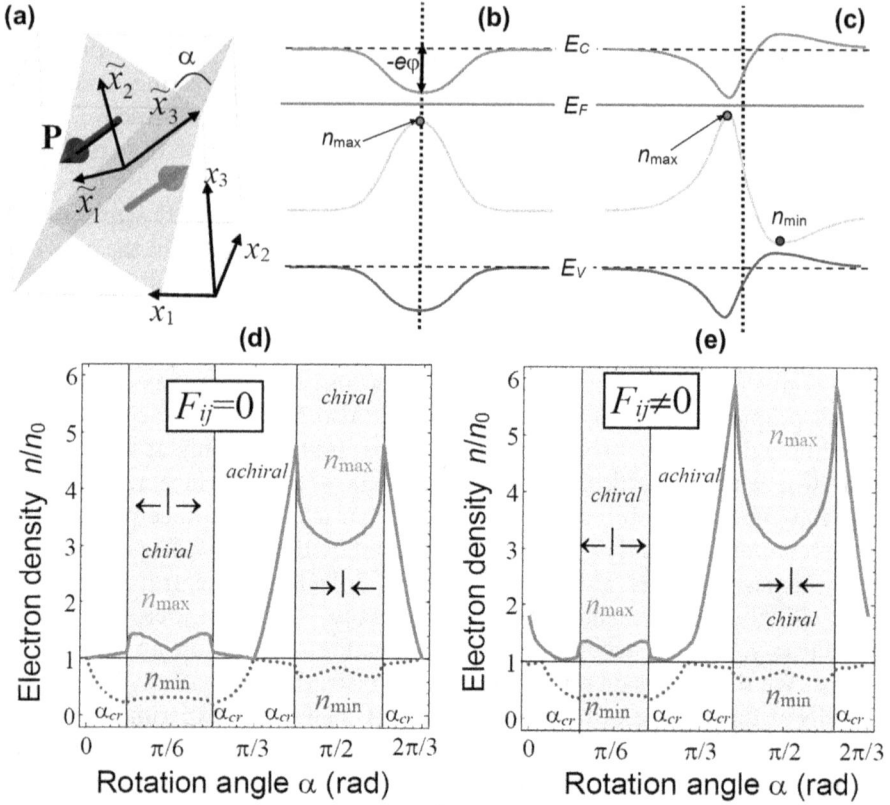

Figure 7.1. (a) Rotated coordinate frame $\{\tilde{x}_1, \tilde{x}_2, \tilde{x}_3\}$ choice for 180-degree nominally uncharged domain walls in the rhombohedral ferroelectric BaTiO$_3$; α is the wall rotation angle counted from crystallographic plane $\langle 101 \rangle$. The distance from the wall plane is $\tilde{x}_1$. Sketches of the local band bending for (b) achiral and (c) chiral domain walls, where the spatial regions with maximal (n_{max}) and minimal (n_{min}) electron density are indicated. Relative maximal n_{max}/n_0 and minimal n_{min}/n_0 electron density versus the domain wall rotation angle α calculated in BaTiO$_3$ at 180 K without flexoelectric coupling (d) $F_{ij} = 0$, and with flexoelectric coupling (e) $F_{ij} \neq 0$. (Reproduced with permission from [18]. Copyright 2013 the American Physical Society.)

the interfacial improper ferroelectricity [22] induced by octahedral rotations in YMnO$_3$ [23], Ca$_3$Mn$_2$O$_7$ [24] and CaTiO$_3$ [6].

The structural and electronic phenomena at the nominally uncharged twin walls in ferroelastics, such as CaTiO$_3$, are conditioned by the elastic strain gradient but also by the chemistry in the form of field-driven segregation of mobile ions [19, 25] and/or oxygen vacancies [26, 27]. In the first approximation of the perturbation method, the surface displacement $u_i^S(x)$, induced by the ferroelastic wall–surface junction in the spatial point x, is given by the convolution of the corresponding elastic Green function, $G_{ij}(x, \xi)$, with the elastic stress field, σ_{jk}^0, unperturbed by the surface [19]:

$$u_i^S(x) = \int_{-\infty}^{\infty} d\xi_1 \int_{-\infty}^{\infty} d\xi_2 \, G_{ij}(x_1 - \xi_1, x_2 - \xi_2, x_3) \sigma_{jk}^0(\xi_1, \xi_2) n_k, \qquad (7.1)$$

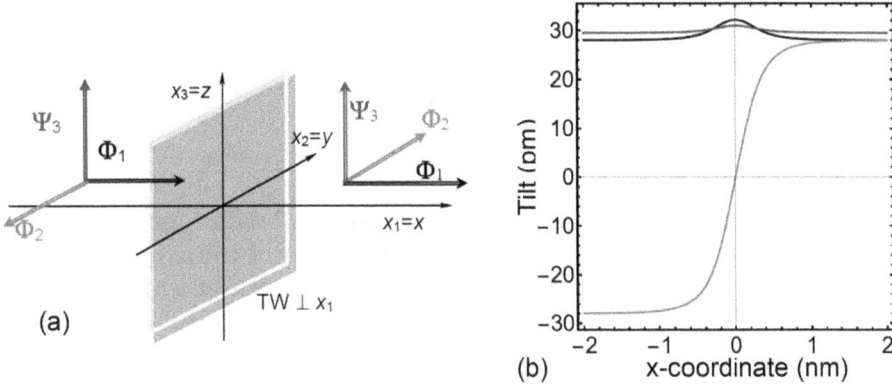

Figure 7.2. (a) Orientation of the structural order parameters near the twin wall–surface junction with respect to pseudo-cubic axes $x_1 = x$, $x_2 = y$ and $x_3 = z$. The gray plane is the head-to-tail twin wall (TW) near the film surface. (b) Distributions of Φ_1, Φ_2 and Ψ_3 across the twin wall. (Reproduced with permission from [19]. Copyright 2012 the American Physical Society.)

where $G_{ij}(x - \xi)$ is known for the elastically isotropic half-space [28] and for the mechanically clamped/free film [29]; n_k is the outer normal to the mechanically free surface $x_3 = 0$. The geometry of the ferroelastic wall–surface junction is shown in figure 7.2(a).

Inhomogeneous elastic stresses $\sigma_{jk}^0(\xi_1, \xi_2)$ originate from the rotostriction coupling with the structural order parameter variation in the vicinity of the ferroelastic twin wall. The long-range order parameter describing oxygen octahedral rotations is the rotation angle or the 'tilt' of the oxygen octahedra [30, 31]. The tilt of tetragonal ferroelastics (e.g. SrTiO$_3$ below 105 K) in the low-symmetry phase is described by a single axial vector $\mathbf{\Phi} = (\Phi_1, \Phi_2, \Phi_3)$ [20, 32] and the tilt in orthorhombic ferroelastics (e.g. CaTiO$_3$ at room temperature) is described by two axial vectors $\mathbf{\Phi} = (\Phi_1, \Phi_2, \Phi_3)$ and $\mathbf{\Psi} = (\Psi_1, \Psi_2, \Psi_3)$ [33, 34].

For 90-degree twins in orthorhombic ferroelastics the minimal number of nonzero tilt components are Φ_1, Φ_2 and Ψ_3 (see figure 7.2(b)). In what follows we use approximate analytical expressions for the structural order parameter components: $\Phi_1 \approx \Phi_S(1 + a \cosh^{-2}(x_1/w))$, $\Phi_2 \approx \Phi_S \tanh(x_1/w)$, $\Psi_3 \approx \Psi_S(1 + b \cosh^{-2}(x_1/w))$, where Φ_S and Ψ_S are spontaneous values, w is the intrinsic width of the twin wall in the bulk, and the amplitudes a and b are much smaller than unity.

Elastic strains $u_{kl}(x, z)$ calculated from equation (7.1) as $u_{kl}(x, z) = u_{kl}^0 + \frac{1}{2}\left(\frac{\partial u_i^S}{\partial x_j} + \frac{\partial u_j^S}{\partial x_i}\right)$, where u_{kl}^0 is the strain field of the twin unperturbed by the surface, have the following form in orthorhombic ferroelastics [19]:

$$u_{xx}^0 \approx R_{11}\Phi_S^2 + R_{12}\Phi_2^2 + V_{111}\Phi_S^4 + 6V_{112}\Phi_S^2\Phi_2^2 + V_{122}\Phi_2^4 + W_{122}\Psi_S^4 +$$

$$\frac{s_{12}(\Phi_S^2 - \Phi_2^2)}{s_{11} + s_{12}}(R_{11} + R_{12} + (V_{111} + V_{122})(\Phi_S^2 + \Phi_2^2) + 6(V_{112} + V_{123})\Phi_S^2), \tag{7.2a}$$

$$u_{yy}^0 \approx (R_{11} + R_{12})\Phi_S^2 + (V_{122} + 6V_{112} + V_{111})\Phi_S^4 + W_{122}\Psi_S^4, \tag{7.2b}$$

$$u_{zz}^0 \approx 2R_{12}\Phi_S^2 + (2V_{122} + 6V_{123})\Phi_S^4 + W_{111}\Psi_S^4. \tag{7.2c}$$

Here elastic compliances, s_{ijkl}, the fourth order rotostriction coefficients, R_{ijkl}, and the sixth order rotostriction coefficients, V_{ijklmn} and W_{ijklmn}, are written in Voigt notation. Since the strains are proportional to the product of corresponding rotostriction coefficients, the second and fourth powers of the oxygen tilt components, their appearance is the typical manifestation of the rotostriction effect. Note, that u_{yy}^0 and u_{zz}^0 are spontaneous strains, which exist in the bulk stress-free homogeneous sample. The strain u_{xx}^0 contains both the spontaneous part and the part proportional to the order parameter variation ($\Phi_S^2 - \Phi_2^2$), that vanishes far from the twin wall.

The trace of the strain tensor, $\mathrm{Tr}\,(u_{ij}) \equiv u_{ii}$, written in the first-order approximation, is [19]

$$\mathrm{Tr}\,(u_{ij}) = u_{xx}^0 + u_{yy}^0 + u_{zz}^0 - \frac{(1+\nu)(1-2\nu)}{Y} \frac{w(2w(w+z)^2 + z(x^2 + (w+z)^2))}{(x^2 + (w+z)^2)^2}\delta\sigma, \tag{7.3}$$

where w is the domain wall thickness and the 'perturbation', stress $\delta\sigma$, is

$$\delta\sigma = \frac{s_{11}\left(R_{12} + (V_{122} + 6V_{112})\Phi_S^2\right) - s_{12}\left(R_{11} + (V_{111} + 6V_{112})\Phi_S^2\right)}{s_{11}^2 - s_{12}^2}\Phi_S^2. \tag{7.4}$$

Equations (7.2–7.4), which are the first-order approximation for strains, do not include polarization-dependent components, e.g. induced electrostriction and flexoelectric coupling. The contribution of the flexoelectric effect leads to the small second order correction in equation (7.3) proportional to the square of the flexoelectric coupling coefficient.

The strain tensor diagonal components (a)–(c) and their trace in the vicinity of the wall–surface junction in CaTiO$_3$ are shown in figure 7.3. Since the strain decrease follows the long-range power law in accordance with equations (7.2)–(7.4), the scale of the strain amplitude decay (about 4 nm) is much higher than the intrinsic width of the domain wall (about 0.4 nm). Naturally, this long-range decay could influence strongly the appearance of the improper polarization and carrier accumulation caused by the wall–surface junction.

7.3 Improper ferroelectricity at the twin wall–surface intersection

7.3.1 Theoretical description

The polarization components induced by the flexoelectric coupling can be estimated as

$$P_i(x, z) \approx \alpha_{ij}^{-1}\left(f_{mnjl}\frac{\partial u_{mn}}{\partial x_l} - \frac{\partial \phi}{\partial x_j}\right). \tag{7.5}$$

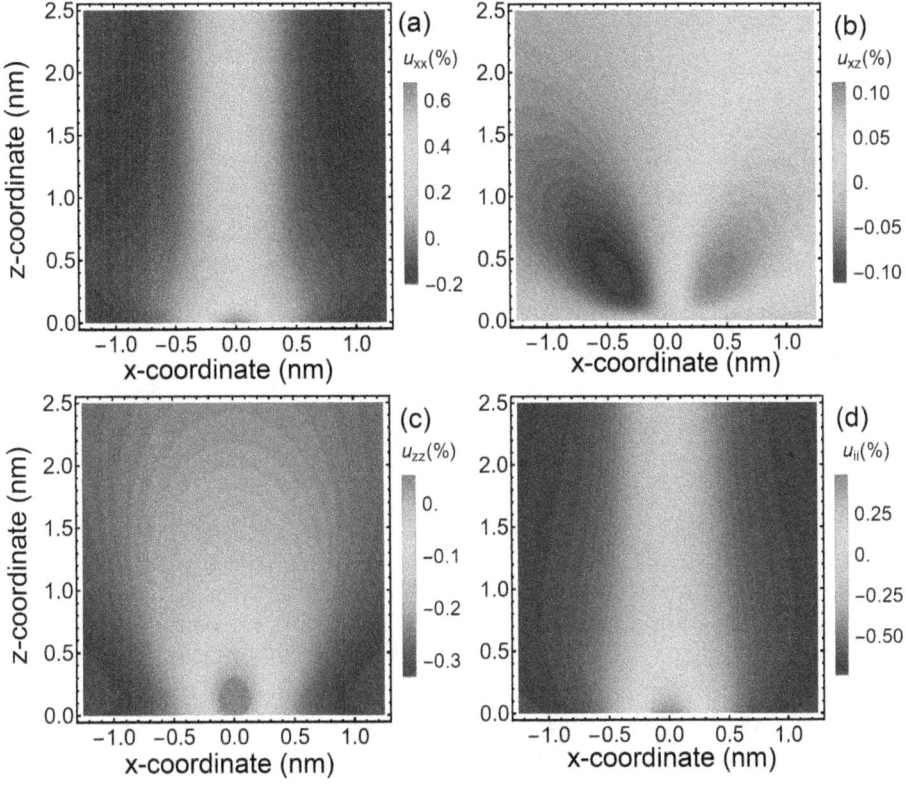

Figure 7.3. (a)–(c) Diagonal components of strain tensor and (d) their trace versus the distance x from the twin wall plane ($x = 0$) and the distance z from the surface ($z = 0$) calculated for CaTiO$_3$ parameters [19] and room temperature. Color scales show the strain components in %.

Flexoelectric tensor coefficients are denoted as f_{mnjl}. Their numerical values were determined experimentally for SrTiO$_3$ [35].

Coefficients α_{ij} in equation (7.5) are affected by elastic strains and biquadratic coupling as the following [20]

$$\alpha_{ij}(T, x, z) = a_{ij}(T) - q_{ijkl}u_{kl}(x, z) - \eta_{ijkl}\Phi_k\Phi_l - \xi_{ijkl}\Psi_k\Psi_l. \quad (7.6)$$

The temperature dependence of the coefficients $a_{ii}(T)$ obey the Barrett law for CaTiO$_3$ [33] and SrTiO$_3$ [36]; q_{ijkl} are electrostriction coefficients, and η_{ijkl} and ξ_{ijkl} are the Houchmanzadeh–Lajzerowicz–Salje biquadratic coupling tensor coefficients between the structural and polar order parameters [37]. Elastic stresses induced by the twin wall–surface junction appear too small to induce ferroelectric polarization in CaTiO$_3$ at room temperature, i.e. $\alpha_{ij}(T, x, z)$ is always positive.

The electrostatic potential ϕ in equation (7.5) should be determined from the Poisson equation,

$$\varepsilon_0\varepsilon_b\frac{\partial^2\phi}{\partial x_i^2} = \frac{\partial P_i}{\partial x_i} - e\left(p - n + N_d^+ - N_a^-\right), \quad (7.7)$$

where $\varepsilon_0 = 8.85 \times 10^{-12}$ F m^{-1} is the universal dielectric constant, ε_b is the background dielectric permittivity unrelated with the soft mode permittivity ε_{ij}^{sm}, which is given by expression $\varepsilon_{ij}^{sm} = \varepsilon_0^{-1}\alpha_{ij}^{-1}$, $e = 1.602 \times 10^{-19}$ C is the electron charge, $p(\phi)$ and $n(\phi)$ are the hole and electron densities, and N_d^+ and N_a^- are the concentrations of corresponding ionized acceptors and donors.

If the depolarization field can be negligibly small (due to, for example, free carrier accumulation) in comparison with the flexoelectric effect, the polarization in equation (7.5) can be approximated as $\delta P_i^{\text{flexo}}(x, z) \sim \alpha_{ij}^{-1} f_{mnjl} \frac{\partial u_{mn}}{\partial x_l} \sim \varepsilon_0 \varepsilon_{ij}^{sm} f_{mnjl} \frac{\partial u_{mn}}{\partial x_l}$. In this case the polarization behavior is primary determined by the convolution of the strain gradient with the flexoelectric effect tensor. Since the strains are proportional to the product of corresponding rotostriction coefficients and the second (or fourth) powers of the oxygen tilt components, it can be concluded that the improper polarization equation (7.5) is caused by the flexo-roto effect. Note that the short-circuit electrical boundary conditions correspond to the minimal depolarization field (e.g. when the surface is covered by the perfectly conductive layer or metallic electrode).

The spatial distributions of the in-plane and out-of-plane polarization components, $P_x(x, z)$ and $P_z(x, z)$, are shown in figures 7.4(a) and 7.4(b), respectively. As expected, the $P_x(x, z)$-profiles are antisymmetric and $P_z(x, z)$-profiles are symmetric with respect to the twin wall plane $x = 0$. $P_x(0, z)$ is identically zero as anticipated from the symmetry consideration, and $P_x(x, z)$ has two maxima in the vicinity of the wall plane, whose amplitude decreases strongly and half-width increases as z increases. The result is in the qualitative agreement with the transmission electron microscopy results on a twin wall of CaTiO$_3$ [5]. $P_z(0, z)$ has a pronounced maximum at the surface $z = 0$, and then it strongly vanishes and diffuses with z increase.

7.3.2 Electron-based techniques to probe polarization at surfaces

The characterization of surface atomic, chemical and electronic structure requires surface sensitive techniques. Low energy electrons or photoelectrons provide a unique tool to explore these surface properties as they are intrinsically sensitive to charge, to chemistry in the case of photoelectrons, and to topography and local surface crystallinity via angularly and spatially resolved imaging in both real space and diffraction modes. Most of the results presented here rely on the use of such techniques.

The very first papers reporting observation of ferroelastic and ferrolectric domain ordering and twin wall formation suggested explicitly the existence of specific surface structures and topography depending on the nature of the twin walls, in particular the angle they formed with the single crystal surface under study. Forsbergh showed that tetragonal (101) twinning in single crystal BaTiO$_3$ gave rise to lamina at 45° to the surface plane [38]. He used optical techniques to observe the domain ordering, while noting that head-to-head or tail-to-tail twins could be expected to generate a surface polarization charge equal to $\sqrt{2}$ times the bulk polarization even though this could not be directly observed by x-ray or optical methods.

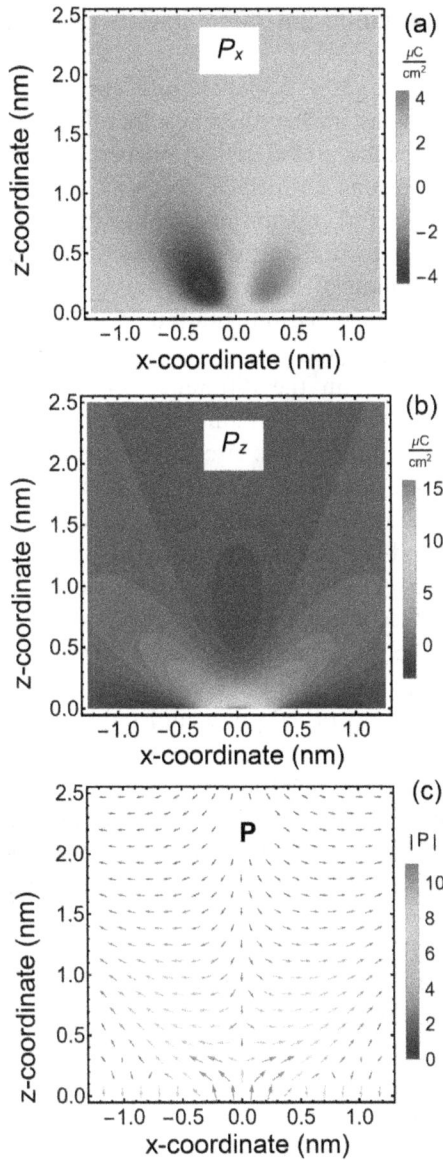

Figure 7.4. (a) The in-plane polarization component $P_x(x, z)$, (b) the out-of-plane polarization component $P_z(x, z)$, and (c) the vector field $\vec{P}(x, z)$ versus the distance x from the twin wall plane $x = 0$ and the distance z from the surface $z = 0$ calculated for $CaTiO_3$ parameters and room temperature. Color scales show the strain components in %.

At the same time, Merz reported optical measurements using crossed Nichols of domain ordering and domain wall migration under electric fields in $BaTiO_3$. Twin walls in the ferroelectric tetragonal phase, following either a/a or c/a ordering, were observable. Furthermore, the characteristic in-plane needle domains creating highly charged head-to-head or tail-to-tail domain wall configurations were observed during ferroelectric switching [39].

However, neither Forsbergh nor Merz could specifically measure the surface structure of ferroelastic twins, rather their results allowed some deduction of the probable twin wall topography and electrical properties at the surface.

The first experimental results on surface electronic imaging were reported by Le Bihan using secondary electron emission in a scanning electron microscope in order to resolve the surface polarization states in different ferroelectric and ferroelastic crystals [40]. Electron imaging of charged (ferroelectric) surfaces was successfully applied to visualize domains in barium titanate, triglycine sulfate, and guanidine aluminum sulfate hexahydrate [40].

Le Bihan's method relied on the different etching rates for domain surface polarized upwards, downwards or in plane, corresponding to c^+, c^- and a type domains. The surface topography gave rise to contrast which could be unambiguously related to the surface polarization state, as shown in figure 7.5.

Scanning probe microscopy is also sensitive to the local surface charge state. The work by Sergei Kalinin and Dawn Bonnell provided a basis for the interpretation of contrast observed on ferroelectric surfaces using electrostatic force microscopy (EFM) and scanning surface potential microscopy (SSPM). Using an image charge model they described the relationship between local polarization and surface potential contrast, notably the complex contrast inversion due to partial or full screening of polarization by adsorbed polar molecules [41] (figure 7.6).

Near field microscopy thus provided unprecedented spatial resolution in the study of domains and domain walls at surfaces, however, it can also be an invasive technique due to tip–surface interactions. A second disadvantage is the intrinsically sequential nature of the data acquisition, preventing simultaneous full-field imaging. Another approach, avoiding tip–surface interactions which may influence the domain polarization, is mirror electron microscopy (MEM) using a low energy electron microscope (LEEM), which allows non-contact, full-field imaging of the surface topography and potential with 20 nm spatial resolution (figure 7.7). The transition from the reflection of the electrons to the back-scattering regime, the MEM–LEEM transition, images the electrostatic potential above the surface, thus mapping the surface charge. The incident electron potential is determined by the sample bias or start voltage. At low start voltages, electrons are reflected by the potential above the surface (MEM), whereas at higher values they penetrate the sample and are backscattered (LEEM). Contrast inversion due to screening of the surface polarization charge by polar adsorbates was confirmed using low energy electron microscopy on single crystal $BaTiO_3$ [43]. A pioneering study using photoemission electron microscopy (PEEM) rather than an incident electron probe was used to probe polarity patterned surfaces of GaN, $PbZrTiO_3$ and $LiNbO_3$, and their screening by polar adsorbates [44].

Interestingly, LEEM can also be used to image fully in-plane domain ordering in perovskite ferroelectrics. Although no net surface charge is expected for in-plane a/a with 90° domain walls perpendicular to the surface, the stray field at the domain walls provides a direct measurement of the polarization direction. Rault *et al* used

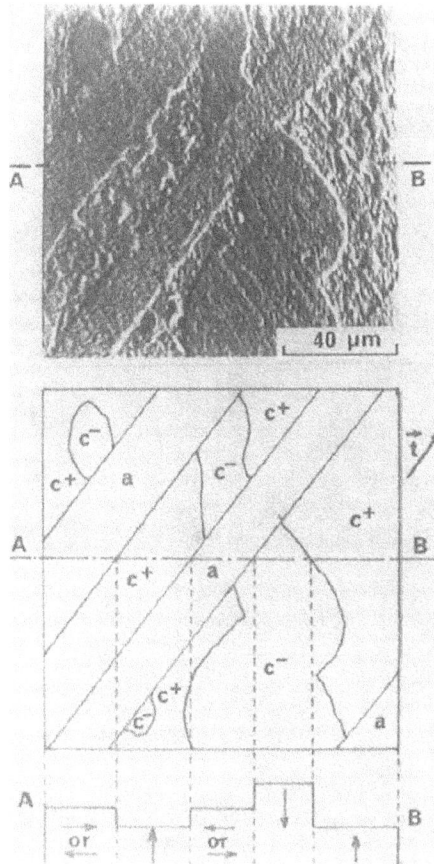

Figure 7.5. Secondary electron image of BaTiO$_3$ surface after polarization-sensitive etching. The schematic identifies the different domain polarizations. (Reproduced with permission from [40]. Copyright 2011 Taylor and Francis.)

LEEM to image in-plane domain switching kinetics at the surface of a BaTiO$_3$ single crystal and were able to identify the direction of the stray field and hence the direction of the a-type polarization from the intensity profile of the reflected electrons across the domain wall [46]. On switching from a^+/a^+ to a^-/a^- the intensity profile also reversed (figure 7.8).

The importance of specific surface sensitive analysis of domain wall structures has been underlined by LEEM and PEEM studies from several groups. PEEM with *in situ* temperature control has been used to investigate the surface polarity of ferroelectric BaTiO$_3$ by probing the local potential modulations at the microscopic scale. PEEM provides parallel imaging in photoemission using electron lenses with a spatial resolution of ~50 nm. Domains with different ferroelectric polarization present different surface charge, which shifts the electronic levels and hence the work function measured by the emitted electrons, as illustrated in figure 7.9. The photoelectrons have a small inelastic mean free path (from a few angstroms to a

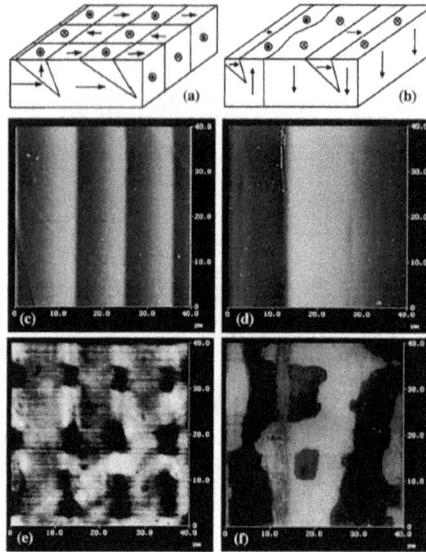

Figure 7.6. (a), (b) Schematic diagrams of domain structure, (c), (d) surface topography and (e), (f) surface potential in the a-domain region with (a), (c), (e) c-domain wedges and in (b), (d), (f) c-domain region with a-domain wedges. The scale is (c), (d) 50 nm and (e), (f) 0.2 V. (Reproduced with permission from [41]. Copyright 2001 the American Physical Society.)

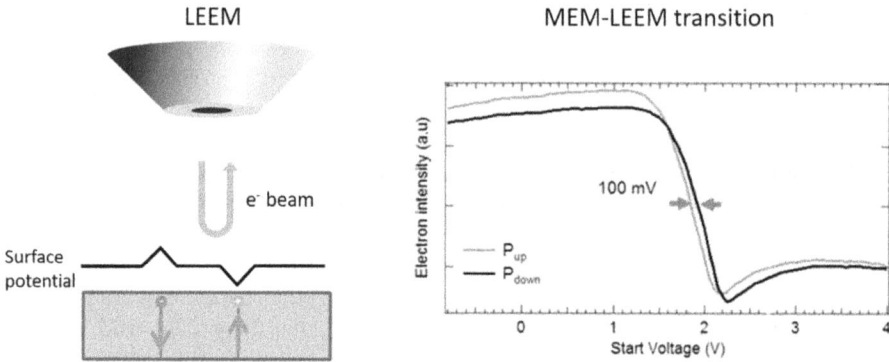

Figure 7.7. Left: Schematic showing how polar walls at the surface of a sample can modulate the surface potential and therefore the potential at which incident electrons reflect from/penetrate into the surface. Right: MEM–LEEM curves measured from upward and downward pointing domains in a LiNbO$_3$ sample, showing a 100 meV shift in the position of the MEM–LEEM transition. (Adapted from [45]. CC BY 4.0.)

few nanometers, depending on the electron kinetic energy) making the technique inherently surface sensitive.

Using laser-induced PEEM, Höfner *et al* provided fascinating evidence of ferroelectric order at the surface of BaTiO$_3$ single crystals well above the Curie temperature. This was linked to the persistence of a tetragonal-like structure due to surface ionic relaxation [47]. Barrett *et al* reported the persistence of domain-like

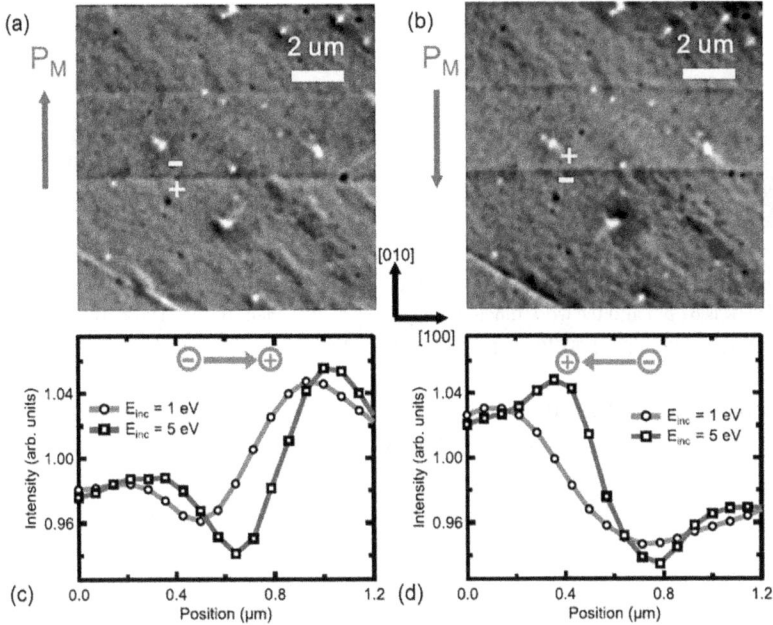

Figure 7.8. LEEM images of sample surface (a) before and (b) after switching, the field of view is 30 mm. Intensity profiles extracted from LEEM images acquired using 1 eV (circles, red line) and 5 eV (squares, blue line) along [010] across a domain wall at the surface of $BaTiO_3$(001) (c) before and (d) after switching. The electric dipole across the wall is shown by the plus and minus signs. (Reproduced with permission from [46]. CC BY-NC-ND 4.0.)

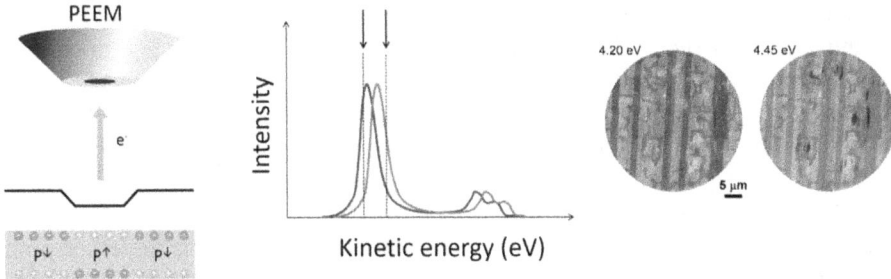

Figure 7.9. Schematic of modulation of the work function by surface polarization charge (left), resulting shift of photoemission spectra from domains with different polarization (center) and contrast inversion in the electron images (right).

structures at $BaTiO_3$ surfaces up to 350 °C using LEEM, related to the well known tweed structure, due to local polarization fluctuations, in this case quasi-static, as a result of the 90° domain walls, i.e. both ferroelectric and ferroelastic. In the same study, the emergence of dynamic ferroelastic domain walls with a polar nature ordered along [010] was observed through the Curie temperature. The directional nature suggests a strong link with elastic energy barrier to be overcome to allow the atomic displacements associated with the ferroelectric to paraelectric structural

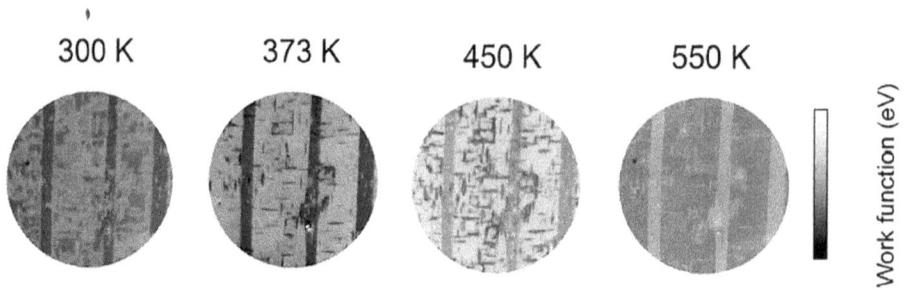

Figure 7.10. Work function maps obtained from image series below TC at 300 K, 373 K, 450 K and 550 K. The field of view is 67 μm and the work function range in each case spans 0.5 eV. Dark gray corresponds to $P\downarrow$, mid-gray to $P_{\text{in-plane}}$ and light gray to $P\uparrow$ polarization directions. (Reproduced from [48]. CC BY 4.0.)

phase transition. Their appearance at a saddle point in the potential energy landscape would also explain their transitory nature. The rapid appearance and disappearance of the transitional ferroelastic domain ordering over approximately 1° at the Curie temperature is reminiscent of Rayleigh waves in geophysics deforming the surface in the vertical direction momentarily, here creating an elastic and polar surface anomaly.

Following the studies of Höfner *et al*, Mathieu *et al* have evidenced in more detail the nature of the surface proximity effect in $BaTiO_3$ [48]. Whereas in the bulk the phase transition to the paraelectric state takes place as expected at the Curie temperature, in the very near surface region, the ferroelectric and ferroelastic domain structure persists up to 550 K. This is related to the outwards relaxation of oxygen ions at the surface, maintaining the tetragonal-like structure well beyond the bulk Curie temperature (figure 7.10). Both Raman and x-ray diffraction confirmed a well-behaved structural transition to the cubic, paraelectric phase, implying that the structural distortion is confined to a thin surface region delimited by an invariant plane. At the invariant plane near the surface, self-reversal of the polarization occurs due to the increasing local depolarization field since the underlying cubic phase cannot screen the surface polarization. The resulting surface domain structure is dominated by ferroelastic twins making a 45° angle with the surface. The polarization of the surface region above the Curie temperature was estimated at 10 μC cm^{-2}. The specific surface domain structure is therefore pinned by the ferroelastic twins, and PEEM mapping of the work function demonstrated that this memory, imprinted elastically, was only lost by heating to higher temperature (975 K).

7.3.3 Experimental investigations at the surface of polar domain walls

A further advantage of LEEM is that by imaging the focal or diffraction plane of the microscope, rather than an intermediate image plane, it is possible to obtain spatially resolved information on the local crystal orientation. This is particularly useful in identifying twin walls where there is no contrast in the electron image of the twin domains due to charge but there is the angular symmetry of the twinned crystal structure. Figure 7.11 shows one such example of the low energy electron diffraction

Figure 7.11. (a) MEM image (field of view 75 μm) of a $CaTiO_3$ (001) surface at a start voltage of -0.13 eV, showing twin walls. No significant domain contrast is observed. (b), (c) LEED images at 17 eV acquired from regions straddling the horizontal and vertical twins in (a), defined by positioning an appropriate field aperture. The double LEED spots directly reflect the twinning. (Reproduced with permission from [49]. Copyright 2019 the American Physical Society.)

(LEED) patterns obtained through microscopic field apertures 1 and 2 positioned over ferroelastic domain walls at the surface of $CaTiO_3(001)$. The splitting of the LEED spots, measured by imaging the diffraction plane, is a direct measure of the twin angle associated with each domain wall.

In 2000, Nepijko *et al* demonstrated experimentally and provided a theoretical basis as to how specific physical topography such as mesas, pits or steps could act as focusing or defocusing elements in electron imaging [50]. By generalizing this idea, quantitative angular anisotropy in the electron images should allow deduction of the surface topography. Most recently, Magagnin *et al* reported the use of PEEM to characterize the ferroelastic twins present at the surface of $CaTiO_3$ from the effect of topology on the photoelectron angular distribution and from symmetry arguments [51]. Low energy electron and photoelectron images of surfaces show intensity modulations due to both electrical and physical topography. In PEEM, electrons cross the diffraction plane on the optical axis for normal emission and off-axis for off-normal emission. By positioning a contrast aperture in the back focal plane, a precise and narrow angular range can be selected.

The effect of the contrast aperture is shown in the schematic of figure 7.12(b). Higher off-centering improves the domain topography contrast dramatically thanks to the angular selection but also induces a shift of the energy scale. This is because off-normal electrons have velocity components perpendicular and parallel to the sample surface; as a result, the kinetic energy measured by the PEEM will be lower, and the threshold deduced for photoemission is shifted to higher energy (within the reference frame of the PEEM).

Emission from domains with different tilt angles are therefore centered at different positions in the diffraction plane, giving rise to intensity variation as shown in figures 7.12(d)–(f) via the angular selection (figure 7.12(b)). We focus on the domains labeled D1, D2, D1$'$ and D3. Domain D1 is used for instrument alignment and its surface normal coincides with the PEEM optical axis. D2, D3, and D1$'$ make finite angles with respect to D1. The twin wall is vertical in the image,

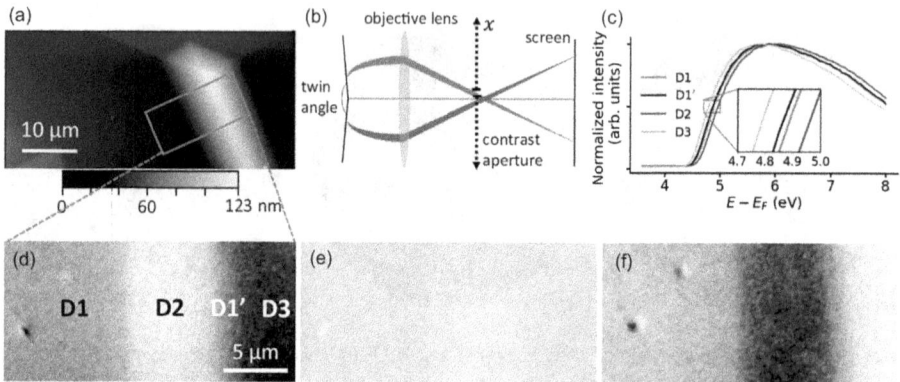

Figure 7.12. (a) Atomic force microscopy topography image of the $CaTiO_3$ surface with a red box highlighting the area of interest containing domains D1, D2, D1' and D3. The surface topography is visible with the twin D2/D3 on the right-hand side. (b) Schematic showing the angular selection by the contrast aperture in the back focal plane of photoelectron emission from twin domains, here the electron emission in red is favored. (c) Photoemission threshold spectra from domains D1, D2, and D3. (d)–(f) PEEM images acquired at $E—E_f = 4.3$ eV for contrast apertures (CA) positions + 140, 0, and −140 μm with respect to the optical axis. (Reproduced with permission from [51]. Copyright 2024 the American Physical Society.)

therefore, by off-centering the aperture horizontally we selectively analyse photo-electrons emitted from domains (figures 7.12(d) and 7.12(f)) on either side of a twin boundary. When the aperture is centered on the optical axis, the contrast between the twin domains is almost zero. In this configuration, the angular difference with respect to D1 is minimized as in figure 7.12(e). Figure 7.12(c) shows the spectra for each domain extracted from the threshold image series.

By modeling the electron optics inside the PEEM with realistic values of tilt angles, contrast aperture opening and PEEM magnification, it was possible to deduce the tilt angles from the PEEM contrast. From the tilt angles, symmetry arguments allowed reconstruction of the ferroelastic surface with clearly identified twin walls.

Figure 7.13 shows the remarkable agreement between the physical topography as deduced from the PEEM measured twin angles and that directly extract from an atomic force microscopy (AFM) map of the same region. The non-invasive, full-field imaging nature of PEEM gives it, in these respects, a decisive advantage over scanning probe techniques.

The variation of surface potential in polar twin walls in $CaTiO_3$ was studied by Nataf *et al* using a combination of resonant piezoelectric spectroscopy (RPS), AFM and LEEM. RPS provided the signature of symmetry breaking in the centrosymmetric $CaTiO_3$. AFM mapped the surface topography induced by the ferroelastic domain walls. MEM–LEEM mapping of the same regions recorded in AFM provided a direct measure of the polarity of the domain walls. From knowledge of the crystal axes and from symmetry arguments, each distinct twin wall could be identified.

The LEEM contrast also provides direct information on the polarity (figure 7.14). The crucial result was that the dark domain walls (R_2, V_1, V_2, and V_3) have an outward-pointing, i.e. positive polarity, whereas the bright R_1 has an inward-pointing polarity, i.e. negative surface charge. Thus, both ridge and valley domain

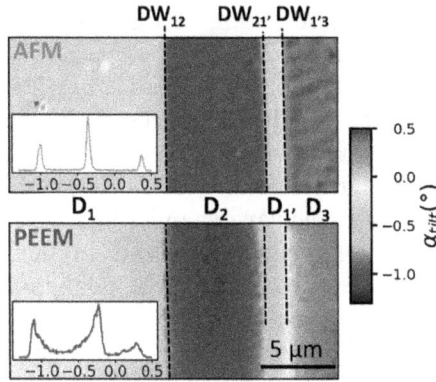

Figure 7.13. PEEM and AFM angular maps of the analysed area, with an indication of the domain walls. Insets show the histograms of the angles extracted from the AFM and PEEM maps and correspondence with D1 to D3. (Reproduced with permission from [51]. Copyright 2024 the American Physical Society.)

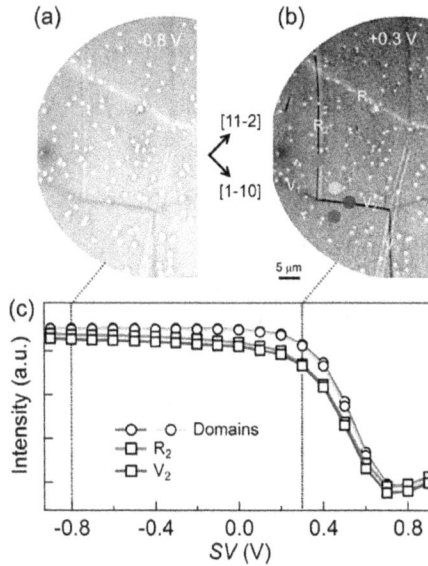

Figure 7.14. MEM images of the domain walls at start voltages of (a) −0.8 and (b) + 0.3 V. (c) Electron intensity as a function of start voltage (incident kinetic energy) measured in domains and at domain walls showing a 100 mV shift in the MEM–LEEM transition to lower values for the in the domain walls. (Reproduced with permission from [52]. Copyright 2017 the American Physical Society.)

walls may adopt the same polarity, in agreement with theoretical calculations of the wall energies [53]. Even more fascinating is that a ridge may have positive or negative polarity. This provides experimental evidence that switching of domain wall polarity might be possible without a topographical change from ridge to valley, although it should be noted that R_1 and R_2, both W' type walls, had complementary azimuthal and inclination angles. Varying the incident electron energy and after additional annealing experiments provided further evidence for the polar nature of the twins at the surface due to screening of the polarization charge by electron injection, trapping and dissipation [52].

This work was then further developed by Zhao *et al* who used a simple model to simulate the contrast expected in LEEM images of different ferroelastic surfaces with polar/non-polar surfaces and twins [49].

Zhao *et al* determined the surface polarity of CaTiO$_3$ in the vicinity of ferroelastic domain boundaries from the molecular dynamics based on an ionic spring model. The calculated interatomic gradient forces lead to the flexoelectricity, which generates the polarity at the surface and twin boundaries. Using LEEM and MEM experiments, it was demonstrated that negatively (or positively) charged surfaces reflect (or attract) incident electrons with low kinetic energy. The electron images reveal the surface valleys and ridges near the intersection of the twin boundary and the surface. The inward 'down' polarity reflects electrons like negative surface charges, and the outward 'up' polarity backscatters electrons like positive surface charges. Both the polarity of the twin boundary and the physical topography scatter electrons in CaTiO$_3$ with (001) and (111) surface terminations. Importantly, three major surface effects were identified from the LEEM/MEM experiments: the surface termination of an ionic crystal (i.e. positive or negative charge), topography of the surface (i.e. the presence of ridges and/or valleys), and surface dipoles (i.e. up or down, or in-plane polarity). Note, that other features are effectively screened for depths more than three layers under the surface and cannot be observed in the LEEM/MEM experiments.

Simulations proved that the density of reflected electrons is mostly influenced by surface charges, and explained how the local work function variations influence the electron reflectivity (figure 7.15). Inclined surfaces scatter the beam preferentially in the direction of the inclination. When two surfaces form a valley, their scattering

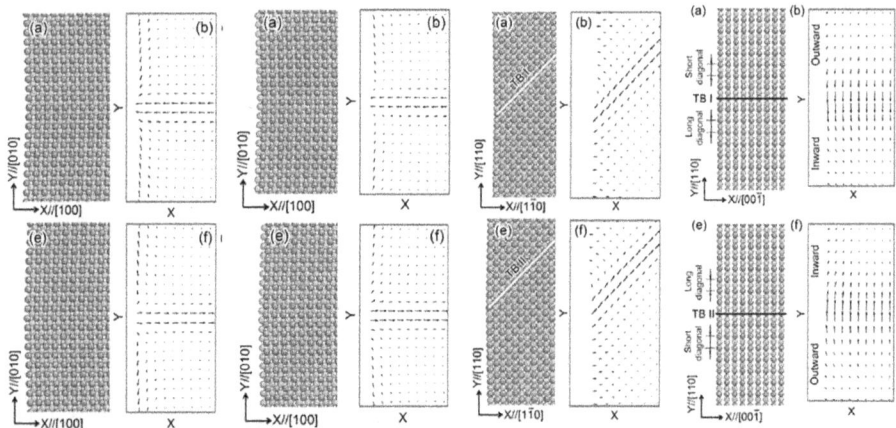

Figure 7.15. Polarity at anion, cation terminated surfaces (the two left-hand sets of images) with valleys and ridges. The atomic configurations and polarizations near the twin boundary (a valley). Polarity of twin walls at neutral surfaces (right-hand images) with topography (45° inclination angle) and flat (twin at 90° with respect to the surface). The dipole displacement is amplified by a factor of 25 for clarity. Cations are in orange and anions are in blue. (Reproduced with permission from [49]. Copyright 2019 the American Physical Society.)

superimposes and enhances near the intersection of surfaces with the domain boundary. When two surfaces form a ridge, the two reflected beams diverge and reduce the overlap. The resulting LEEM/MEM pattern shows dark lines in the first case and bright lines in the second. The absence of a net surface charge allows polarity due to the surface dipoles to play a dominant role in determining the electron reflectivity. In the case of 45° domain boundaries, they generate patterns with dark and bright lines accompanied by a strong change of the LEEM intensity between the adjacent domains. The third case is 45° twins without topographical peculiarities at the surface corresponding to the purely in-plane polarization. The stray field laterally deviates electrons and gives rise to an asymmetric intensity profile in MEM perpendicular to the wall and to the appearance of vortex-like polarization structures on either side of the twin wall.

7.4 Carrier accumulation and electric conductivity at the twin wall–surface junction

The first theoretical predictions of domain wall conductivity in ferroelectrics were made in early 1969 [54], and the proposed 'direct' mechanism is based on the compensation of polarization bound charge divergency by mobile carriers accumulated at the 'charged' walls. According to the direct mechanism, the electrical conductance of the ferroelectric domain walls is a continuous function of the domain wall orientation, which can vary from the head-to-head to uncharged (or 'parallel') and then to the tail-to-tail domains walls (see figures 7.16(a) and 7.16(b)). Corresponding change of the conductivity is due to the carrier accumulation and the band-structure changes at the walls. The physical origin of the charged domain walls conductivity is the bound charge variation across the wall, which induces the depolarization (or stray) electric field. The stray field attracts screening carriers which conduct and causes the band-structure changes at and near the wall.

Following the above logic, the 'nominally uncharged' Ising-type 180° ferroelectric domain walls should be energetically preferable in uniaxial ferroelectrics. However,

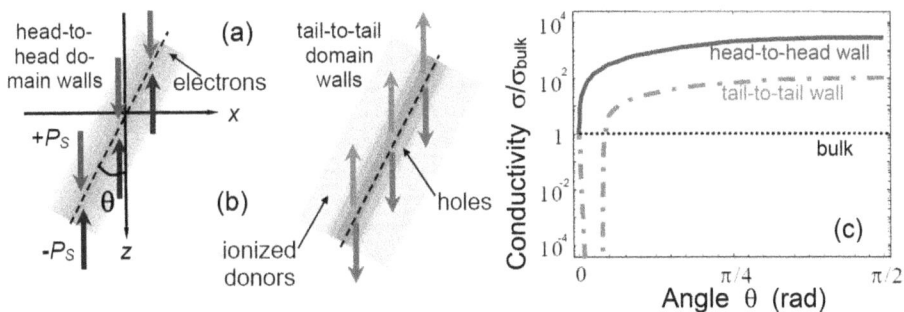

Figure 7.16. Sketch of the (a) head-to-head and (b) tail-to-tail charged domain walls in the uniaxial ferroelectrics-semiconductors of *n*-type. The gradient colors indicate the free carrier concentration (electrons in the case (a) and donors + holes in the case (b)) increase at the domain wall. (c) Dependence of the relative static conductivity at the domain wall plane on the wall incline angle θ calculated for head-to-head (solid curves) and tail-to tail domain walls (dashed curves). (Adapted with permission from [55]. Copyright 2012 Taylor and Francis.)

this is not always true. For instance, Cherifi–Hertel *et al* [56] probed the internal structure of the 180-degree ferroelectric domain walls in a uniaxial ferroelectric $LiTaO_3$ using the second-harmonic generation microscopy and detected a pronounced second-harmonic signal at the walls. Local polarimetry analysis of this signal revealed the existence of a planar polarization within the walls with the Bloch-like configurations [56]. Other results [57] demonstrate an evident deviation from the ideal Ising configuration of domain walls that is expected in uniaxial ferroelectrics, supporting the hypothesis of the formation of much more complex chiral structures (see e.g. [18] and figure 7.17).

The domain wall structure near the surface of multiaxial ferroelectrics can be even more complex, as discussed in the previous section. Numerous attempts have been made to describe polarization vector structure at the domain walls in multiaxial ferroelectrics since the early days of ferroelectricity [58–60], but progress has been limited by technical abilities. With more mature calculation possibilities, the combination of the LGD approach and the phase-field modeling emerged as a powerful tool for the simulation of domain wall structure and their electronic properties in uniaxial [61], multiaxial [62–64] and improper [65] ferroelectrics. Specifically, Gureev *et al* [61] considered the structure and calculated the energy of a charged 180° domain wall and revealed that the scales controlling the wall structure can be very different from the Debye screening radius. The scale depends on the spontaneous polarization and the concentration of free carriers. Hlinka *et al* [62–64] revealed the almost uncharged Ising–Bloch–Neel-type 180° domain walls in $BaTiO_3$ for definite rotation angles. Using density functional theory, the LGD approach and molecular dynamics simulations, Lee *et al* [66], Rakesh *et al* [67] and Gu *et al* [68] showed that 180° domain walls in $PbTiO_3$, $LiNbO_3$ and thin strained films of $BaTiO_3$ manifest a mixture of Ising-like (predominantly), Bloch-like and Néel-like characters.

As a part of the computational procedure, the direct mechanism of the domain wall conduction was incorporated in the finite element modeling (FEM) in uniaxial [69] and multiaxial ferroelectrics [70], which allows one to simulate the dependence of the wall conductivity on the bound charge. However, experimental results also report conductivity of the nominally uncharged domain walls in different

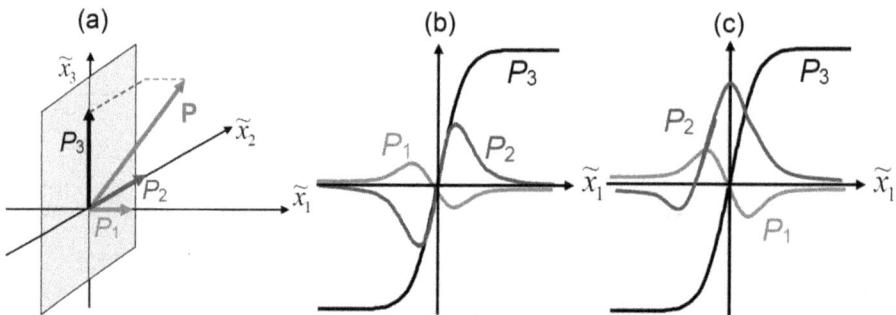

Figure 7.17. (a) Polarization vector geometry for the mixed-type walls: P_3 is the Ising component, P_2 is the Bloch component and P_1 is the Neel component. The distance from the wall plane is $\tilde{x}_1$. Schematics of the polarization components distribution inside the (b) achiral and (c) chiral domain walls. (Adapted with permission from [18]. Copyright 2013 the American Physical Society.)

multiferroics [71, 72] and multiaxial ferroelectrics [73, 74]. In this case more 'indirect' mechanisms of the domain wall conductivity should exist.

Using a deformation potential concept [75], the strain-induced shift of the conduction and valence band edges, E_C and E_V, caused by the wall–surface junction, is proportional to the strain variation $u_{ij}^S(x, z) = u_{ij}(x, z) - u_{ij}^0(\infty, z)$. Here u_{ij}^0 is the corresponding spontaneous strain due to the surface discontinuity far from the wall, where $\Phi_2^2 \to \Phi_S^2$. The band shift is

$$E_C\left(u_{ij}^S(x, z)\right) = E_{C0} + \Xi_{ij}^C u_{ij}^S(x, z), \quad E_V\left(u_{ij}^S(x, z)\right) = E_{V0} + \Xi_{ij}^V u_{ij}^S(x, z). \quad (7.8)$$

Ξ_{ij}^C and Ξ_{ij}^V are deformation potential tensors of electrons in the conduction band and holes in the valence bands, respectively. The values E_{C0} and E_{V0} are the corresponding energetic positions far from the wall.

Electric field, $E_i = -\partial\phi/\partial x_i$, and electrostatic potential φ, are determined self-consistently from the Poisson equation, in which the flexo-polarization, $\delta P_i^{\text{flexo}}(x, z) \cong \alpha_{ij}^{-1} f_{mnjl} \frac{\partial u_{mn}}{\partial x_l}$, can be used as the zero-order approximation [73]. Changes of the electric field and potential due to the flexoelectric effect are $\delta E_j^{\text{flexo}} \cong -f_{mnjl} \frac{\partial u_{mn}}{\partial x_l}$ and $\delta\phi_{\text{flexo}}(x, z) \cong f_{mn33} u_{mn}^S(x, z) \sim f_{ij33} R_{ijkl} \Phi_k \Phi_l$, respectively.

Allowing for the deformation potential variation and the flexoelectric mechanism, local band bending caused by the twin wall–surface junction can be estimated as $\Delta E_n(x, z) = \Xi_d^C u_{ii}^S(x, z) + e\delta\phi_{\text{flexo}}(x, z)$ for electrons and $\Delta E_p(x, z) = -\Xi_d^V u_{ii}^S(x, z) - e\delta\phi_{\text{flexo}}(x, z)$ for holes. The change in band structure and the changes in electrochemical potentials modulate the densities of free electrons $n(x, z)$ and holes $p(x, z)$ accumulated by the wall–surface junction. The effect can be estimated in the Boltzmann approximation as

$$n(x, z) \approx n_0 \exp\left(\frac{\Xi_d^C u_{ii}^S(x, z) + ef_{ij33} u_{ij}^S(x, z)}{k_B T}\right), \quad (7.9a)$$

$$p(x, z) \approx p_0 \exp\left(\frac{-\Xi_d^V u_{ii}^S(x, z) - ef_{ij33} u_{ij}^S(x, z)}{k_B T}\right). \quad (7.9b)$$

The terms $\Xi_d^C u_{ii}^S(x, z)$ and $\Xi_d^V u_{ii}^S(x, z)$ in equation (7.9a), related with the deformation potentials, make the *direct contribution* to the carrier accumulation at twin domain walls in ferroelastics. The trace of the strain tensor $u_{ii}(x, z)$ is given by equation (7.3). The term $\pm ef_{ij33} u_{ij}^S(x, z)$ in equation (7.9a) is related to the flexo-electric coupling with the antiferrodistortive tilt, because $f_{mn33} u_{mn}^S(x, z) \sim f_{ij33} R_{ijkl} \Phi_k \Phi_l$. It is responsible for the *indirect contribution* to the carrier accumulation at twin domain walls in ferroelastics. Hence, the roto-flexo field at the twin wall–surface junctions is the *indirect mechanism* of the carrier accumulation.

The static conductivity $\rho(x, z)$ is proportional to the free carrier density:

$$\rho(x, z) = -e\mu_n n(x, z) + e\mu_p p(x, z), \quad (7.10)$$

where μ_n and μ_p are the electron and hole mobilities. We refer to the changes of electrochemical potential due to the flexoelectric effect, $\pm e f_{ij33} u_{ij}^S(x, z)$ in equation (7.8), as the indirect mechanism of domain walls conductivity.

The carrier density induced by the local bend bending and flexoelectric effect at the twin wall–surface junction in CaTiO$_3$ parameters is shown in figure 7.18(a) (for electrons) and figure 7.18(b) (for holes). The profiles are symmetric with respect to

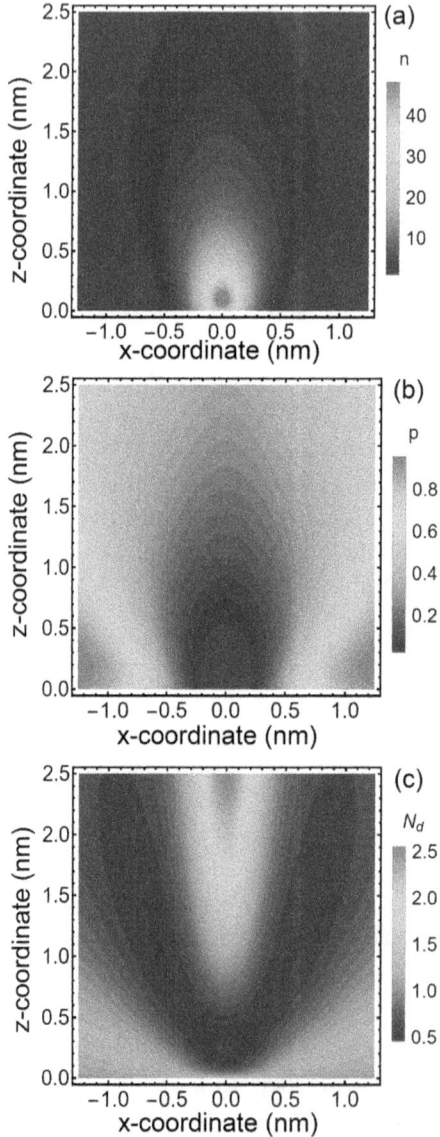

Figure 7.18. Color maps of the (a) electron n, (b) hole p, and (c) oxygen vacancy N_d^+ concentration (in relative units) versus the distance x from the twin wall plane $x = 0$ and the distance z from the surface $z = 0$ calculated for CaTiO$_3$ parameters and room temperature.

the wall plane $x = 0$ and have a sharp maximum at $x = 0$. The maximal amplitude decreases and its half-width increases with an increase in z.

Similarly to the electrons accumulation and holes depletion, the twin wall–surface junction in $CaTiO_3$ (or $SrTiO_3$) can accumulate donors (oxygen vacancies) or acceptors (titanium vacancies). Here, the point charge defect (other name is 'dilation center also known as elastic dipole' [76]) tensor $\beta_{jk}^{a,\,d}$ plays the same role in the vacancies segregation as the deformation potential tensor in the electron accumulation but of course with totally different mobilities. The structure of the chemical strain tensor is controlled by the crystalline symmetry of the material. It is diagonal and reduces to scalar, $\beta_{ij}^{a,\,d} = \beta_{ii}^{a,\,d}\delta_{ij}$, for isotropic or cubic media.

Using analytical expressions for the single-ionized donor concentration, $N_d^+ \approx N_{d0}^+\exp\left((\beta_{jk}^d u_{jk} - e\phi)/k_B T\right)$, and $N_a^- \approx N_{a0}^-\exp\left((\beta_{jk}^a u_{jk} + e\phi)/k_B T\right)$ for the single-ionized acceptor concentration, and the expression for potential variation, $\delta\phi_{\text{flexo}}(x, z) = f_{mn33}u_{mn}^S(x, z)$, approximate expressions for N_d^+ and N_a^- have the form

$$N_d^+(x, z) \approx N_{d0}^+\exp\left(\frac{\beta_{ii}^d u_{ii}^S(x, z) - ef_{ij33}u_{ij}^S(x, z)}{k_B T}\right),$$ (7.11a)

$$N_a^-(x, z) \approx N_{a0}^-\exp\left(\frac{\beta_{ii}^a u_{ii}^S(x, z) + ef_{ij33}u_{ij}^S(x, z)}{k_B T}\right).$$ (7.11b)

Here N_{d0}^+ and N_{a0}^- are the concentrations of single-ionized defects in the bulk. For neutral defects the corresponding expressions are $N_d^0(x, z) \approx N_{d0}\exp\left(\frac{\beta_{ii}^d u_{ii}^S(x,z)}{k_B T}\right)$ and $N_a^0(x, z) \approx N_{a0}\exp\left(\frac{\beta_{ii}^a u_{ii}^S(x,z)}{k_B T}\right)$.

The spatial distribution of oxygen vacancy concentration N_d^+ in the vicinity of the twin wall–surface junction in $CaTiO_3$ is shown in figure 7.18(c). It is seen from the figure that the distribution of the vacancies is rather complex and affected by the surface. The conductivity response of the twin wall–surface junction can be verified by SSPM, since the junction is the part of the conduction path involved in the probing.

Beyond pure ferroelastics, Hunnestad *et al* have published a review of the characterization of ferroelectric domain walls, including charged–conductive domain walls, using scanning electron microscopy [77]. The different acquisition modes all rely on the interaction of the primary electrons with surface charges, whether it be a fixed polarization charge, screening charge or free charge. Furthermore, different physical mechanisms can shed light on the charged nature of the domain walls. For example, the pyroelectric effect due to local electron beam heating of the sample surface can result in a reduction of the ferroelectric polarization and therefore overscreening of the polarization charge, giving rise to a change in image contrast. A second example is the domain wall orientation. 180°

domain walls can be neutral (antiparallel polarization), positively charged (head-to-head) or negatively charged (tail-to-tail). The mobile charge in the ferroelectric will screen these three domain wall configurations differently, providing distinctive contrast in the electron imaging. This is similar to the mechanism reported by Kalinin and Bonnell using SSPM [41]. Finally, the classical trade-off in SEM between primary energy and secondary electron emission can result in positive or negative sample charging, giving rise to a distinct signal from a domain wall with respect to the adjacent polarized domains.

XPEEM, used in x-ray absorption mode, can also be used to probe domain wall charge and even to deduce domain wall conductivity qualitatively. Schaab *et al* studied the XPEEM signal of the secondary electrons at the Mn L_3 edge from domains and domain walls in $ErMnO_3$ [78]. By using the *x*-cut of the $ErMnO_3$, it was possible to simultaneously image the charge neutral domain surfaces, head-to-head and tail-to tail domain walls. Fascinatingly, only tail-to-tail domain walls showed a secondary electron contrast. The mechanism is related to the additional screening possible at the domain walls due to the higher free electron density. The screening compensated the photoemission induced sample charging and resulted in emission at higher kinetic energy in the laboratory reference frame (figure 7.19).

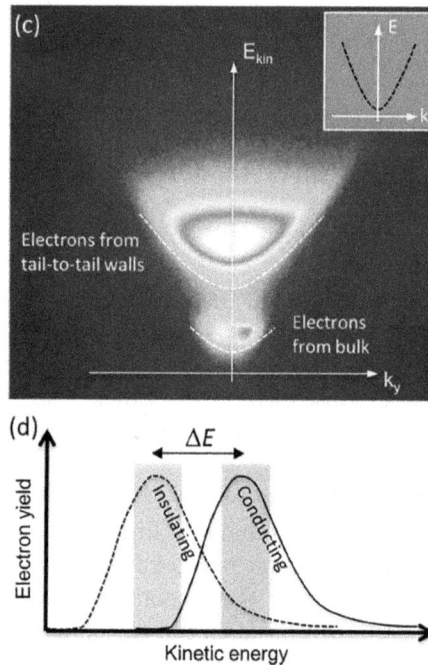

Figure 7.19. Reciprocal space E versus k image of the photoelectron parabola excited at the Mn L_3 edge of $ErMnO_3$. The high electron density at tail-to-tail domain walls allow one to screen the positive holes resulting from the photoemission process more efficiently, and the detected secondary electrons have a higher kinetic energy in the laboratory reference frame. (Reproduced with permission from [78]. Copyright 2014 AIP Publishing.)

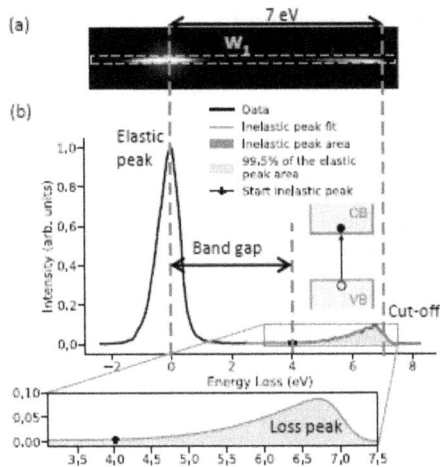

Figure 7.20. Electron energy loss spectrum acquired from a twin wall at the $CaTiO_3(001)$ surface in a LEEM.

Nevertheless, these are indirect measurements or deductions of domain wall conductivity, providing evidence for probable conductivity based on the distinctive charged nature of the domain walls. Furthermore, the experimental contrasts observed depend on the discontinuity in the bulk polarization at the domain wall. In the case of ferroelastics, there is no bulk polarization, so domain wall conductivity must therefore be related to the intrinsic domain wall electronic structure, possibly enhanced by the surface discontinuity [49] or strong interaction between twin walls and charged defects such as oxygen vacancies.

One particularly promised technique to explore more domain wall conductivity directly is energy loss spectroscopy in LEEM. The onset of the secondary electron losses can be used to probe the local band gap narrowing at twin walls in $CaTiO_3$ (unpublished research). This requires clean surfaces in order to be sure that the secondary electron onset is not due to surface contamination (figure 7.20).

7.5 Summary

Since the first experimental observations of ferroelectric and ferroelastic domains by Forsbergh, Merz and others, domain ordering has been a constant source of study to reveal the underlying physical phenomena, particularly from the perspective of applications in a wide variety of fields, ranging from microelectronics to photovoltaics, sensors and electromechanical devices. The realization that domain walls could carry novel functionalities absent in the bulk ferroelectrics and ferroelastics provided both new impetus but also demanded the use of specific surface sensitive techniques to unravel the connections between structure, chemistry and ordering. A particularly attractive, but perhaps the least studied, field is the functionality of ferroelastic domain walls and the possibility to engineer their physical properties. Inhomogeneous elastic strains, which exist due to the rotostriction in the vicinity of the twin wall–surface junctions in ferroelastics, can strongly affect their electronic properties. In particular, the strains change the band structure at the wall–surface junction via the deformation potential, rotostriction and flexoelectric coupling mechanisms. The calculated decrease of the local band gap can be considered as the direct mechanism of the uncharged domain wall conductivity increase in the ferroelastics ($CaTiO_3$, $EuTiO_3$, $SrTiO_3$, etc), and by extension in other multiferroics ($BiFeO_3$) and semiconducting ferroelectrics ($PbTiO_3$, $BaTiO_3$, etc). Flexoelectric and rotostriction couplings lead to the appearance of the inhomogeneous electric field, called the roto-flexo field [20, 32], proportional to the structural order parameter gradient across the wall. The fields, which are localized at the domain wall plane and lead to the carrier accumulation in the wall region, are considered as indirect mechanism of the uncharged domain walls conductivity (see figure 7.21). Comparison of the contribution of the direct and indirect mechanisms to the twin wall conductivity demonstrated their complex interplay in antiferrodistortive ferroelastics, such as $SrTiO_3$ and $CaTiO_3$. Experimentally, polarity and/or enhanced conductivity are now routinely observed in the bulk and also at the surface. In particular, the use of low energy electrons or photoelectrons has proven to be a valuable tool thanks to their intrinsic surface, charge and chemical sensitivity.

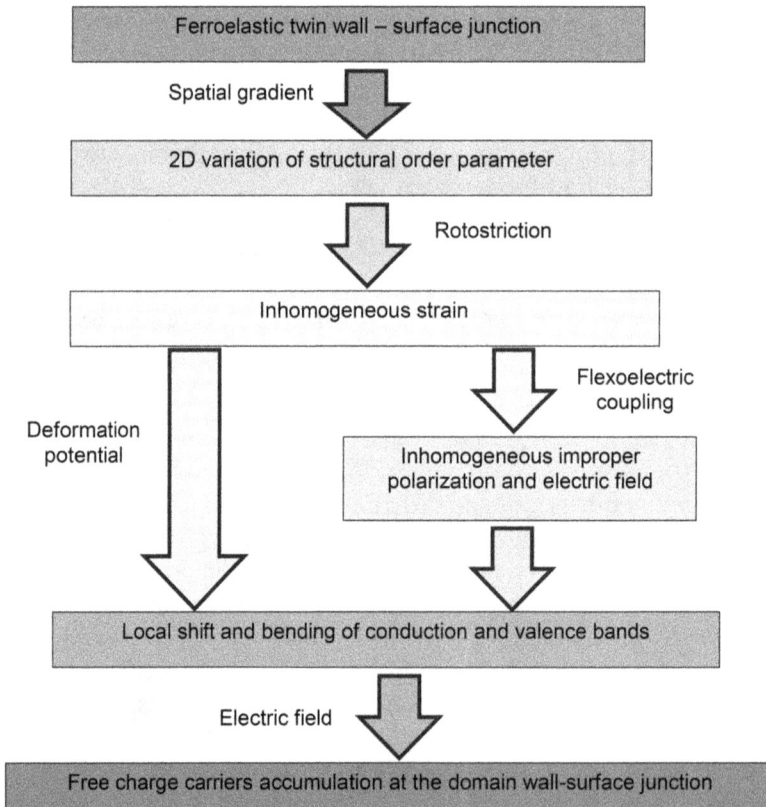

Figure 7.21. Origin of the twin wall–surface junction spontaneous polarization and electro-conductivity: direct and indirect mechanisms. (Adapted with permission from [19]. Copyright 2012 the American Physical Society.)

Acknowledgments

E.A.E. and A.N.M. gratefully acknowledge the DOE Software Project on 'Computational Mesoscale Science and Open Software for Quantum Materials' under Award Number DE-SC0020145 as part of the Computational Materials Sciences Program of the US Department of Energy, Office of Science, Basic Energy Sciences. A.N.M. also acknowledges support from the Horizon Europe Framework Programme (HORIZON-TMA-MSCA-SE), project no 101131229, 'Piezoelectricity in 2D-materials: materials, modeling, and applications' (PIEZO 2D). Results were visualized in Mathematica Wolfram Research, v.14.0. N.B. gratefully acknowledges the National Research Fund (Luxemburg) FNR/P12/4853155/Kreisel, Ile-de-France SESAME project MesoXcopy, ANR-10-LABX-0035, ANR-12-IS04–0001–01 CHEM-SWITCH, Leverhulme Trust (EM−2016–004), EPSRC (EP/K009702/1), ANR-DFG-HREELM (ANR−14-CE35–0019–03), INTER/MOBILITY/13/ 6572599, Programme Key Research Shaanxi Province (2018ZDXM-GY−145), Scientific Research Programme of Shaanxi Education Department (Grant No. 16JK1561), doctoral starting grant Xi'an University of Technology (Grant No. 101-4541115016), and the National Natural Science Foundation of China (Grant No. 51602255).

Bibliography

[1] Schlom D G, Chen L-Q, Fennie C J, Gopalan V, Muller D A, Pan X, Ramesh R and Uecker R 2014 Elastic strain engineering of ferroic oxides *MRS Bull.* **39** 118

[2] Nataf G F, Guennou M, Gregg J M, Meier D, Hlinka J, Salje E K H and Kreisel J 2020 Domain-wall engineering and topological defects in ferroelectric and ferroelastic materials *Nat. Rev. Phys.* **2** 634

[3] Hayward S A, Chrosch J, Salje E K H and Carpenter M A 1997 Thickness of pericline twin walls in anorthoclase: an x-ray diffraction study *Eur. J. Mineral.* **8** 1301

[4] Aird A and Salje E K H 1998 Sheet superconductivity in twin walls: experimental evidence of WO_{3-x} *J. Phys. Condens. Matter* **10** L377

[5] Van Aert S, Turner S, Delville R, Schryvers D, Van Tendeloo G and Salje E K H 2012 Direct observation of ferrielectricity at ferroelastic domain boundaries in $CaTiO_3$ by electron microscopy *Adv. Mater.* **24** 523

[6] Goncalves-Ferreira L, Redfern S A T, Artacho E and Salje E K H 2008 Ferrielectric twin walls in $CaTiO_3$ *Phys. Rev. Lett.* **101** 097602

[7] Salje E and Zhang H 2009 Domain boundary engineering *Phase Transit.* **82** 452

[8] Viehland D D and Salje E K H 2014 Domain boundary-dominated systems: adaptive structures and functional twin boundaries *Adv. Phys.* **63** 267

[9] Bhattacharya K and James R D 2005 The material is the machine *Science* **307** 53

[10] Catalan G, Seidel J, Ramesh R and Scott J F 2012 Domain wall nanoelectronics *Rev. Mod. Phys.* **84** 119

[11] Tagantsev A K 1986 Piezoelectricity and flexoelectricity in crystalline dielectrics *Phys. Rev. B* **34** 5883

[12] Catalan G, Sinnamon L J and Gregg J M 2004 The effect of flexoelectricity on the dielectric properties of inhomogeneously strained ferroelectric thin films *J. Phys. Condens. Matter* **16** 2253

[13] Maranganti R, Sharma N D and Sharma P 2006 Electromechanical coupling in nonpiezo-electric materials due to nanoscale nonlocal size effects: Green's function solutions and embedded inclusions *Phys. Rev.* B **74** 014110

[14] Eliseev E A, Morozovska A N, Glinchuk M D and Blinc R 2009 Spontaneous flexoelectric/flexomagnetic effect in nanoferroics *Phys. Rev.* B **79** 165433

[15] Morozovska A N, Eliseev E A, Svechnikov G S and Kalinin S V 2011 Nanoscale electromechanics of paraelectric materials with mobile charges: size effects and nonlinearity of electromechanical response of $SrTiO_3$ films *Phys. Rev.* B **84** 045402

[16] Tagantsev A K, Meunier V and Sharma P 2009 Novel electromechanical phenomena at the nanoscale: phenomenological theory and atomistic modeling *MRS Bull.* **34** 643

[17] Yudin P V, Tagantsev A K, Eliseev E A, Morozovska A N and Setter N 2012 Bichiral structure of ferroelectric domain walls driven by flexoelectricity *Phys. Rev.* B **86** 134102

[18] Eliseev E, Yudin P, Kalinin S, Setter N, Tagantsev A and Morozovska A 2013 Structural phase transitions and electronic phenomena at 180-degree domain walls in rhombohedral $BaTiO_3$ *Phys. Rev.* B **87** 054111

[19] Eliseev E A, Morozovska A N, Gu Y, Borisevich A Y, Chen L-Q, Gopalan V and Kalinin S V 2012 Conductivity of twin-domain-wall/surface junctions in ferroelastics: interplay of deformation potential, octahedral rotations, improper ferroelectricity, and flexoelectric coupling *Phys. Rev.* B **86** 085416

[20] Morozovska A N, Eliseev E A, Kalinin S V, Qing Chen L and Gopalan V 2012 Surface polar states and pyroelectricity in ferroelastics induced by flexo-roto field *Appl. Phys. Lett.* **100** 142902

[21] Gopalan V and Litvin D B 2011 Rotation-reversal symmetries in crystals and handed structures *Nat. Mater.* **10** 376

[22] Bousquet E, Dawber M, Stucki N, Lichtensteiger C, Hermet P, Gariglio S, Triscone J-M and Ghosez P 2008 Improper ferroelectricity in perovskite oxide artificial superlattices *Nature* **452** 732

[23] Fennie C J and Rabe K M 2005 Ferroelectric transition in $YMnO_3$ from first principles *Phys. Rev.* B **72** 100103

[24] Benedek N A and Fennie C J 2011 Hybrid improper ferroelectricity: a mechanism for controllable polarization–magnetization coupling *Phys. Rev. Lett.* **106** 107204

[25] Novak J and Salje E K H 1998 Surface structure of domain walls *J. Phys. Condens. Matter* **10** L359

[26] Goncalves-Ferreira L, Redfern S A T, Artacho E, Salje E and Lee W T 2010 Trapping of oxygen vacancies in the twin walls of perovskite *Phys. Rev.* B **81** 024109

[27] Rauschen R and De Souza R A 2022 Oxygen transport at a ferroelastic domain wall in $CaTiO_3$ *Phys. Rev. Mater.* **6** L111401

[28] Lur'e A I 1964 *Three-Dimensional Problems of the Theory of Elasticity* (New York: Interscience)

[29] Morozovska A N, Eliseev E A and Kalinin S V 2007 The piezoresponse force microscopy of surface layers and thin films: effective response and resolution function *J. Appl. Phys.* **102** 074105

[30] Cao W and Barsch G R 1990 Landau–Ginzburg model of interphase boundaries in improper ferroelastic perovskites of D^{18}_{4h} symmetry *Phys. Rev.* B **41** 4334

[31] Tagantsev A K, Courtens E and Arzel L 2001 Prediction of a low-temperature ferroelectric instability in antiphase domain boundaries of strontium titanate *Phys. Rev.* B **64** 224107

[32] Morozovska A N, Eliseev E A, Glinchuk M D, Chen L-Q and Gopalan V 2012 Interfacial polarization and pyroelectricity in antiferrodistortive structures induced by a flexoelectric effect and rotostriction *Phys. Rev.* B **85** 094107

[33] Gu Y, Rabe K, Bousquet E, Gopalan V and Chen L Q 2012 Phenomenological thermodynamic potential for $CaTiO_3$ single crystals *Phys. Rev.* B **85** 064117

[34] Carpenter M A, Becerro A I and Seifert F 2001 Strain analysis of phase transitions in (Ca,Sr) TiO_3 perovskites *Am. Mineral.* **86** 348

[35] Zubko P, Catalan G, Buckley A, Welche P R L and Scott J F 2007 Strain-gradient-induced polarization in $SrTiO_3$ single crystals *Phys. Rev. Lett.* **99** 167601

[36] Barrett J H 1952 Dielectric constant in perovskite type crystals *Phys. Rev.* **86** 118

[37] Houchmandzadeh B, Lajzerowicz J and Salje E 1991 Order parameter coupling and chirality of domain walls *J. Phys. Condens. Matter* **3** 5163

[38] Forsbergh P W 1949 Domain structures and phase transitions in barium titanate *Phys. Rev.* **76** 1187

[39] Merz W J 1954 Domain formation and domain wall motions in ferroelectric $BaTiO_3$ single crystals *Phys. Rev.* **95** 690

[40] Le Bihan R 1989 Study of ferroelectric and ferroelastic domain structures by scanning electron microscopy *Ferroelectrics* **97** 19

[41] Kalinin S V and Bonnell D A 2001 Temperature dependence of polarization and charge dynamics on the $BaTiO_3(100)$ surface by scanning probe microscopy *Appl. Phys. Lett.* **78** 1116

[42] Kalinin S V and Bonnell D A 2001 Local potential and polarization screening on ferroelectric surfaces *Phys. Rev.* B **63** 125411

[43] Wang J L, Vilquin B and Barrett N 2012 Screening of ferroelectric domains on $BaTiO_3(001)$ surface by ultraviolet photo-induced charge and dissociative water adsorption *Appl. Phys. Lett.* **101** 092902

[44] Yang W-C, Rodriguez B J, Gruverman A and Nemanich R J 2005 Photo electron emission microscopy of polarity-patterned materials *J. Phys. Condens. Matter* **17** S1415

[45] Nataf G F, Grysan P, Guennou M, Kreisel J, Martinotti D, Rountree C L, Mathieu C and Barrett N 2016 Low energy electron imaging of domains and domain walls in magnesium-doped lithium niobate *Sci. Rep.* **6** 33098

[46] Rault J E, Mentes T O, Locatelli A and Barrett N 2014 Reversible switching of in-plane polarized ferroelectric domains in $BaTiO_3(001)$ with very low energy electrons *Sci. Rep.* **4** 6792

[47] Höfer A, Fechner M, Duncker K, Hölzer M, Mertig I and Widdra W 2012 Persistence of surface domain structures for a bulk ferroelectric above T_C *Phys. Rev. Lett.* **108** 087602

[48] Mathieu C, Lubin C, Le Doueff G, Cattelan M, Gemeiner P, Dkhil B, Salje E K H and Barrett N 2018 Surface proximity effect, imprint memory of ferroelectric twins, and tweed in the paraelectric phase of $BaTiO_3$ *Sci. Rep.* **8** 13660

[49] Zhao Z, Barrett N, Wu Q, Martinotti D, Tortech L, Haumont R, Pellen M and Salje E K H 2019 Interaction of low-energy electrons with surface polarity near ferroelastic domain boundaries *Phys. Rev. Mater.* **3** 043601

[50] Nepijko S A, Sedov N N, Ziethen C, Schönhense G, Merkel M and Escher M 2000 Peculiarities of imaging one and two-dimensional structures in an emission electron microscope. 1. Theory *J. Microsc.* **199** 124

[51] Magagnin G, Lubin C, Escher M, Weber N, Tortech L and Barrett N 2024 Ferroelastic twin angles at the surface of $CaTiO_3$ quantified by photoemission electron microscopy *Phys. Rev. Lett.* **132** 56201

[52] Nataf G F *et al* 2017 Control of surface potential at polar domain walls in a nonpolar oxide *Phys. Rev. Mater.* **1** 074410

[53] Zykova-Timan T and Salje E K H 2014 Highly mobile vortex structures inside polar twin boundaries in $SrTiO_3$ *Appl. Phys. Lett.* **104** 082907

[54] Guro G I, Ivanchik I I and Kovtoniuk N F 1969 C-domain $BaTiO_3$ crystal in a short-circuited capacitor *Sov. Sol. St. Phys.* **11** 1956–64

[55] Morozovska A N 2012 Domain wall conduction in ferroelectrics *Ferroelectrics* **438** 3

[56] Cherifi-Hertel S, Bulou H, Hertel R, Taupier G, Dorkenoo K D, Andreas C, Guyonnet J, Gaponenko I, Gallo K and Paruch P 2017 Non-Ising and chiral ferroelectric domain walls revealed by nonlinear optical microscopy *Nat. Commun.* **8** 15768

[57] Cherifi-Hertel S, Voulot C, Acevedo-Salas U, Zhang Y, Crégut O, Dorkenoo K D and Hertel R 2021 Shedding light on non-Ising polar domain walls: insight from second harmonic generation microscopy and polarimetry analysis *J. Appl. Phys.* **129** 081101

[58] Zhirnov V A 1959 A contribution to the theory of domain walls in ferroelectrics *Sov. Phys. JETP* **35** 822

[59] Darinskii B and Fedosov V 1971 Structure of 90° domain walls in $BaTiO_3$ *Sov. Phys. Solid State* **13** 17–22

[60] Cao W and Cross L E 1991 Theory of tetragonal twin structures in ferroelectric perovskites with a first-order phase transition *Phys. Rev.* B **44** 5

[61] Gureev M Y, Tagantsev A K and Setter N 2011 Head-to-head and tail-to-tail 180° domain walls in an isolated ferroelectric *Phys. Rev.* B **83** 184104

[62] Hlinka J and Márton P 2006 Phenomenological model of a 90° domain wall in $BaTiO_3$-type ferroelectrics *Phys. Rev.* B **74** 104104

[63] Marton P, Rychetsky I and Hlinka J 2010 Domain walls of ferroelectric $BaTiO_3$ within the Ginzburg–Landau–Devonshire phenomenological model *Phys. Rev.* B **81** 144125

[64] Stepkova V, Marton P and Hlinka J 2012 Stress-induced phase transition in ferroelectric domain walls of $BaTiO_3$ *J. Phys. Condens. Matter* **24** 212201

[65] Mostovoy M, Nomura K and Nagaosa N 2011 Theory of electric polarization in multi-orbital Mott insulators *Phys. Rev. Lett.* **106** 047204

[66] Lee D, Behera R K, Wu P, Xu H, Li Y L, Sinnott S B, Phillpot S R, Chen L Q and Gopalan V 2009 Mixed Bloch–Néel–Ising character of 180° ferroelectric domain walls *Phys. Rev.* B **80** 060102

[67] Behera R K, Lee C-W, Lee D, Morozovska A N, Sinnott S B, Asthagiri A, Gopalan V and Phillpot S R 2011 Structure and energetics of 180° domain walls in $PbTiO_3$ by density functional theory *J. Phys. Condens. Matter* **23** 175902

[68] Gu Y, Li M, Morozovska A N, Wang Y, Eliseev E A, Gopalan V and Chen L-Q 2014 Flexoelectricity and ferroelectric domain wall structures: phase-field modeling and DFT calculations *Phys. Rev.* B **89** 174111

[69] Eliseev E A, Morozovska A N, Svechnikov G S, Gopalan V and Shur V Y 2011 Static conductivity of charged domain walls in uniaxial ferroelectric semiconductors *Phys. Rev.* B **83** 235313

[70] Eliseev E A, Morozovska A N, Svechnikov G S, Maksymovych P and Kalinin S V 2012 Domain wall conduction in multiaxial ferroelectrics *Phys. Rev.* B **85** 045312

[71] Balke N *et al* 2012 Enhanced electric conductivity at ferroelectric vortex cores in BiFeO$_3$ *Nat. Phys.* **8** 81

[72] Alikin D *et al* 2020 Strain-polarization coupling mechanism of enhanced conductivity at the grain boundaries in BiFeO$_3$ thin films *Appl. Mater. Today* **20** 100740

[73] Vasudevan R K, Wu W, Guest J R, Baddorf A P, Morozovska A N, Eliseev E A, Balke N, Nagarajan V, Maksymovych P and Kalinin S V 2013 Domain wall conduction and polarization-mediated transport in ferroelectrics *Adv. Funct. Mater.* **23** 2592

[74] Sharma P, Morozovska A N, Eliseev E A, Zhang Q, Sando D, Valanoor N and Seidel J 2022 Specific conductivity of a ferroelectric domain wall *ACS Appl. Electron. Mater.* **4** 2739

[75] Ashcroft N W and Mermin N D 1976 *Solid State Physics* (New York: Harcourt)

[76] Freedman D A, Roundy D and Arias T A 2009 Elastic effects of vacancies in strontium titanate: short- and long-range strain fields, elastic dipole tensors, and chemical strain *Phys. Rev.* B **80** 064108

[77] Hunnestad K A, Roede E D, van Helvoort A T J and Meier D 2020 Characterization of ferroelectric domain walls by scanning electron microscopy *J. Appl. Phys.* **128** 191102

[78] Schaab J *et al* 2014 Imaging and characterization of conducting ferroelectric domain walls by photoemission electron microscopy *Appl. Phys. Lett.* **104** 232904

IOP Publishing

Ferroelastic Materials

Guillaume F Nataf, Blai Casals and Ekhard K H Salje

Chapter 8

Martensitic materials

Marcel Porta, Eduard Vives and Antoni Planes

8.1 Introduction

Martensitic materials are those that show a martensitic transition (MT). At this transition, the lattice undergoes a virtually diffusionless distortion that gives rise to a change of its symmetry that can be accounted for by a shear (or a combination of shears) mechanism to a good approximation [1]. These transitions are thus classified as structural displacive transitions and are commonly first-order phase transitions (FOPTs), thus taking place through nucleation and growth. The strain that determines the lattice distortion is the primary order parameter and, thus, the transition can be induced by either changing the temperature or by application of a mechanical stress, which is the thermodynamic force conjugated to the strain. In the current chapter we will assume that the group symmetry of the product phase is a subgroup of the group symmetry of the parent phase. This is the case of MTs taking place in the so-called shape-memory materials [2]. Note that these MTs are in fact a particular class of ferroelastic transitions.[1] Therefore, the symmetry properties of the high-temperature phase (or parent phase) result in quite a large degeneracy g of the low-temperature phase (or martensitic phase), which is determined by the ratio of the symmetry operations of the parent and martensitic phases. This means that the transition can give rise to g variants of the martensitic phase, which only differ in their crystallographic orientation with respect to the parent phase, but are energetically equivalent in the absence of an externally applied stress, which breaks the degeneracy. These variants tend to form a complex pattern or polyvariant structure (see figure 8.1). The morphology of this pattern and the kinetics of its formation process are mainly determined by the minimization of the elastic strain energy induced by the lattice misfit between the martensite and parent phases and (to a lesser extent) between differently oriented variants. During the growth process, when

---

[1] Usually, a transition with a group–subgroup change of symmetry is denoted as martensitic instead of ferroelastic when it occurs in metals or alloys.

Figure 8.1. Optical image obtained with polarized light of the surface of a Cu–Zn–Al single crystalline alloy in the martensitic phase revealing the coexistence of different energetically equivalent variants. (Reproduced with permission from [3]. Copyright 2017 the American Physical Society.)

the stored elastic energy balances the transition driving force that is provided by the difference between the stress-free free energies of the martensite and parent phases, the transition is said to occur in thermoelastic equilibrium. This condition may occur in a given temperature range and level of applied stress, which define the transformation coexistence region (see figure 8.2(a) and (b)). In this region the transition can only take place along an optimal transition path at which an almost complete accommodation of the transformation-induced elastic strains is achieved. In other words, the corresponding arrangement of variants (self-accommodating) tends to eliminate the long-range character of the strain field (volume-dependent energy) which would result from a coherent single-variant martensitic region embedded in the parent phase. It is worth noting that under ideal thermoelastic conditions the transition should occur with null hysteresis (see figures 8.2(a) and (b)).

From a kinetic point of view, these transitions often show athermal character, which means that, contrary to thermally activated systems, thermal fluctuations do not play a very relevant role and, consequently, do not control the dynamics of the transition. In these circumstances, the condition of thermoelasticity may not be strictly satisfied. In this quasi-thermoelastic case, as temperature or stress vary continuously, the transition shows hysteresis due the existence of energy barriers that cannot be overcome due to the fact that thermal fluctuations play an irrelevant role. The existence of disorder in the materials (vacancies, dislocations, or even the sample surface) creates large heterogeneities in the nucleation barriers. The transition then occurs intermittently through a sequence of avalanches. The system evolves through a sequence of metastable minima of the complex free energy landscape that characterizes the parent–martensite two-phase coexisting region (see figures 8.2(c) and (d)). Actually, avalanches are associated with strain discontinuities and are related to fast relaxations from one metastable minimum to another. They involve energy dissipation which is ultimately responsible for the hysteresis observed in this class of transitions.

Associated with thermoelastic (and quasi-thermoelastic) behaviour, shape-memory materials display very interesting thermomechanical properties known

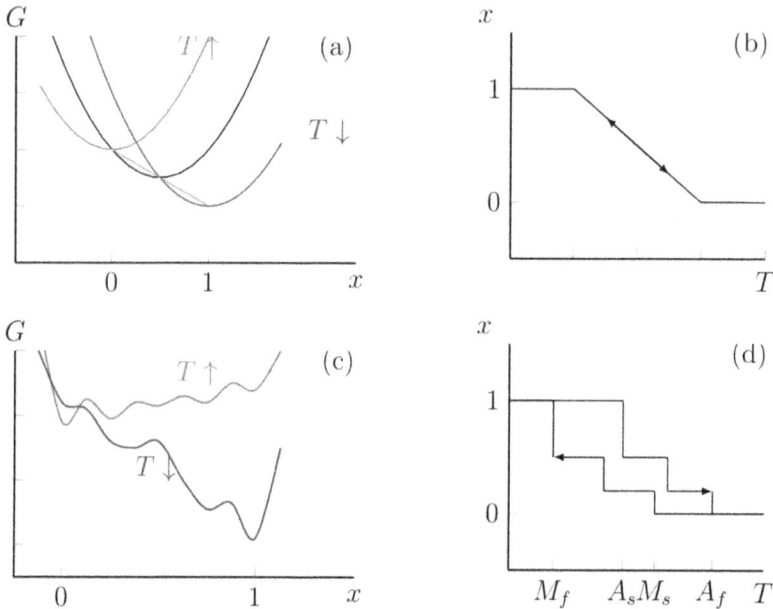

Figure 8.2. (a) Schematic representation of the free energy functional G as a function of the transformed fraction x at different temperatures for a system in thermoelastic equilibrium. (b) The corresponding behaviour of the transformed fraction x (order parameter) as a function of temperature obtained by minimization of $G(x)$. (c) Behaviour of $G(x)$ for an athermal system with quenched disorder, showing quasi-thermoelastic equilibrium. (d) Corresponding behaviour of the order parameter exhibiting avalanches and hysteresis.

generically as shape-memory properties. They are related to a highly nonlinear response to an applied stress that provides to these materials the capacity of remembering an original shape after a severe deformation process. Specifically, the shape-memory effect refers to the fact that when these materials are deformed in the low-temperature martensitic phase, they are able to recover their original shape by the reverse transition upon heating. In the high-temperature phase, the same systems display another unique property called superelasticity that consists of the possibility of recovering, upon loading and unloading, a large strain (in many cases >10%) associated with the stress-induced transition. These two properties make this class of materials very attractive from a technological point of view, since they may function as sensors as well as actuators, and are promising candidates as smart materials. Moreover, the combination of superelasticity and the significant latent heat of the MT give rise to interesting elastocaloric properties [4] that provide shape-memory alloys an enormous potential for future developments of a new, environmentally friendly solid-state refrigeration technology.

The present chapter is organized as follows. In the next section, experimental techniques suitable to detect avalanches in MTs are introduced, and avalanche features are discussed in relation to kinetic characteristics of this class of structural transitions. The particular case of ferromagnetic shape-memory alloys is also briefly discussed. Next, models are introduced that conveniently reproduce transition features, its intermittent dynamics through avalanches and the morphology of the martensitic phase. In the last section, the same models are used to analyse elastocaloric properties in shape-memory alloys and the results are compared with experimental data. Finally, we provide our conclusions.

8.2 Avalanche response

Avalanches can be used to describe many different phenomena. This kind of intermittent response was first proposed by Bak in his seminal theory for self-organized criticality (SOC) [5]. It is the expected response of spatially extended, externally driven, out-of-equilibrium systems with disorder that evolve in athermal conditions. The general SOC framework has been applied to many problems not only in physics but also in geophysics (earthquakes), biology, and social sciences. Within physics, avalanches have been shown to occur in magnetism, ferrolectricity, superconductivity, condensation, boiling, failure of materials, and the MTs of interest in this chapter. In many cases these kinds of intermittent responses have also been called 'crackling noise' [6] or 'jerks' when referring to the movement of the interfaces [7].

The word 'criticality' was motivated by the comparison of avalanche phenomena with equilibrium critical phenomena for which fluctuations are very large and display no characteristic scales. Close to classical equilibrium critical points the description of the thermodynamic properties of the system (order parameter, susceptibility, specific heat, etc) as a function of the 'distance' of the driving parameters (temperature, field, etc) to the critical point is given by power-law functions which reflect the existence of scale invariance.

Similarly, in out-of-equilibrium, avalanche properties such as avalanche energies, avalanche sizes (length, surface, volume), avalanche durations, etc, are stochastic in nature. They can be described by probability densities that, rather than being Gaussian as is expected for standard physical properties, extend for many decades with fat tails, and usually show power-law probability densities, which also reflect the scale invariance of avalanche properties. One of the immediate consequences is that, in many cases, the average values of avalanche properties (i.e. the expected or typical values) are not well defined.

As explained in the introduction, for the case of systems displaying an FOPT such as MTs, the existence of metastable states (and hysteresis) is the usual situation. Since the systems have spatially distributed disorder, heterogeneous random barriers separating metastable states appear. When such systems are driven through the FOPT by a smooth and slow ramp of the control parameter, the trajectories on the free energy landscape consist of a sequence of avalanches.

The experimental detection of avalanche response has inherent difficulties for two main reasons. First, its intermittent character: avalanches are fast and sharp events separated by long waiting times without activity. This requires experimental techniques that are able to resolve very short time scales but at the same time able to wait for long times without saturating storage devices. Second, the absence of a characteristic size requires devices be able to measure physical magnitudes spanning many decades, which is only possible with instruments with many digits.

Fortunately, in some physical systems, the sharp variation of physical magnitudes associated with avalanches induces electromagnetic or elastic waves that can be used as indirect methods to monitor their occurrence and understand some of their properties. The paradigmatic cases are Barkhausen noise [8], associated with magnetic avalanches, and acoustic emission (AE), associated with structural avalanches in solids. For the case of MTs, the existence of AE has been known for many decades [9, 10]. The almost instantaneous collective motion of a large number of atoms that reorganize in a new crystalline structure, resulting from the jump from one metastable state to another, produces an elastic wave that propagates through the solid and can be acquired by appropriate transducers attached on the sample surface. The detection of AE is typically done in the ultrasonic range (100 kHz–3 MHz) with piezoelectric transducers, but the transformation events can be so sharp and so energetic that, in some cases, sound waves are even emitted in the audible range. The transducers convert the micrometric surface vibrations to voltages that can be amplified and analysed with appropriate electronic circuits. The detection of AE although being an indirect method to study the dynamics of the MT, has an unbeatable temporal resolution since typical sampling rates are of the order of 40 MHz. On the other hand, voltage resolution is rather good, typically reaching six decades (18 bits). The technique, as will be detailed below, has been used for the study of MTs under thermal driving conditions, in uniaxial tensile experiments with force control or displacement control, in flexure experiments, etc.

Moreover, AE measurements have been shown to correlate rather well with other direct and indirect experimental techniques: calorimetry [11], optical imaging [12], strain field measurements by the grid method [13], and infrared emission detection [14, 15].

8.2.1 Experimental detection of acoustic emission

In typical AE detection experiments one or more piezoelectric sensors are placed in acoustic contact with the sample surface. The signal from the sensors is typically amplified $\times 100$ (40 dB) or $\times 1000$ (60 dB), digitized at high frequency ($\sim$40 MHz) and analysed by a proper electronic device. In contrast to the classical spectral methods for continuum noise analysis, in the last few decades a different analysis method more adequate for the study of intermittent signals has been developed. The philosophy behind it is that the burst signals that carry the information from the atomic rearrangements in the material must be separated from the long pauses that correspond to the regime in which the sample remains silent in one metastable state and stores elastic deformation energy (see figure 8.3(a)). The avalanche analysis technique proceeds as follows: a threshold $\pm V_{th}$ (for both positive and negative voltages) that will separate signals from noise is set by performing blank experiments when the sample is not externally driven. There are unavoidable sources of electric noise that will overlap with real signals that arise from the small transformation events. A good electric ground is required and some experiments are carried out inside

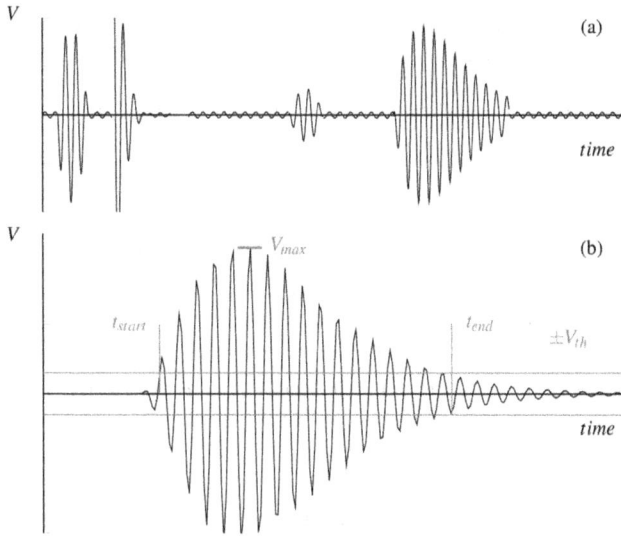

Figure 8.3. (a) Schematic representation of the continuum voltage signal recorded and amplified from an AE transducer. (b) The definition of an individual AE hit and some of its properties.

Faraday cages to minimize noise. During a real experiment the sample is driven through the transition with an imposed driving rate (force, temperature, displacement, etc). Then, when the amplified voltage signal crosses the threshold, (either in the positive or negative direction) one sets the starting time t_{start} of a so-called AE hit (see figure 8.3(b)). Since recorded signals oscillate with a frequency that depends mostly on the sensor properties, in order to decide the end of the hit, one needs to use a temporal parameter called the hit definition time (HDT). Only when the voltage remains within the $\pm V_{th}$ thresholds for a time longer than the HDT (typically in the range of 100 μs), one considers that the hit has ended in the last instant t_{end} that the threshold was crossed. With this procedure, AE hits are identified and can then be analysed and/or stored. Typically the experiments render a catalogue in which, for every hit, one records the time of occurrence t_{start}, the hit maximum voltage V_{max} that is transformed into hit amplitude $A = 20\log_{10}(V_{max}/V_0)$, (in dB, V_0 being a reference voltage), the hit duration $D = t_{end} - t_{start}$ and the hit energy E (typically in aJ) which is defined as

$$E = \frac{1}{R_0} \int_{t_{start}}^{t_{end}} V(t)^2 dt, \tag{8.1}$$

where $R_0 = 10$ kΩ, is a reference resistance.

The relation of the AE-hit properties with the properties of the physical phenomenon it originates from is know as the 'source problem'. It cannot be fully understood without assuming a model for the nucleation and the movement of the martensite–austenite and martensite–martensite interfaces. The initially stable interfaces accelerate from zero velocity, reach a maximum velocity, and decelerate until arresting again. Within the simplest 1D models, the voltage detected by the transducers is assumed to be proportional to the interface speed and therefore the hit energy can be considered to be proportional to the total kinetic energy released (dissipated) in the source event. Of course, elastic waves are not emitted isotropically and can be attenuated and distorted during the propagation through the material and in the sample/transducer contact. Thus, it is very difficult to perform absolute estimations of the physical properties of the real transformation events. Instead, measurements should only be used to establish relative differences between them.

When several transducers are attached on the sample surface it is possible to implement location techniques (similar to those used in geophysics) to determine the source position, from the difference in times of arrival of the signal to the different sensors. The group of hits in the different sensors that are assumed to correspond to the same physical origin allow one to characterize the so-called AE events. From the individual hit properties corresponding to the same event it is also possible to estimate event properties. For instance, with two sensors (1 and 2) one can locate the position of the source in elongated samples that can be approximated as one-dimensional objects [16]. Knowing the speed of ultrasounds in the material c, if sensor 1 is at the origin of the x axis and sensor 2 is at $x = L$, the position of the AE event will be given by $x = \frac{L + c(t_1 - t_2)}{2}$ and the event energy can be estimated as $E = \frac{E_1 + E_2}{2}$, except for a proportionality factor that depends on the attenuation of ultrasounds in the material.

8.2.2 Results

The discontinuous character of the recorded AE signal during smooth ramps of temperature or force is indeed a clear indication of the athermal character of the transition. The kinetics consists of a sequence of sudden transformation events between metastable states with associated AE hits followed by waiting times in which the system simply stores elastic energy and does not emit elastic waves.

A result known since more than 40 years ago [11] is that, in MTs, the acoustic emission rate (number of AE hits recorded per second) correlates extremely well with the signal detected from calorimetric measurements which, after a subtraction of an appropriate baseline, is proportional to the latent heat emitted or absorbed per unit time, which in turn is proportional to the rate of transformation, to a good approximation (see figure 8.4(a)). This correlation can be used in order to determine the transformation temperatures with a much higher precision than that allowed by calorimetric techniques given the sensitivity and time resolution of the AE. [17]. If the driving parameters (temperature, force, etc) are reversed, AE can also be used to characterize the hysteretic behaviour associated with the lack of thermoelasticity (hysteresis width, partial hysteresis cycles, return point memory, etc).

The AE/calorimetry correlation is, in fact, rather surprising. On the one hand, it is not clear at all why avalanches (which are dissipative events) are related to latent heat, which is a reversible effect. But experience shows that this is indeed the case. Moreover, even assuming that for every metastable jump, the energy of the relaxation events is proportional to the free energy difference between the initial and final state (before and after the avalanche) one would then expect a good correlation of the calorimetric signal, not with the number of events but with the

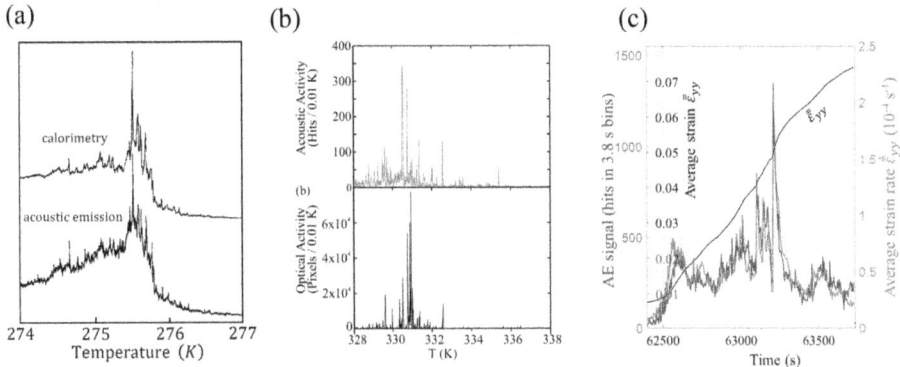

Figure 8.4. (a) Correlation between the AE activity (number of hits per temperature interval) and the calorimetric curve in a temperature driven martenstic transition. (b) Correlation between AE activity and optical activity in a Ni–Mn–Ga temperature driven martensitic transition. (c) Correlation between the AE activity and the strain rate in a stress-driven martensitic transition. ((a) Reproduced with permission from [11]. Copyright 1981 John Wiley and Sons. (b) Reproduced with permission from [12]. Copyright 2014 the American Physical Society. (c) Reproduced from [13]. CC BY 4.0.)

total energy of the events recorded per unit time. Anyway, experiments show that the number of events typically correlates much better than the total energy. The reason for this apparent contradiction is probably hidden behind the fact that AE-energies are, as expected for most avalanche properties, described by power-law (or at least fat tail-like) probability distributions [18]. The sum (or the average) of avalanche energies in a certain time interval is essentially dominated by the largest ones, which depend on microscopic details such as sample defects or other kinds of disorder, and thus does not inform on the overall free energy difference between two metastable states. The number of barriers that separate two given metastable states in a trajectory seems to be a good estimation of the reversible energy difference between them.

More recently it has been shown that the AE activity also correlates very well with the optical activity measured with polarized light in thermally driven MT in Ni–Mn–Ga [12] or with the strain rate measured with the grid method in stress-driven MT in Cu–Zn–Al single crystals [13] (see figure 8.4(b) and (c)).

8.2.2.1 Thermal driving experiments

The careful study of AE activity in thermally driven MT has allowed the understanding of important physical aspects of the complexity of these phase transitions in which the system typically transforms from a high symmetry phase to many coexisting variants of the low symmetry phase. By performing cooling/heating ramps at different rates back and forth across the transition between the same two fixed temperatures (with samples that have previously been cycled many times) one can separate the athermal avalanches that will show a good scaling with the rate (typically most of them) from the avalanches which are thermally activated and that do not show scaling. For instance, using these ideas it has been shown that the cubic-monoclinic II (18R) transition in Cu–Zn–Al alloys is much more athermal than the cubic to orthorhombic (2H) transition in Cu–Al–Ni alloys [19].

A still unsolved question is related to the asymmetry that is often observed between the total number of AE events recorded during heating and cooling ramps. This asymmetry could be related to purely experimental facts, such as the fact that attenuation in martensite is different from the attenuation in austenite but also could be a consequence of the fact that, while nucleation is required for the transition from the cubic to martensitic phase to take place, the reverse transition occurs by fast shrinkage of martensitic domains of the different variants that already exist [20]. It seems, however, that the problem is more complex since in some materials a larger number of AE events in the forward than in the reverse transition has been reported [21].

Moreover, if the AE studies are performed by consecutively cycling virgin samples [22, 23] that have been cast at high temperatures and have never been through the transition, one can understand how, during the first cycles, the samples rearrange some of the metastable barriers in order to find an optimal path linking the parent phase to the martensitic phase. In this learning process that is typically finished after the first ~10–12 cycles, the probability distribution of the AE hit energies evolves from being subcritical (with an excess of large avalanches) towards a final situation with a power-law distribution. This is, therefore, a paradigmatic example of a driven system that rearranges until reaching a 'critical' state, with lack of characteristic scales, in which a certain path connecting the high-temperature phase with a combination of variants of the low-temperature phase has been selected.

The final probability densities of avalanche energies ($g_E(E)$ for the energies E, $g_V(V)$ for the maximum voltages V, and $g_D(D)$ for the durations D) can then be written as

$$g_E(E) \equiv \frac{dP(E)}{dE} = \frac{E_0^{1-\varepsilon}}{\varepsilon - 1} E^{-\varepsilon}$$

$$g_V(V) \equiv \frac{dP(V)}{dV} = \frac{V_0^{1-\alpha}}{\alpha - 1} V^{-\alpha} \qquad (8.2)$$

$$g_D(D) \equiv \frac{dP(D)}{dD} = \frac{D_0^{1-\delta}}{\delta - 1} D^{-\delta},$$

where E_0, V_0 and D_0 are the lower thresholds required for the power-law density to be normalizable and ε, α and δ are the so-called critical exponents [18, 24, 25] characterizing the distributions of each of the three different avalanche properties. These laws are displayed as straight lines in log–log scales with negative slopes that correspond to the critical exponents (figure 8.5). Note that when estimating these laws by representing histograms of the measured data given by the catalogues, it is important to take into account that the graphical slopes in log–log scale are $-\varepsilon$, $-\alpha$ and $-\delta$ when linear bins are used, but become $-(\varepsilon - 1)$, $-(\alpha - 1)$ and $-(\delta - 1)$ when logarithmic bins are used.

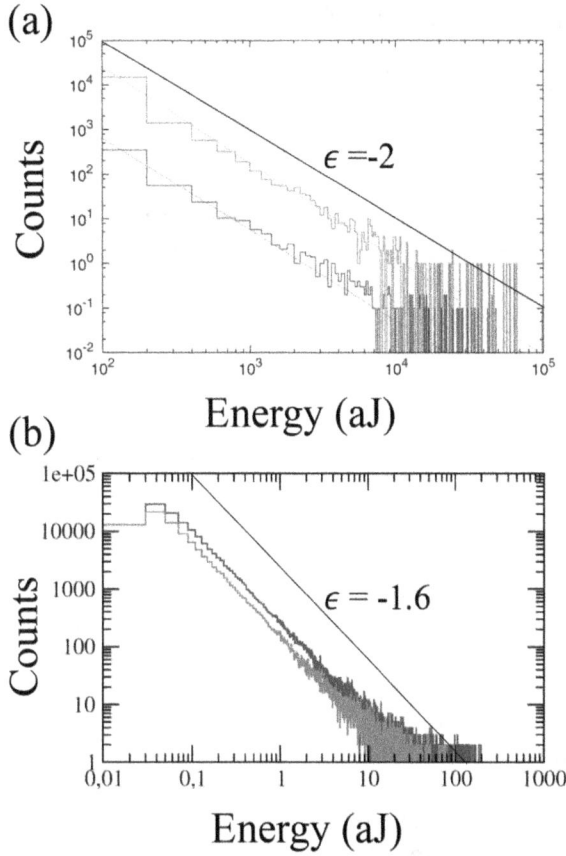

Figure 8.5. Log–log plots of the histograms revealing the power-law distribution of AE avalanche energies for (a) CuZnAl (transforming to monoclinic structure) and (b) FePd (transforming to tetragonal structure).

It should also be noted that, when avalanche sizes are characterized from the amplitudes A (typically in dB), which corresponds to a logarithmic measure of the maximum voltage V, the corresponding distribution is an exponential of the form

$$g_A(A)dA \sim e^{-(\alpha-1)A}dA, \qquad (8.3)$$

which again is revealed by a linear decaying behaviour in linear–log scales.

The experimental determination of the critical exponents is difficult and usually subjected to large error bars. The first requirement to obtain good estimations is to record a large number of events. In this sense it must be noted that reasonable catalogues should contain about 10^5–10^6 hits. These catalogues can be easily obtained when studying the more athermal samples driven at low-temperature rates. When the driving rate is too fast, avalanches overlap and then one obtains fewer signals and deformed histograms with an excess of large avalanche energies, and durations.

In addition, the measurement of the avalanche properties shows different problems [26]:

- When studying the amplitude distribution $g_V(V)$ (or $g_A(A)$) it should be taken into account that low-amplitude signals are sharply cut because of the experimental threshold. High amplitudes can also be affected by the saturation of the amplifier.

- Avalanche energies, E, usually display cleaner power-law distributions extending more decades. This is because energies are time integrated quantities. Consequently, they are less affected by the wave deformations during the propagation through the sample than avalanche amplitudes. The effect of the unavoidable experimental threshold on the avalanche energy distribution $g_E(E)$ is not a sharp cut-off, but a bending in the low energy region of the histograms.

- The most difficult measurements are avalanche durations D since the exact definition of t_{start} and t_{end} is very sensitive to the threshold and to the details of the avalanche temporal profiles also called avalanche shapes. Theoretical works on avalanche criticality postulate a universal function $\hat{\Phi}(x)$ that allows to scale the shape of the individual avalanches $V(t) \sim V_{max}\hat{\Phi}(t/T)$, but distortions due to the response of the AE sensors make it difficult to determine [26]. Recent studies [27, 28] suggest that the measurement of the rise time of the signals (time from t_{start} to the time at which V_{max} is reached) would be more appropriate to determine the true avalanche duration than the hit duration D.

Many years ago [10, 17, 29] it was already revealed that the values of the exponents have a certain degree of universality and take values that depend on the symmetry of the martensitic phase, as shown in table 8.1. For instance, alloys Cu–Al–Zn, Cu–Al–Be and Ni–Al transforming to a monoclinic structure show the values $\varepsilon = 2.0 \pm 0.2$ and $\alpha = 3.0 \pm 0.2$. Cu–Al–Ni and Cu–Al–Mn alloys displaying transitions to an orthorhombic structure show a smaller $\alpha = 2.4 \pm 0.1$ exponent, and the Fe–Pd alloys transforming to a tetragonal structure show even smaller ones $\varepsilon = 1.6 \pm 0.1$ and $\alpha = 2.0 \pm 0.3$. In general, when the low-temperature structure has less variant multiplicity, the exponents are smaller.

This general result suggests that the critical exponents reflect the number of available paths in the metastable free energy landscape connecting the high-

Table 8.1. Table showing the values of the critical exponents corresponding to alloys transforming to different martensitic structures

M-symmetry	Multiplicity	ε	α
Monoclinic	12	2.0 ± 0.2	3.0 ± 0.2
Orthorhombic	6	—	2.4 ± 0.1
Tetragonal	3	1.6 ± 0.1	2.0 ± 0.3

temperature phase with the low temperature phase. The more available paths (the more variant multiplicity), the smaller is the ratio of large to small avalanches thus increasing the power-law exponent.

An important problem, common to almost any avalanche analysis is related to the fitting method used to determine the power-law exponents from the hit catalogues. Least square fitting of a straight line to the log–log plot of the histogram is known to be a very bad choice. Instead, maximum likelihood methods should be used [30]. This is the reason why some of the exponents determined in early experiments with least square fitting methods with very large error bars were later revised and/or corrected.

In addition to the univariate distributions $g_E(E)$, $g_V(V)$ and $g_D(D)$, one can also study the existence of statistical correlations between these three quantities. This is usually done [24, 25] by representing the so-called $V-E$ or $V-D$ maps. These are plots in log–log scale of the recorded amplitudes V against the corresponding energies E or the amplitudes V against the duration D [26]. These cloud maps (in fact they are representations of bivariate probability densities $g(V, E)$ and $g(V, D)$) exhibit sharp crest-like behaviour indicating the existence of strong statistical correlations. Moreover the crests tend to be clearly linear in log–log scales thus allowing new critical exponents z and x to be defined from

$$E \sim V^z$$
$$V \sim D^x. \tag{8.4}$$

By imposing that the univariate power-law distributions correspond to integration of the bivariate distributions, the exponents z and x can be related to the ε, α and δ exponents by the scaling relations

$$\alpha - 1 = z(\varepsilon - 1) = x(\delta - 1). \tag{8.5}$$

Typically one can easily find $z \gtrsim 2$ which is compatible, within error bars, with the scaling relations and the measured values of α and ε. As has been already mentioned, the determination of the avalanche duration is much more sensitive to experimental deviations. Thus, the values of the exponents δ and x have not been systematically determined for the different MTs, especially for those in which alloys have higher attenuation and, thus, the statistics of avalanches is poorer.

Some of the early values δ and x, determined long ago still using the least squares fitting methods (for instance $\delta = 3.5 \pm 0.8$ and $x = 1.0 \pm 0.1$ for monoclinic Cu–Zn–Al [24]), were, in fact, compatible with the scaling relations. However, it was realized later [10] that these experimental exponent values were incompatible with the universal shape scaling hypothesis. In fact a simple integration of the avalanche shape renders a scaling relation $z = 2 + \frac{1}{x}$. The value $z \gtrsim 2$ would imply very large values of x that are never found. The problem was suggested to be related to the poor determination of avalanche duration, as discussed above [26]. This yet unsolved contradiction is in fact part of the recently called energy–duration–size enigma [27]. The authors of [27] have proposed that true avalanche duration would be better estimated from the AE raising times rather than from AE duration. Systematic measurement of avalanche duration correcting for the deformations due to the convolution with the AE sensors [26] will help in order to elucidate the validity of the universal shape scaling hypothesis for avalanches in martensitic transformations.

It should also be mentioned that in more recent works, other sources of distortions of this general criticality scenario have been found. For instance in reference [31] it was shown that the existence of elastic hardening in Cu–Al–Be reduces the ε critical exponent to $\varepsilon \sim 1.6$ despite the fact that the transformation occurs from cubic to monoclinic symmetries. This change has also been suggested to be accompanied by a reduction of the number of allowed martensitic variants, thus reducing effectively the number of transformation paths through the free energy landscape.

On the other hand, E–D maps can be used to discriminate avalanches originating from different sources, which is a frequent problem in AE avalanche analysis. For instance, in porous NiTi samples the MT was studied under compression. This resulted in a rather difficult analysis since AE signals corresponding to the pore collapses were overlapping with those generated from the MT. In this particular case of porous NiTi, the study of the $E-D$ maps allowed one to design specific filters to separate the true MT signals. When the two sources cannot be easily separated one then finds probability distributions that correspond to damped power-laws or mixtures of power-laws [32].

Finally, within this section on thermal driven MTs, it should be mentioned that the existence of Omori-like temporal correlations between consecutive AE avalanches [33] have also found in MT. This has been achieved by performing statistical analysis of the waiting times separating consecutive avalanches above a certain amplitude or energy threshold. Interestingly, the result is very clear for AE avalanches but it is also observed (less clearly) for calorimetric avalanches detected using high sensitivity calorimetry.

This kind of temporal correlations is parallel to the existence of aftershocks found in the study of earthquakes. This means that avalanches do not occur randomly as expected in a Poisson process but rather there is a certain attraction that favours small avalanches to occur closer (in time and temperature) to large avalanches. This suggests a sort of avalanche arrangement in which after the growth of a big martensitic domain, smaller domains compatible with the crystallographic orientations grow to fill in the nearby yet not transformed spaces.

8.2.2.2 Mechanical driving experiments

Martensitic transition can be induced at room temperature by applying a mechanical external force (when the transition temperature is very low the stress necessary can be very large). This includes uniaxial extension, uniaxial compression and non-homogeneous stresses such as those in flexure or torsion. In such experiments the austenitic cubic phase transforms to a phase with a combination of martenstic variants which are favoured by the direction of the applied stress. This selection of variants will not be very relevant in the case of polycrystals in which grains are randomly oriented with respect to the external applied force, but the orientation will be crucial in single crystals.

The fact that MTs are athermal to a good approximation and, thus, do not occur in equilibrium has the consequence that specific features of the driving mechanism influence the AE avalanches. For instance, in uniaxial loading experiments, it is expected that it will not be equivalent to drive the transition by imposing an elongation ramp with a fixed velocity than imposing a force ramp with a fixed force rate. The differences, of course, reveal the non-equilibrium nature of the metastable transformation path. In addition, the effect of the driving rate should also be taken into account. Very fast driving can have two different effects. On the one hand, there might be an overlap of the AE avalanches. On the other hand, if the latent heat released during the transition cannot be dissipated, local temperature changes are expected to occur. For very slow driving rates a different problem will arise: thermally activated transformation events might occur in the case of samples which are not very athermal, which may overlap affecting the statistics of the avalanches induced by loading.

Experiments with Cu–Zn–Al single crystals revealed that the metastable path is indeed different under stress (force) and strain (elongation) control and also depends on the driving rate [34]. Further results prove that the AE energy and amplitude avalanche distributions were also dependent on the driving mechanism (figure 8.6)

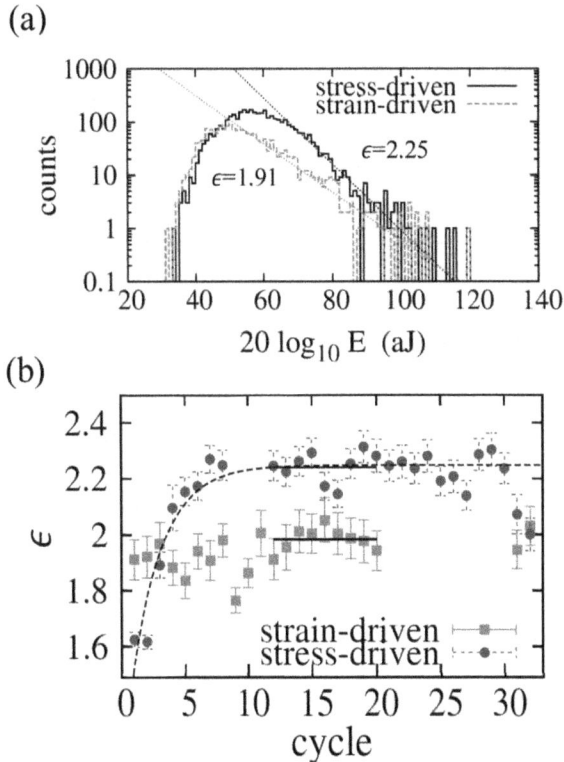

Figure 8.6. (a) Example of power-law energy distributions in stress-driven and strain driven experiments in a Cu–Zn–Al samples after the sample has been cycled many times. (b) Evolution of the fitted ε exponent with cycling. (Reproduced with permission from [36]. Copyright 2009 the American Physical Society.)

[35, 36]. For stress (force) driven experiments the AE distributions reveal the existence of a learning process during the first 5–10 cycles. This is very similar to what was found in thermal driving experiments [35], but the learning process is not present in the case of strain driven experiments. This is due to the fact that one is imposing a control in the sample length that is constraining the freedom of the system to find for some minima in the metastable landscape.

With both driving mechanisms the amplitude and energy probability distributions show a power-law (critical) shape, but the exponents for the strain driven case are smaller. This is in agreement with the hypothesis that the elongation constraint limits the allowed variants (or its allowed size), preventing one from finding an optimal transformation path. This favours the relative occurrence of large events compared to small events, decreasing the critical exponent.

In addition to the uniaxial tensile experiments, AE avalanches have also been studied under other driving mechanical fields. More recently, AE avalanches have also been studied in flexure experiments [37]. In this case, as the sample is bent the transition takes place in a small region that displaces along the sample driven by the bending displacement. In such a situation, energy distribution of AE avalanches has been found to loose its fat tail character and becomes a log-normal distribution.

8.2.2.3 Location experiments

As explained in section 8.2.1, by using several AE sensors it is possible to estimate the location of the source positions. In the case of elongated samples under uniaxial tensile driving (or under flexure) the problem can be simplified to a 1D problem. Location analysis then can be performed with only two (or three) sensors. For instance, in the Cu–Zn–Al alloy [16] the location of the AE events confirmed the fact that the transformation paths were different under elongation (hard device) or force control (soft device). In the case of elongation control, it was shown that martensite fronts exhibit dynamic interactions that prevent them from advancing simultaneously. When the most energetically favourable front advances, the others are arrested due to the driving constraint. In force driven experiments (soft driving) this phenomenon does not occur.

More recently efforts have been directed to correlate the AE location with other methods that allow one to monitor where the transition is taking place. For instance, 1D location experiments have been performed on long single crystal wires of CuAlNi (see figure 8.7(a) and (b)) [14]. In this particular case the located AE events were correlated with the front paths detected by infrared methods. Similarly, for elongated CuZnAl the location of the AE events was also highly correlated with the local strain avalanches detected on the surface by an optical method that measures distortions of a fine grid attached to the sample surface [13]. Results similar to those found in uniaxial tensile experiments have also been found in three-point bending experiments of a CuZnAl sample under flexure [37] (see figure 8.7 (c)).

Finally we should remark that 2D location in small samples presents many more difficulties. Using four sensors 2D location was achieved in the case of thermally induced MT in Ni–Mn–Ga exhibiting qualitative positive correlation with the activity detected by optical means [12].

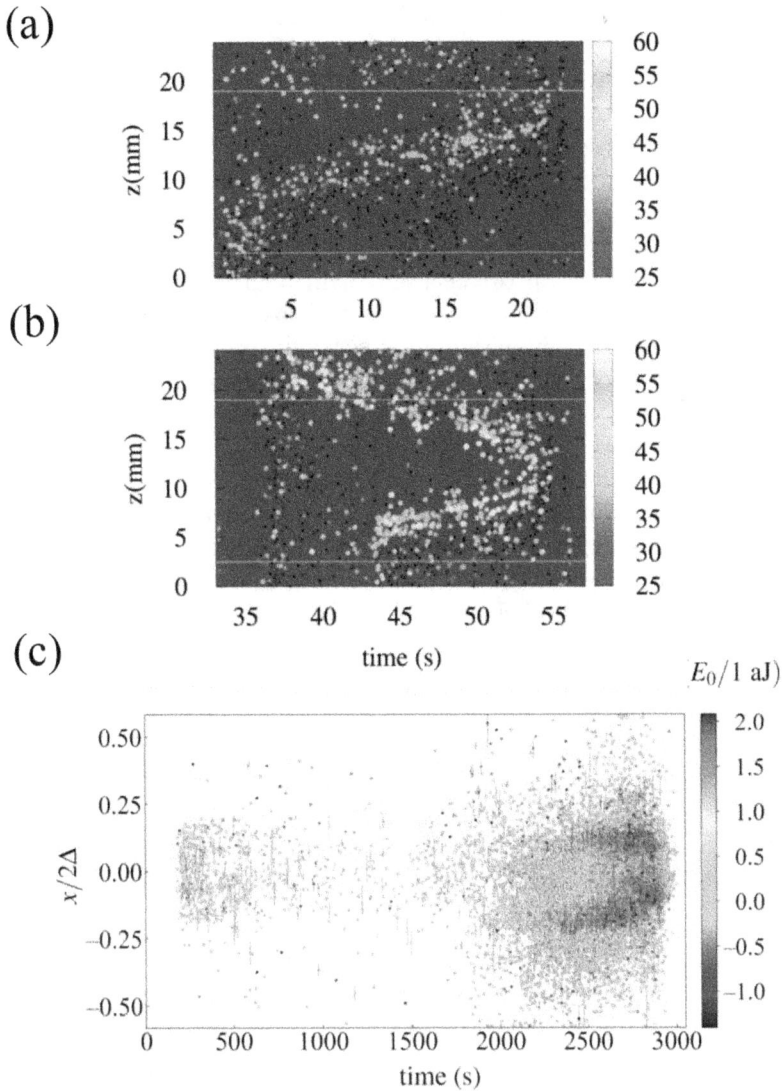

Figure 8.7. (a) Location of AE events as a function of time in a vertical cylindrical sample of Cu–Al–Ni subjected to loading at 0.1 mm s^{-1}. The plot reveals the formation of a martensitic front in the bottom part of the sample that crosses it from bottom to top. The colour indicates the amplitude of every event. (b) The same data but obtained during unloading when the sample transforms back to cubic austenite. In this case, two fronts form in the upper and lower edge of the sample that move towards the centre and merge. (c) Location of AE events in a horizontal sample of Cu–Al–Ni that is subjected to three-point bending and unbending. The colour indicates the source energy of every event.

8.2.2.4 Magnetic shape-memory alloys

A number of shape-memory alloys are ferromagnetic in the region where they undergo the MT. In these materials the transition is accompanied by a change of their magnetic properties [38, 39] and, depending on the nature of this change, two main families are usually considered. In the so-called magnetic shape-memory alloys the magnetocrystalline anisotropy strongly increases while the magnetization only changes slightly at the transition. Instead, in metamagnetic shape-memory alloys the magnetocrystalline anisotropy remains essentially unchanged while the magnetization undergoes a big change. This big change of the magnetization is a consequence of the fact that the transition is magnetostructural, which means that martensitic and magnetic transitions occur simultaneously.

Magnetic shape-memory alloys show a complex martensitic microstructure constituted of a combination of the usual twin related martensitic variants and magnetic domains. In general, 180°-domains form inside martensitic variants. Then, magnetic moments in neighbouring twin-variants are perpendicular to a good approximation (90°-domain walls). This is illustrated in figure 8.8.

Interestingly, in this class of alloys the shape-memory effect can be induced not only by means of an applied stress but also by means of an applied magnetic field since, due to the high magnetic anisotropy, Zeeman energy is minimized by rotation of martensitic variant instead of the usual mechanism based on the rotation of magnetic moments. Therefore, this magnetostrictive mechanism permits magnetically inducing very large deformations that can be erased by warming up the material above the MT. This behaviour defines the magnetic shape-memory effect.

In contrast, metamagnetic shape-memory alloys show magnetic superelasticity, which is essentially based on the possibility of magnetically inducing the magneto-structural MT. Therefore, in both magnetic and metamagnetic shape-memory alloys huge deformations can be induced by an applied magnetic field. From a practical perspective, metamagnetic materials have the advantage that the deformation mechanism based on magnetic superelasticity can support a much larger work output [41]. Overall, magnetic and metamagnetic shape-memory alloys are multifunctional materials that open up new opportunities in the field of applications.

Figure 8.8. Twin-related variants with magnetic domains inside the variants in a Ni–Mn–Ga magnetic shape-memory alloy. 180°- and domains oriented at 90° to each other are indicated. (Adapted with permission from [40]. Copyright 2004 AIP Publishing.)

Most magnetic and metamagnetic shape-memory alloys are Heusler materials that display an L2$_1$ structure at high temperatures. The stoichiometric Ni$_{50}$Mn$_{25}$Ga$_{25}$ alloy is the prototypical magnetic shape-memory alloy. Within the family of Heusler alloys, metamagnetic shape-memory alloys are non-stoichiometric compounds of composition N$_{50}$Mn$_{25+x}$Z$_{25-x}$ with $x > 5$ and Z being period 5 non-magnetic elements such as In, Sn or Sb. In both classes of alloys to a first approximation the magnetism arises from magnetic moments that are localized at Mn-atoms and are coupled through an oscillatory RKKY-type exchange mediated by the conduction electrons. The particular characteristics of this exchange mechanism allow us to explain that in alloys with an excess of Mn-atoms, closer-neighbouring Mn–Mn pairs can become antiferromagnetically coupled in the martensitic phase, which is the essential feature that permits understanding the metamagnetic behaviour of non-stoichiometric alloys. The magnetic state of the martensitic low temperature phase has been suggested to be sensitive to the Z-element. It has been most often proposed that it is paramagnetic but a mixed ferro-antiferromagnetic state or even some kind of ferrimagnetic order have also been reported.

Due to the fact that in ferromagnetic alloys the MT involves a change of the magnetic properties of the material, elastic and magnetic avalanches are expected to occur during the transition in both magnetic and metamagnetic shape-memory alloys. These avalanches can be detected separately using AE and Barkhausen detectors. This class of measurements has been reported in [42] for a Ni$_{50.5}$Mn$_{29.5}$Ga$_{20.0}$ magnetic shape-memory alloy and in [43] for a Ni$_{45}$Co$_5$Mn$_{36.6}$In$_{13.4}$ metamagnetic shape-memory alloy. For the two alloys, both elastic and magnetic avalanches show power-law behaviour. In the magnetic shape-memory alloy the magnetic exponent seems to be smaller than the elastic exponent, which might be consistent with the fact that magnetic domains occur inside martensitic variants. This could explain as well the fact that magnetic avalanches seems to be affected by an exponential cut-off imposed by the length scale associated with variant boundaries. Instead, elastic and magnetic exponents are the same in the metamagnetic alloy, which could indicate that both structural and magnetic transitions have the same origin with avalanche events occurring independently.

8.3 Modelling

The problem of the nucleation of a crystallographic phase into a matrix with a given symmetry dates back to the work of Eshelby [44, 45], Roitburd [46, 47] and Khachaturyan [48], who considered inclusions in a parent matrix with different shapes and properties [49]. In this problem, strain accommodation and energy minimization lead to elastic heterogeneities such as habit planes between parent and product phases or twin domain walls between product phases. A number of approaches have been developed to study the microstructure in martensitic materials, all based in the minimization of a free energy with the constraint of lattice coherency.

In the approach of Khachaturyan [48] the order parameters (OPs) of the phase transition are assumed to be the amplitudes of the deformation modes leading to the martensitic phase, using different order parameters for each one of the variants of the product phase. The total deformation is written as the summation of the transformation strain and an elastic strain that is introduced to guarantee lattice coherency.

Other approaches use the total strain or deformation gradient as the OP. The differences lie in whether the calculation uses displacements or strains as the variables of the model, and how lattice coherency is enforced. They also differ in whether the approach holds for finite deformation or only in the small strain limit.

The approach due to Ball and James [50] holds for finite deformations. It uses the deformation gradient as the OP and lattice coherency is enforced using the Hadamard jump condition. Another approach is to use the total strain as the OP and enforce continuity across the interfaces via the Saint-Vénant compatibility equation [51–53]. In the latter case most of the results have been obtained in the approximation of geometric linear elasticity. However, it has been shown how finite deformations may be treated within the scope of the strain-only-based approach by ensuring compatibility of displacement gradients [54].

8.3.1 Deformation gradient, strain and compatibility

In this subsection, we summarize different approaches to describe the lattice deformation in a proper MT. The physical magnitude that describes such a deformation is the strain tensor, ε. As this tensor is defined in terms of displacement gradients one can use the displacement field, $\mathbf{u}$, as the variables of the model. In this case, its components, u_i, are independent variables. Instead, one can take the displacement gradients as the variables of the model, or equivalently, the components of the deformation gradient, defined as

$$F_{ij} = \frac{\partial u_i}{\partial X_j} + \delta_{ij},$$

(8.6)

where X_j is the jth component of the position vector of a volume element in the reference configuration and δ_{ij} is the Kronecker delta. In such a case, the components of the deformation gradient are not independent variables as they must satisfy an integrability condition leading to a single-valued continuous displacement field with no cracks or voids.

As the curl of a gradient must vanish, the compatibility condition simply reads,

$$\nabla \times F = 0.$$

(8.7)

Compatibility also yields the Hadamard jump condition [1], which is a matching condition that needs to be satisfied at the interface between different variants of the martensitic phase,

$$QF_1 - F_2 = \mathbf{v} \otimes \mathbf{n},$$

(8.8)

where F_1 and F_2 are the deformation gradient at the two variants, Q is a rotation matrix, $\mathbf{v} = (v_1, v_2)$ is a vector and $\mathbf{n} = (x, y)$ is the normal to the interface.

In the approximation of geometric linear elasticity, compatibility yields the Saint-Vénant condition which in three dimensions reads [53, 54]

$$\nabla \times (\nabla \times \varepsilon)^T = 0. \tag{8.9}$$

In two dimensions this condition reduces to the single equation [54]

$$\frac{\partial^2 \varepsilon_{xx}}{\partial y^2} + \frac{\partial^2 \varepsilon_{yy}}{\partial x^2} = 2\frac{\partial^2 \varepsilon_{xy}}{\partial x \partial y}, \tag{8.10}$$

where ε_{ij} are components of the strain tensor. Moreover, compatibility relates the rotation of the lattice,

$$\omega = \frac{1}{2}\left(\frac{\partial u_x}{\partial y} - \frac{\partial u_y}{\partial x}\right), \tag{8.11}$$

to the linear strain fields through the equation [54]

$$\nabla^2 \omega + \left(\frac{\partial^2}{\partial x^2} - \frac{\partial^2}{\partial y^2}\right)\varepsilon_{xy} = \frac{\partial^2(\varepsilon_{xx} - \varepsilon_{yy})}{\partial x \partial y}. \tag{8.12}$$

In the present chapter we use the total strain as the order parameter of martensitic phase transitions and present results using both displacements or strains as the variables of the model. In the latter case, compatibility is enforced using the Saint-Vénant equation.

8.3.2 Modelling the free energy of martensites

The characteristic features of a martensitic transformation are the symmetries of the parent and martensitic phases, in particular their group–subgroup relationship which leads to the shape-memory effect. Ginzburg–Landau models are thus appropriate to describe this kind of phase transitions.

Ginzburg–Landau models are expansions of the free energy density of the parent phase in terms of deformations leading to the martensitic phase. The components or linear combinations of the components of the strain tensor that describe these deformations are denoted as the OPs of the phase transition. The free energy associated with these deformations needs to be equal for all variants of the martensitic phase that are equivalent by symmetry. Thus, the expansion of the free energy density is written in terms of polynomials of components of the strain tensor which are invariant to the symmetry operations of the parent phase. Usually, the invariant polynomials are written in terms of the so-called symmetry adapted strains. In two dimensions, except for the rectangle-to-oblique transformation, these are [53]

$$e_1 = \frac{c_1}{2}(\varepsilon_{xx} + \varepsilon_{yy}),$$

$$e_2 = \frac{c_2}{2}(\varepsilon_{xx} - \varepsilon_{yy}), \qquad (8.13)$$

$$e_3 = \frac{c_3}{2}(\varepsilon_{xy} + \varepsilon_{yx}) = c_3\varepsilon_{xy},$$

where c_1, c_2 and c_3 are constants which depend on the symmetry of the phase transformation.

In what follows, we develop models for the square-to-oblique and square-to-rectangle transformations that will serve as prototypes to study strain microstructure and avalanches in martensitic transformations.

8.3.2.1 The square-to-oblique transformation

The square-to-oblique transformation has two independent one component OPs: the deviatoric strain, e_2, and shear strain, e_3, (figure 8.9(a)) whereas the compression strain, e_1, is a non-OP and $c_1 = c_2 = \sqrt{2}$, and $c_3 = 1$.

For this first-order phase transition the free energy density including all symmetry allowed terms up to sixth order in the OPs takes the form,

$$f_{\rm OP}^{\rm SO} = f_2 + f_3 + f_{23}, \qquad (8.14)$$

with

$$f_2 = \frac{1}{2}A_2e_2^2 + \frac{1}{4}\beta_2e_2^4 + \frac{1}{6}\gamma_2e_2^6,$$

$$f_3 = \frac{1}{2}A_3e_3^2 + \frac{1}{4}\beta_3e_3^4 + \frac{1}{6}\gamma_3e_3^6, \qquad (8.15)$$

$$f_{23} = \frac{1}{4}\beta_{23}e_2^2e_3^2 + \frac{1}{6}\gamma_{23}e_2^4e_3^2 + \frac{1}{6}\gamma_{32}e_2^2e_3^4,$$

(a) (b)

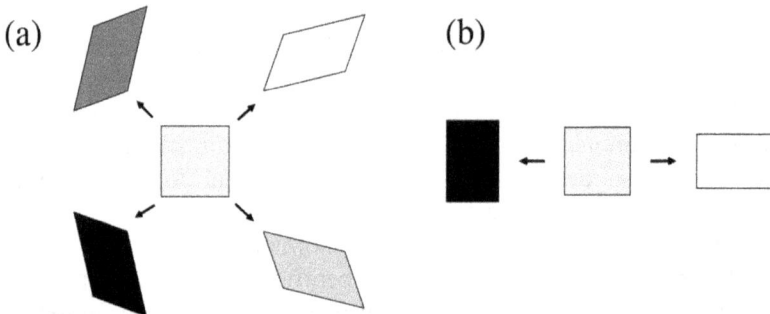

Figure 8.9. Schematic representation of (a) the square-to-oblique transformation with four possible variants of the low symmetry phase and (b) the square-to-rectangle transformation with two possible variants of the low symmetry phase.

where the second order elastic constants, $A_2 = C_{11} - C_{12}$ and $A_3 = 4C_{44}$ are assumed to vary linearly with temperature, T,

$$A_2 = a_2(T - T_c),$$
$$A_3 = a_3(T - T_c),$$
(8.16)

and T_c is the stability limit of the austenite phase.

In addition to the expansion of the free energy density in terms of the OPs of the phase transformation, we also include the lowest order terms in the symmetry adapted strains that are not OPs of the phase transformation. Thus we include the free energy term

$$f_{\text{non-OP}}^{\text{SO}} = \frac{1}{2} A_1 e_1^2.$$
(8.17)

Because compatibility needs to be satisfied, in the approximation of geometric linear elasticity the volumetric strain, e_1, can be written entirely in terms of the deviatoric, e_2, and shear, e_3, strains. From equation (8.10) and taking $c_1 = c_2 = \sqrt{2}$ and $c_3 = 1$ we obtain

$$\nabla^2 e_1 - \left(\frac{\partial^2}{\partial x^2} - \frac{\partial^2}{\partial y^2} \right) e_2 = \sqrt{8} \frac{\partial^2}{\partial x \partial y} e_3.$$
(8.18)

Thus, in Fourier space the volumetric strain can be written as

$$\hat{e}_1(\mathbf{k}) = \frac{k_x^2 - k_y^2}{k_x^2 + k_y^2} \hat{e}_2(\mathbf{k}) + \frac{\sqrt{8} k_x k_y}{k_x^2 + k_y^2} \hat{e}_3(\mathbf{k}),$$
(8.19)

where $\hat{e}_m(\mathbf{k})$ stands for the Fourier transform of $e_m(\mathbf{r})$. Using this relation, the contribution of the volumetric strain to the free energy becomes,

$$\mathscr{F}_{\text{non-OP}}^{\text{SO}} = \int d\mathbf{r} f_{\text{non-OP}}^{\text{SO}} = \frac{1}{(2\pi)^2} \frac{1}{2} A_1 \int d\mathbf{k} \hat{e}_1(\mathbf{k}) \hat{e}_1(-\mathbf{k})$$

$$= \frac{1}{(2\pi)^2} \frac{1}{2} \int d\mathbf{k} \left[\hat{e}_2(\mathbf{k}) \hat{U}_{22}(\mathbf{k}) \hat{e}_2(-\mathbf{k}) + \hat{e}_2(\mathbf{k}) \hat{U}_{23}(\mathbf{k}) \hat{e}_3(-\mathbf{k}) \right.$$
(8.20)
$$\left. + \hat{e}_3(\mathbf{k}) \hat{U}_{32}(\mathbf{k}) \hat{e}_2(-\mathbf{k}) + \hat{e}_3(\mathbf{k}) \hat{U}_{33}(\mathbf{k}) \hat{e}_3(-\mathbf{k}) \right],$$

with

$$\hat{U}_{22}(\mathbf{k}) = A_1 \frac{\left(k_x^2 - k_y^2\right)^2}{\left(k_x^2 + k_y^2\right)^2},$$

$$\hat{U}_{23}(\mathbf{k}) = \hat{U}_{32}(\mathbf{k}) = A_1 \frac{\left(k_x^2 - k_y^2\right)\sqrt{8} k_x k_y}{\left(k_x^2 + k_y^2\right)^2},$$
(8.21)

$$\hat{U}_{33}(\mathbf{k}) = A_1 \frac{8 k_x^2 k_y^2}{\left(k_x^2 + k_y^2\right)^2}.$$

This free energy term can be written as

$$\mathcal{F}_{\text{non-OP}}^{\text{SO}} = \frac{1}{2} \int d\mathbf{r} \int d\mathbf{r}' \left[e_2(\mathbf{r}) U_{22}(\mathbf{r}-\mathbf{r}') e_2(\mathbf{r}') \right.$$
$$\left. + 2e_2(\mathbf{r}) U_{23}(\mathbf{r}-\mathbf{r}') e_3(\mathbf{r}') + e_3(\mathbf{r}) U_{33}(\mathbf{r}-\mathbf{r}') e_3(\mathbf{r}') \right], \tag{8.22}$$

where $U_{nm}(\mathbf{r})$ is the inverse Fourier transform of $\hat{U}_{nm}(\mathbf{k})$. Thus, this term introduces long-range anisotropic interactions between the OPs at different locations of the system and is the cause of the orientation of strain modulations such as interfaces in specific directions.

Finally, we also include a dependence of the free energy density on the gradients of the OPs,

$$f_{\text{grad}}^{\text{SO}} = \frac{1}{2}\kappa_2 |\nabla e_2|^2 + \frac{1}{2}\kappa_3 |\nabla e_3|^2 + \frac{1}{2}\kappa_{23}\left(\frac{\partial e_2}{\partial x}\frac{\partial e_3}{\partial y} + \frac{\partial e_2}{\partial y}\frac{\partial e_3}{\partial x} \right). \tag{8.23}$$

Thus, the total free energy density is written as the summation of three terms,

$$f^{\text{SO}} = f_{\text{OP}}^{\text{SO}} + f_{\text{non-OP}}^{\text{SO}} + f_{\text{grad}}^{\text{SO}}. \tag{8.24}$$

8.3.2.2 The square-to-rectangle transformation

The square-to-rectangle transformation is driven by the deviatoric strain component, e_2, (figure 8.9(b)) whereas the compression, e_1, and shear, e_3, strains are non-OPs and $c_1 = c_2 = \sqrt{2}$ and $c_3 = 1$.

Expanding the elastic free energy up to sixth order in the OP we obtain

$$f_{\text{OP}}^{\text{SR}} = \frac{1}{2}a_2(T - T_c)e_2^2 + \frac{1}{4}\beta e_2^4 + \frac{1}{6}\gamma e_2^6, \tag{8.25}$$

and the non-OP and gradient terms are

$$f_{\text{non-OP}}^{\text{SR}} = \frac{1}{2}A_1 e_1^2 + \frac{1}{2}A_3 e_3^2,$$
$$f_{\text{grad}}^{\text{SR}} = \frac{1}{2}\kappa_2 |\nabla e_2|^2. \tag{8.26}$$

Integrating the non-OP free energy density in Fourier space we obtain,

$$\mathcal{F}_{\text{non-OP}}^{\text{SR}} = \frac{1}{(2\pi)^2} \int d\mathbf{k} \left[\frac{1}{2}A_1 \hat{e}_1(\mathbf{k})\hat{e}_1(-\mathbf{k}) + \frac{1}{2}A_3 \hat{e}_3(\mathbf{k})\hat{e}_3(-\mathbf{k}) \right]. \tag{8.27}$$

Using compatibility (equation (8.19)) this free energy can be written as

$$\mathcal{F}_{\text{non-OP}}^{\text{SR}} = \frac{1}{(2\pi)^2}\frac{1}{2} \int d\mathbf{k} \left\{ \hat{e}_2(\mathbf{k})\hat{U}_{22}(\mathbf{k})\hat{e}_2(-\mathbf{k}) + \hat{e}_2(\mathbf{k})\hat{U}_{23}(\mathbf{k})\hat{e}_3(-\mathbf{k}) \right.$$
$$\left. + \hat{e}_3(\mathbf{k})\hat{U}_{32}(\mathbf{k})\hat{e}_2(-\mathbf{k}) + \hat{e}_3(\mathbf{k})\left[\hat{U}_{33}(\mathbf{k}) + A_3 \right]\hat{e}_3(-\mathbf{k}) \right\}, \tag{8.28}$$

where $\hat{U}_{nm}(\mathbf{k})$ are given in equation (8.21). This free energy can be minimized analytically with respect to the non-OP component e_3. This yields an expression for each non-OP strain component as a function of the OP [51],

$$\hat{e}_1(\mathbf{k}) = \frac{A_3\left(k_x^2 - k_y^2\right)\left(k_x^2 + k_y^2\right)}{A_1 8k_x^2 k_y^2 + A_3\left(k_x^2 + k_y^2\right)^2}\hat{e}_2(\mathbf{k}),$$

$$\hat{e}_3(\mathbf{k}) = -\frac{A_1\left(k_x^2 - k_y^2\right)\sqrt{8}\,k_x k_y}{A_1 8k_x^2 k_y^2 + A_3\left(k_x^2 + k_y^2\right)^2}\hat{e}_2(\mathbf{k}),$$

(8.29)

and the contribution of these strain components to the free energy can be written as

$$\mathscr{F}_{\text{non-OP}}^{\text{SR}} = \frac{1}{(2\pi)^2}\frac{1}{2}\int d\mathbf{k}\frac{A_1 A_3\left(k_x^2 - k_y^2\right)^2}{A_1 8k_x^2 k_y^2 + A_3\left(k_x^2 + k_y^2\right)^2}\hat{e}_2(\mathbf{k})\hat{e}_2(-\mathbf{k}),$$

(8.30)

where it becomes clear that this energy is minimized when the OP variations are oriented in $\langle 11 \rangle$ directions.

8.3.2.3 Disorder in the Ginzburg–Landau model
As important as the symmetries of the phase transition, disorder is essential to model avalanches in martensitic transformations. Disorder is introduced into the model through a spatial variation of the stability limit of the austenite phase, T_c, which is written as

$$T_c(\mathbf{r}) = \overline{T}_c[1 + \eta(\mathbf{r})],$$

(8.31)

where $\overline{T}_c$ is the volume averaged value of $T_c(\mathbf{r})$, and $\eta(\mathbf{r})$ is a Gaussian-distributed random variable with zero mean, with standard deviation ζ, and spatially correlated with an exponential pair correlation function with correlation length ξ,

$$\langle \eta(\mathbf{r})\eta(\mathbf{r}') \rangle = \zeta^2 \exp(-|\mathbf{r} - \mathbf{r}'|/\xi).$$

(8.32)

The disorder arises from the strong dependence of the temperature of the phase transformation and the associated stability limits on the composition of these materials. Consequently, the unavoidable fluctuations of the composition lead to a spatial variation of the transition temperature. Alternatively, disorder can be introduced as quenched stress arising from point defects that couples to the OPs of the phase transformation or simply to the compression strain, e_1 [52, 55].

8.3.3 Model dynamics
The dynamical equation of a mechanical system is Newton's second law, which in a continuum is written as

$$\rho \ddot{u}_i = \rho g_i + \sum_j \frac{\partial \sigma_{ij}}{\partial x_j},$$

(8.33)

where the dots stand for time derivative, ρ is the density of the deformed configuration, g_i is the ith component of an external force (per unit mass), σ_{ij} are the components of the Cauchy stress tensor, and x_j is the jth component of the position vector of a volume element in the deformed configuration. Using more appropriate variables, this dynamical equation can be written as [56]

$$\rho_0 \ddot{u}_i = \rho_0 g_i + \sum_j \frac{\partial \tau_{ij}}{\partial X_j}, \qquad (8.34)$$

where ρ_0 is the density of the undistorted medium, τ_{ij} are the components of the first Piola–Kirchhoff stress tensor and X_j is the jth component of the position vector of a volume element in the reference undistorted configuration. The first Piola–Kirchhoff stress tensor is the work conjugate of the deformation gradient. Thus, this stress tensor can be obtained as the functional derivative of the free energy with respect to the components of the deformation gradient,

$$\tau_{ij} = \frac{\delta \mathscr{F}}{\delta F_{ij}}. \qquad (8.35)$$

Using that the free energy has been written as a functional of the symmetry adapted strains we have

$$\tau_{ij} = \sum_{n=1}^{3} \frac{\delta \mathscr{F}}{\delta e_n} \frac{\partial e_n}{\partial F_{ij}}. \qquad (8.36)$$

Moreover, as the free energy is a functional of the symmetry adapted strains of the form

$$\mathscr{F} = \int f(e(\mathbf{r}), \nabla e(\mathbf{r})) d\mathbf{r}, \qquad (8.37)$$

we have

$$\frac{\delta \mathscr{F}}{\delta e_n} = \frac{\partial f}{\partial e_n} - \nabla \cdot \frac{\partial f}{\partial \nabla e_n}. \qquad (8.38)$$

If one uses the displacement field as the variables of the model, its time evolution can be obtained by numerical integration of equation (8.34). A damping stress tensor, $\bar{\tau}$, arising from a Rayleigh potential, R, is also included to dissipate the excess free energy during relaxation to equilibrium,

$$\bar{\tau}_{ij} = \frac{\delta}{\delta \dot{F}_{ij}} \int R \, d\mathbf{X} = \frac{\partial R}{\partial \dot{F}_{ij}}, \qquad (8.39)$$

with

$$R = \frac{1}{2} \sum_{n=1}^{3} \bar{A}_n \dot{e}_n^2, \qquad (8.40)$$

which leads to the damping force,

$$h_i = \frac{1}{\rho_0} \sum_j \frac{\partial \bar{\tau}_{ij}}{\partial X_j}. \tag{8.41}$$

If one uses strains or the deformation gradient as the variables of the model, an appropriate inertial dynamics needs to be derived from equation (8.34). To this end, we differentiate this equation with respect to X_k and obtain a dynamical equation for the displacement gradients,

$$\rho_0 \frac{\partial \ddot{u}_i}{\partial X_k} = \rho_0 \frac{\partial g_i}{\partial X_k} + \sum_j \frac{\partial^2 \tau_{ij}}{\partial X_j \partial X_k}, \tag{8.42}$$

from which we also obtain the dynamic equation for the linear strain tensor. In two dimensions we have

$$\rho_0 \ddot{\varepsilon}_{xx} = \rho_0 \frac{\partial g_x}{\partial X} + \frac{\partial^2 \tau_{xx}}{\partial X^2} + \frac{\partial^2 \tau_{xy}}{\partial X \partial Y},$$

$$\rho_0 \ddot{\varepsilon}_{xy} = \frac{1}{2} \left[\rho_0 \left(\frac{\partial g_x}{\partial Y} + \frac{\partial g_y}{\partial X} \right) + \frac{\partial^2 \tau_{xx}}{\partial X \partial Y} + \frac{\partial^2 \tau_{xy}}{\partial Y^2} + \frac{\partial^2 \tau_{yx}}{\partial X^2} + \frac{\partial^2 \tau_{yy}}{\partial X \partial Y} \right], \tag{8.43}$$

$$\rho_0 \ddot{\varepsilon}_{yy} = \rho_0 \frac{\partial g_y}{\partial Y} + \frac{\partial^2 \tau_{yx}}{\partial X \partial Y} + \frac{\partial^2 \tau_{yy}}{\partial Y^2},$$

where $X = X_x$ and $Y = X_y$, and τ_{ij} is interpreted as the linear stress tensor, which is symmetric. Linear combinations of these equations yield the time evolution equation of the symmetry adapted strains.

An alternative to the inertial dynamics is the relaxational dynamics,

$$\frac{\partial e_n}{\partial t} = -\Gamma_n \frac{\delta \mathscr{F}}{\delta e_n}, \tag{8.44}$$

where Γ_n is a relaxational coefficient. In this case only the time evolution equations for the OPs are considered, and the relationship between the OPs and the non-OP components of the symmetry adapted strains is taken into account in the functional derivative of the free energy. Thus, the relaxational dynamics is a simple steepest descent minimization of the free energy with the constraint of compatibility.

8.3.4 Microstructure of martensite

The nucleation of the martensitic phase within a matrix of the high-temperature phase has the energy cost associated with the elastic distortion of the matrix that occurs due to the mismatch between both phases. Minimization of this energy is the origin of the microstructure of the martensite. The mechanism to reduce the elastic distortion of the high-temperature phase is twinning of the nucleated martensite, which is the simultaneous nucleation of different variants of the martensitic phase [57, 58].

Strain compatibility (equation (8.18)) means that spatial variations of the OP of a phase transition (twinning) require variations of non-OP strain components unless these modulations occur in very specific directions. Thus, minimization of the elastic energy associated with non-OP strain components determines the orientation of the twin boundaries, and to a high extent the morphology of the microstructure of the martensite.

In the square-to-rectangle phase transition, this energy term is proportional to $(k_x^2 - k_y^2)^2$ which is minimized when the twin boundaries are oriented along $\langle 11 \rangle$ directions. Similarly, the energy cost associated with modulations of the shear strain is minimized when they are oriented in $\langle 10 \rangle$ directions, as the non-OP free energy is proportional to $k_x^2 k_y^2$. When modulations of the deviatoric and shear strains occur simultaneously, as in the square-to-oblique transformation, the analysis is more complex. In this case, we obtain the orientation of the twin boundaries from the Hadamard jump condition, which is a matching condition between two homogeneous domains which is equivalent to compatibility.

The deformation gradient in a given domain of the oblique phase can be written as

$$F_1 = \begin{pmatrix} 1 + \mu & \nu \\ \nu & 1 - \mu \end{pmatrix} \tag{8.45}$$

with $\mu = e_2/\sqrt{2}$ and $\nu = e_3$. In another orientational domain with different deviatoric and shear strains the deformation gradient will be

$$F_2 = \begin{pmatrix} 1 - \mu & -\nu \\ -\nu & 1 + \mu \end{pmatrix}. \tag{8.46}$$

In 2D, the rotation matrix Q can be written in a general form as

$$Q = \begin{pmatrix} \cos\theta & \sin\theta \\ -\sin\theta & \cos\theta \end{pmatrix}, \tag{8.47}$$

where θ is the rotation angle of the lattice. In the limit of a small rotation angle the rotation matrix becomes

$$Q = \begin{pmatrix} 1 & \theta \\ -\theta & 1 \end{pmatrix}. \tag{8.48}$$

Thus, in the limit of small strains the Hadamard jump condition (equation (8.8)) can be written as

$$\begin{pmatrix} 2\mu & 2\nu + \theta \\ 2\nu - \theta & -2\mu \end{pmatrix} = \begin{pmatrix} v_1 x & v_1 y \\ v_2 x & v_2 y \end{pmatrix}, \tag{8.49}$$

Figure 8.10. Strain microstructure of an oblique phase at $T = 0.8T_c$ with four orientational domains. The strain field $2e_2(\mathbf{r}) + e_3(\mathbf{r})$ is plotted using a grey scale. Selected domain boundaries are highlighted with lines (see text).

which has the solutions

$$\frac{y}{x} = \frac{\nu \pm \sqrt{\mu^2 + \nu^2}}{\mu}. \tag{8.50}$$

Using the model parameters $a_2 = a_3$, $\beta_2 = \beta_3$, $\gamma_2 = \gamma_3$ and $\beta_{23} = \gamma_{23} = \gamma_{32} = 0$ yields $\mu = \pm\nu/\sqrt{2}$. Thus, the possible values of the orientation angle of the twin boundaries are

$$\Phi = \arctan\left(\frac{y}{x}\right) = \arctan(\pm\sqrt{2} \pm \sqrt{3}) = \begin{cases} \pm 17.6° \\ \pm 72.4° \end{cases}. \tag{8.51}$$

In figure 8.10 we show a strain microstructure of an oblique phase with several boundaries between domains highlighted with colours. Boundaries between domains with the same value of the shear strain correspond to a variation of the deviatoric strain only and are thus oriented along $\langle 11 \rangle$ directions (highlighted with green dashed lines). Boundaries where there is only a variation of the shear strain are oriented along $\langle 10 \rangle$ directions (highlighted with blue dashed lines), and boundaries where there is a variation of both the deviatoric and shear strains have orientation angles $\Phi = -72.4°$ or $\Phi = 17.6°$ for $\mu = -\nu/\sqrt{2}$ (highlighted with red dashed lines) and $\Phi = -17.6°$ or $\Phi = 72.4°$ for $\mu = +\nu/\sqrt{2}$ (highlighted with orange lines).

8.3.5 Avalanches in the Ginzburg–Landau model

In the presence of disorder the Ginzburg–Landau model does not have a smooth response to temperature or applied stress changes but evolves through jumps in both

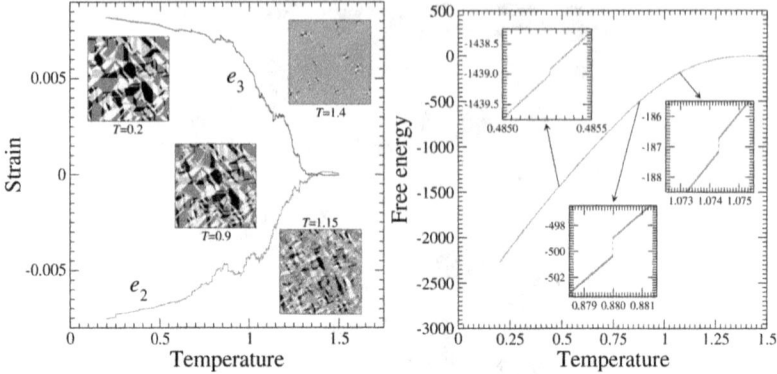

Figure 8.11. (Left panel) Strain versus temperature during a cooling process through the square-to-oblique phase transition. The jumps in the strain are clearly visible. Snapshots of the strain microstructure at selected temperatures are shown in the insets. (Right panel) Free energy versus temperature during the cooling process shown in the left panel. Selected jumps of the free energy are enlarged in the insets.

the strain configuration and its corresponding free energy. The reason is the existence of energy barriers that hinders the nucleation of the martensitic phase and the dynamics of the interfaces. Avalanches occur when an energy barrier pinning the strain microstructure can be overcome by the application of the appropriate temperature or stress change.

As an example, in figure 8.11 we show the temperature evolution of the strain fields and the free energy during cooling through a square-to-oblique phase transition. The model parameters used in these simulations are $A_1 = a_2 = a_3 = 1$, $\beta_2 = \beta_3 = -400$, $\gamma_2 = \gamma_3 = 2 \times 10^5$ and $\beta_{23} = \gamma_{23} = \gamma_{32} = 0$, which yield a transition temperature $T_0 = 1.15 T_c$. The jumps in the strain are clearly visible (left panel) and the corresponding jumps in the free energy (right panel) are enlarged in the insets. Snapshots of the strain microstructure are also shown in the insets of the left panel at $T = 1.4 T_c$, $T = 1.15 T_c$, $T = 0.9 T_c$ and $T = 0.2 T_c$.

The study of the statistical properties of the avalanches for cooling processes through both the square-to-oblique phase transition and the square-to-rectangle transformation has confirmed the existence of a power-law distribution of the number of avalanches in a strain and energy variation regime, $P \sim (\Delta e)^{-\alpha}$, $P \sim (\Delta \mathscr{F})^{-\varepsilon}$, where the strain variation, Δe, was defined as $\Delta e = \sqrt{(\Delta e_2)^2 + (\Delta e_3)^2}$ for the square-to-oblique transformation and $\Delta e = |\Delta e_2|$ for the square-to-rectangle transformation (figure 8.12). In addition, the dependence of the critical exponents on the multiplicity of the martensitic phase was also obtained, as summarized in table 8.2. This dependence of the critical exponents on the symmetry change is in good agreement with the experimental results discussed previously.

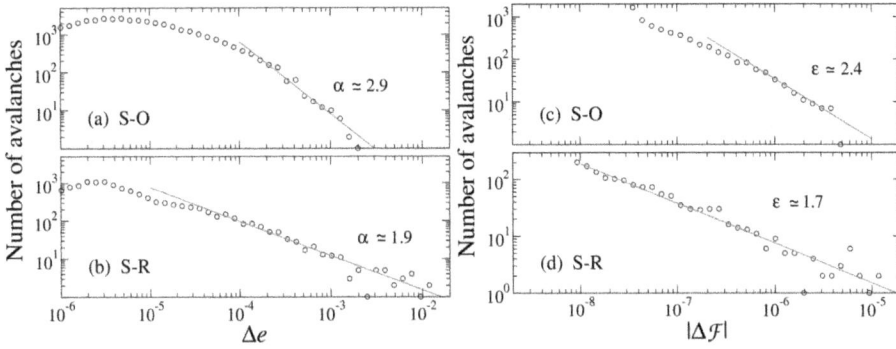

Figure 8.12. Log–log plot of the number of avalanches versus the strain variation Δe and versus the avalanche's energy $|\Delta\mathcal{F}|$ in (a), (c) the square-to-oblique transformation, and (b), (d) the square-to-rectangle transformation. The exponents are obtained using the maximum likelihood method, and lines are guides for the eye.

Table 8.2. Critical exponents obtained from the simulations for the square-to-oblique and the square-to-rectangle transformations.

M-symmetry	Multiplicity	α	ε
Oblique	4	2.9± 0.3	2.4 ± 0.2
Rectangular	2	1.9 ± 0.2	1.7 ± 0.1

8.4 Caloric effects

From the point of view of applications, in addition to those arising directly from shape-memory properties, which include the design of sensors and actuators, applications related to their large and reversible thermal response induced by stress have raised a lot of interest in recent years. Thanks to this large thermal response these materials show a giant elastocaloric effect that provides them with great potential for solid-state refrigeration and energy harvesting applications.

The elastocaloric effect depends on the reversible thermal response of a given material to an applied uniaxial stress or strain [4, 59]. When the mechanical field is applied or removed isothermally, the elastocaloric effect is quantified by a change of entropy, while it is measured by a change of temperature when it is applied adiabatically. On the other hand, energy harvesting is based on the inverse effect and exploits the possibility of using shape-memory materials to convert temperature gradients into mechanical work. Here, we will focus on the elastocaloric properties.

In shape-memory alloys, a huge thermal response occurs associated with the superelastic property, which is a consequence of the possibility of quasi-reversibly recovering the huge deformation associated with the MT upon loading and unloading, which is thus accompanied by the transition latent heat. In fact, this behaviour can also be induced by non-homogeneous loading that gives, for instance, rise to bending or twisting. In the first case the corresponding caloric effect is denoted as the flexocaloric effect, while in the second it is denoted as the twistocaloric effect [37, 60]. Compared to the homogeneous situation, in these cases large strain gradients occur in the regions of maximum curvature, which permits the caloric effect to be concentrated in these regions with low applied forces [61].

If, for instance, a stress $\sigma(\mathbf{r})$ is applied isothermally, the induced entropy change will be given as the difference of the entropies of the deformed and non-deformed states

$$\Delta S = S[T, e(\mathbf{r})] - S(T, 0), \tag{8.52}$$

where the entropy S can be obtained as

$$S[T, e(\mathbf{r})] = -\frac{\partial \mathscr{F}_t}{\partial T} \tag{8.53}$$

with the free energy function $\mathscr{F}_t$ given by

$$\mathscr{F}_t = \mathscr{F}[e(\mathbf{r}), \nabla e(\mathbf{r})] + \mathscr{F}_{\text{vib}}(T), \tag{8.54}$$

where the first term is the elastic free energy that will depend on the actual change of symmetry at the transition and the second is a vibrational free energy that takes into account the thermodynamic behaviour outside the transition region, which is assumed to occur at a high enough temperature so that the classical limit to this contribution applies. It can be expressed as

$$\mathscr{F}_{\text{vib}} = k_{\text{B}} T \sum_{i=1}^{3N} \ln \left(\frac{\hbar \omega_i}{k_{\text{B}} T} \right). \tag{8.55}$$

As an example, in figure 8.13(a) we show entropy curves obtained at a given temperature associated with the MT induced by bending a 2D-beam caused by application of a continuous distribution of forces for selected values of a parameter f_0 that characterizes the distribution. Here, f_0 measures the force applied perpendicularly in the middle of the beam. From these curves, ΔS and ΔT curves are obtained that characterize the resulting flexocaloric effect. They are shown in figures 8.13(b) and (c).

When these results are compared to the elastocaloric effect induced by uniaxial application of an homogeneous stress, it is concluded that a larger caloric response can be obtained in the range of low exchanged work in the case of bending. This is a consequence of the fact that, in the superelastic region above the MT temperature, a larger threshold must be overcome to induce the transformation by application of a uniaxial stress than by bending.

It is worth noting that metamagnetic shape-memory alloys can show, in addition to the elastocaloric effect, the magnetocaloric effect associated with the magnetic superelasticity [62]. Therefore, this class of materials is intrinsically multicaloric, which means that caloric effects can be induced simultaneously by application of mechanical and magnetic fields.

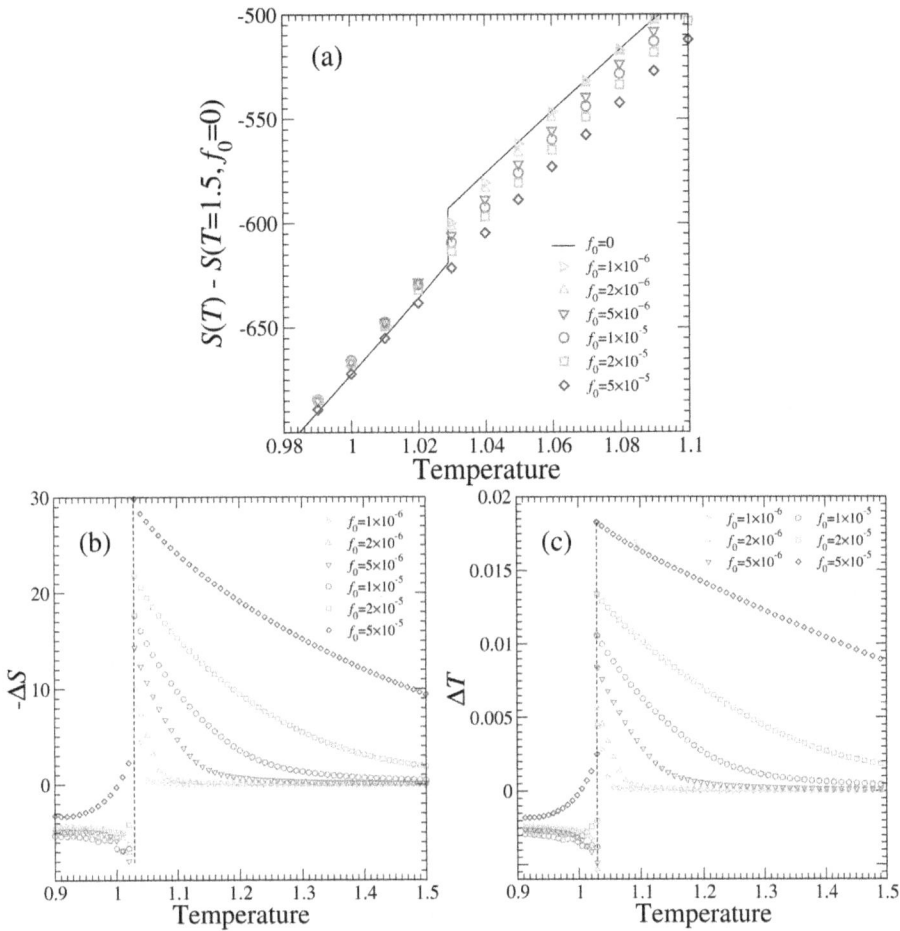

Figure 8.13. (a) Entropy curves, (b) isothermal entropy change versus temperature, and (c) adiabatic temperature change versus initial temperature.

8.5 Summary

In this chapter we have summarized the recent approach to MTs in shape-memory alloys based on the idea that in the athermal limit, deviations from ideal thermo-elastic behaviour are manifested by the fact that the transition occurs intermittently through a sequence of avalanches that carry out the dissipated energy, which is at the origin of the transition hysteresis. The fact that avalanche properties such as duration, size, or energy exhibit a power-law distribution reflects the fact that these local events occur without time and length scales. Therefore, martensitic and in general, ferroelastic transitions, in spite of usually being first-order transitions, show 'crackling noise' with associated avalanche criticality. In MTs, similarly to equilibrium critical phenomena, universality classes exist governed to a large extent by the change of symmetry taking place at the transition. Experimental and model results confirm that critical exponents depend on the degeneracy g given by the ratio of symmetry operations of the high- and low-temperature phases.

Acknowledgments

This publication is part of the R&D project PID2020-113549RB-I00 financed by the Agencia Estatal de Investigación from the Ministerio de Ciencia e Innovación (Spain) (MCIN/AEI/10.13039/501100011033).

References

[1] Bhattacharya K 2003 Microstructure of martensite *Why it Forms and How it Gives Rise to the Shape-Memory Effect* (Oxford: Oxford University Press)

[2] 1998 *Shape Memory Materials* ed K Otsuka and C M Wayman (Cambridge: University Press)

[3] Torrents G, Illa X, Vives E and Planes A 2017 Geometrical model for martensitic phase transitions: understanding criticality and weak universality during microstructure growth *Phys. Rev.* E **95** 013001

[4] Bonnot E, Romero R, Mañosa L, Vives E and Planes A 2008 Elastocaloric effect associated with the martensitic transition in shape-memory alloys *Phys. Rev. Lett.* **100** 125901

[5] Bak P 1996 *How Nature Works: the Science of Self Orgnized Criticality* (Berlin: Springer)

[6] Sethna J P, Dahmen K A and Myers C R 2001 Crackling noise *Nature* **410** 242–50

[7] Nataf G F and Salje E K H 2020 Avalanches in ferroelectric, ferroelastic and coelastic materials: phase transition, domain switching and propagation *Ferroelectrics* **569** 82–107

[8] Durin G and Zapperi S 2006 The Barkhausen effect *The Science of Hystherisis* vol 2 G Bertotti and I Mayergoyz (New York: Academic) pp 181–267

[9] Clapp P C 1995 How would we recognize a martensitic transformation if it bumped into us on a dark & austy night? *J. Phys.* IV **5** 11

[10] Planes A, Mañosa L and Vives E 2013 Acoustic emission in martensitic transformations *J. Alloys Compd.* **577S** S699–704

[11] Planes A, Macqueron J L, Morin M and Guénin G 1981 Study of martensitic transformation of Cu–Zn–Al alloy by coupled enthalpy and acoustic emission measurements *Phys. Stat. Sol.* **66** 717–24

[12] Niemann R, Kopeček J, Heczko O, Romberg J, Schultz L, Fähler S, Vives E, Mañosa L and Planes A 2014 Localizing sources of acoustic emission during the martensitic transformation *Phys. Rev.* B **89** 214118

[13] Blaysat B, Balandraud X, Grédiac M, Vives E, Barrera N and Zanzotto G 2020 Concurrent tracking of strain and noise bursts at ferroelastic phase fronts *Commun. Mater.* **99** 224101

[14] Ianniciello L, Romanini M, Mañosa L, Planes A, Engelbrecht K and Vives E 2020 Tracking the dynamics of power sources and sinks during the martensitic transformation of a Cu–Al–Ni single crystal *Appl. Phys. Lett.* **116** 183901

[15] Capellera G, Ianniciello L, Romanini M and Vives E 2021 Heat sink avalanche dynamics in elastocaloric Cu–Al–Ni single crystal detected by infrared calorimetry and Gaussian filtering *Appl. Phys. Lett.* **119** 151905

[16] Vives E, Soto-Parra D E, Mañosa L and Planes A 2011 Imaging the dynamics of martensitic transitions using acoustic emission *Phys. Rev.* B **84** 060101

[17] Bonnot E, Mañosa L, Planes A, Soto-Parra D, Vives E, Ludwig B, Strothkaemper C, Fukuda T and Kakeshita T 2008 Acoustic emission in the fcc-fct martensitic transition of $Fe_{68.8} Pd_{31.2}$ *Phys. Rev.* B **78** 184103

[18] Vives E, Ortín J, Mañosa L, Ràfols I, Pérez-Magrané R and Planes A 1994 Distributions of avalanches in martensitic transformations *Phys. Rev. Lett.* **72** 1694

[19] Pérez-Reche F J, Vives E, Mañosa L and Planes A 2004 Athermal character of structural phase transitions *Phys. Rev. Lett.* **89** 195701

[20] Planes A and Vives E 2017 Avalanche criticality in thermal-driven martensitic transitions: the asymmetry of the forward and reverse transitions in shape-memory materials *J. Phys.: Condens. Matter* **29** 334001

[21] Beke D L, Bolgár M, Tóth L Z and Daróczi L 2018 On the asymmetry of the forward and reverse martensitic transformations in shape memory alloys *J. Alloys Compd.* **741** 106–15

[22] Pérez-Reche F J, Stipcich M, Vives E, Mañosa L, Planes A and Morin M 2004 Kinetics of martensitic transitions in Cu-Al-Mn under thermal cycling: analysis at multiple length scales *Phys. Rev.* B **69** 64101

[23] Bonnot E, Vives E, Mañosa L and Planes A 2008 Learning through cycling in martensitic phase transitions *Mater. Sci. Eng.* A **481-482** 223–6

[24] Vives E, Ràfols I, Mañosa L, Ortín J and Planes A 1995 Statistics of avalanches in martensitic transformations. I. Acoustic emission experiments *Phys. Rev.* B **52** 12644

[25] Ràfols I and Vives E 1995 Statistics of avalanches in martensitic transformations. II. Modeling *Modeling Phys. Rev.* B **52** 12651

[26] Baró J 2019 Avalanches in out of equilibrium systems: statistical analysis of experiments and simulations *PhD Thesis* Universitat de Barcelona

[27] Casals B, Dahmen K A, Gou B, Rooke S and Salje E K H 2021 The duration-energy-size enigma for acoustic emission *Sci. Rep.* **11** 5590

[28] Chen Y, Gou B, Yuan B, Ding X, Sun J and Salje E K H 2022 Multiple avalanche processes in acoustic emission spectroscopy: multibranching of the energy-amplitude scaling *Phys. Status Solidi* b **259** 2100465

[29] Carrillo L, Mañosa L, Ortín J, Planes A and Vives E 1998 Experimental evidence for universality of acoustic emission avalanche distributions during structural transitions *Phys. Rev. Lett.* **81** 1889

[30] Clauset A, Rohilla-Shallizi C and Newman M E J 2009 Power-law distributions in empirical data *SIAM Rev.* **51** 661–703

[31] Romero F J, Martín-Olalla J M, Gallardo M C, Soto-Parra D E, Salje E K H, Vives E and Planes A 2019 Scale-invariant avalanche dynamics in the temperature-driven martensitic transition of a Cu-Al-Be single crystal *Phys. Rev.* B **99** 224101

[32] Salje E K H, Planes A and Vives E 2017 Analysis of crackling noise using the maximum-likelihood method: power-law mixing and exponential damping *Phys. Rev.* E **96** 224101

[33] Baró J, Martín-Olalla J M, Romero F J, Gallardo M C, Salje E K H, Vives E and Planes A 2014 Avalanche correlations in the martensitic transition of a Cu–Zn–Al shape memory alloy: analysis of acoustic emission and calorimetry *J. Phys. Condens. Matter* **26** 125401

[34] Bonnot E, Romero R, Illa X, Mañosa L, Planes A and Vives E 2007 Hysteresis in a system driven by either generalized force or displacement variables: martensitic phase transition in single-crystalline Cu–Zn–Al *Phys. Rev.* B **70** 064105

[35] Bonnot E, Vives E, Mañosa L and Planes A 2008 Acoustic emission and energy dissipation during front propagation in a stress-driven martensitic transition *Phys. Rev.* B **78** 094104

[36] Vives E, Soto-Parra D, Mañosa L, Romero R and Planes A 2009 Driving-induced crossover in the avalanche criticality of martensitic transitions *Phys. Rev.* B **80** 180101R

[37] Pérez-Junyent C, Porta M, Valdés E, Mañosa L, Planes A, Saxena A and Vives E 2022 Flexocaloric effect in superelastic materials *APL Mater.* **10** 121103

[38] Acet M, Mañosa L and Planes A 2013 Magnetic-field-induced effects in martensitic Heusler-based magnetic shape memory alloys *Handbook of Magnetic Materials* (Amsterdam: Elsevier) pp 231–89

[39] Planes A, Mañosa L and Acet M 2013 Recent progress and future perspectives in magnetic and metamagnetic shape-memory Heusler alloys *Mater. Sci. Forum* **738-739** 391–9

[40] Ge Y, Heczko O, Söderberg O and Lindroos V K 2004 Various magnetic domain structures in a Ni–Mn–Ga martensite exhibiting magnetic shape memory effect *J. Appl. Phys.* **96** 2159–63

[41] Karaca H E, Karaman I, Basaran B, Ren Y, Chumlyakov Y I and Maier H J 2009 Magnetic field-induced phase transformation in NiMnCoIn magnetic shape-memory alloys—a new actuation mechanism with large work output *Adv. Func. Mater.* **19** 983–98

[42] Baró J, Dixon S, Edwards R S, Fan Y, Keeble D S, Mañosa L, Planes A and Vives E 2013 Simultaneous detection of the acoustic emission and Barkhausen noise during the martensitic transition of a Ni-Mn-Ga shape-memory alloy *Phys. Rev.* B **88** 174108

[43] Samy N M, Bolgár M K, Barta N, Daróczi L, Tóth L Z, Chumlyakov Y I, Karaman I and Beke D L 2018 Thermal, acoustic and magnetic noises emitted during martensitic transformation in single crystalline $Ni_{45}Co_5Mn_{36.6}In_{13.4}$ meta-magnetic shape memory alloy *J. Alloys Comp.* **778** 669–80

[44] Eshelby J D 1957 The determination of the elastic field of an ellipsoidal inclusion, and related problems *Proc R. Soc. Lond.* A **241** 376–96

[45] Eshelby J D 1959 The elastic field outside an ellipsoidal inclusion *Proc. R. Soc. Lond.* A **252** 561–9

[46] Roitburd A L 1978 Martensitic transformation as a typical phase transformation in solids *Solid State Phys.* **33** 317–90

[47] Roitburd A L 1993 Elastic domains and polydomain phases in solids *Phase Transit.* **45** 1–33

[48] Khachaturyan A G 1983 *Theory of Structural Transformations in Solids* (New York: Wiley)

[49] Mura T 1987 *Micromechanics of Defects in Solids* (Dordrecht: Kluwer)

[50] Ball J M and James R D 1987 Fine phase mixtures as minimizers of energy *Arch. Ration. Mech. Anal.* **100** 13–52

[51] Kartha S, Krumhansl J A, Sethna J P and Wickham L K 1995 Disorder-driven pretransitional tweed pattern in martensitic transformations *Phys. Rev.* B **52** 803–22

[52] Shenoy S R, Lookman T, Saxena A and Bishop A R 1999 Martensitic textures: multiscale consequences of elastic compatibility *Phys. Rev.* B **60** R12537

[53] Lookman T, Shenoy S R, Rasmussen K Ø, Saxena A and Bishop A R 2003 Ferroelastic dynamics and strain compatibility *Phys. Rev.* B **67** 024114

[54] Porta M and Lookman T 2013 Heterogeneity and phase transformation in materials: energy minimization, iterative methods and geometric nonlinearity *Acta Mater.* **61** 5311–40

[55] Yang Y, Xue D, Yuan R, Zhou Y, Lookman T, Ding X, Ren X and Sun J 2019 Doping effects of point defects in shape memory alloys *Acta Mater.* **176** 177

[56] Howell P, Kozyreff G and Ockendon J 2009 *Applied Solid Mechanics* (Cambridge: Cambridge University Press)

[57] Horovitz B, Barsch G R and Krumhansl J A 1991 Twin bands in martensites: statics and dynamics *Phys. Rev.* B **42** 1021–33

[58] Porta M, Castán T, Lloveras P, Lookman T, Saxena A and Shenoy S R 2009 Interfaces in ferroelastics: fringing fields, microstructure, and size and shape effects *Phys. Rev.* B **79** 214117

[59] Mañosa L and Planes A 2017 Materials with giant mechanocaloric effects: cooling by strength *Adv. Mater.* **29** 1603607

[60] Porta M, Castán T, Saxena A and Planes A 2021 Flexocaloric effect near a ferroelastic transition *Phys. Rev.* E **104** 094108

[61] Porta M, Castán T, Saxena A and Planes A 2023 Caloric effects induced by uniform and non-uniform stress in shape-memory materials *Shape Memory Superelast.* **9** 345–52

[62] Planes A, Mañosa L and Acet M 2009 Magnetocaloric effect and its relation to shape-memory properties in ferromagnetic Heusler alloys *J. Phys. Cond. Matter* **21** 232201

IOP Publishing

Ferroelastic Materials

Guillaume F Nataf, Blai Casals and Ekhard K H Salje

Chapter 9

Ferroelastic hybrid organic–inorganic perovskites

Wei Li

Hybrid organic–inorganic perovskites (HOIPs) are an emerging subclass of materials with the perovskite architecture, which are chemically abundant and structurally diverse. They include halides, formates, azides, cyanides, hypophosphites, borohydrides, dicyanamides as well as dicynametallates. Compared with perovskite oxides, HOIPs show significant more structural flexibility due to the existence of molecular components. This virtue allows HOIPs to show abundant phase transitions and many of them are ferroelastic. In contrast to perovskite oxides, the ferroelastic phase transitions in HOIPs can be driven by more complex mechanisms which include not only conventional octahedral tilting and atomic displacements, but also order–disorder and rearrangement of molecular interactions. In this chapter, we introduce the ferroelastic transitions of several prototypical HOIPs, from the mechanism, evolution of ferroelastic domains to related functionalities. Meanwhile, a few other representative ferroelastic hybrid crystals are also briefly discussed.

9.1 Introduction

Perovskite, the name of a famous mineral that was discovered in the Ural Mountains in 1839 by German mineralogist Gustav Rose [1], has been widely used to describe the ABX_3-type crystal structure derived from the prototypical mineral $CaTiO_3$, nowadays independent of the chemical and compositional variations [2]. When the A-, B-, and X-sites of the perovskite architecture are replaced partially by organic cations and/or linkers while the lattice match and charge balance are maintained, a subclass of perovskites emerge, namely hybrid organic–inorganic perovskites (HOIPs) (figure 9.1). Specifically, by introducing organic cations on the A-site and/or molecular anions on the X-site, several families of HOIPs can be formed, which include halides (X = Cl^-, Br^-, I^-), formates (X = $HCOO^-$), azides (X = N_3^-), cyanides (X = CN^-), hypophosphites

(X = H$_2$PO$_2^-$), borohydrides (X = BH$_4^-$), dicyanamides (X = dca$^-$), as well as dicynametallates (X = [Ag(CN)$_2$]$^-$, [Au(CN)$_2$]$^-$), which are usually named by the X-sites [3]. In terms of B-sites, most of them are divalent metal ions or mixed monovalent/trivalent metal ions, although a few are monovalent metal ions because of the existence of organic diamine cations on the A-site. Additionally, NH$_4^+$ can also serve as the B-site to construct metal-free perovskites by octahedrally connecting with six halogen ions with the templating organic diamine cations. For the most diverse A-sites, there are tens of protonated ammonium cations that can be used to construct HOIPs, including ammonium, alkylammonium, aromatic ammonium, heterocyclic ammonium, and so on.

The Goldschmidt tolerance factor (TF, t) is a general criterion to judge if the selected ions can structurally form a three-dimensional (3D) perovskite lattice [4]. Because the A-sites and/or X-sites in HOIPs are not spherical ions but molecular groups, the following formula expresses the calculation of TF for HOIPs,

$$t = \frac{(r_{Aeff} + r_{Xeff})}{\sqrt{2}(r_B + 0.5h_{Xeff})} \tag{9.1}$$

where r_{Aeff}, r_{Xeff}, r_B, and h_{Xeff} represent the effective radius of the A-site molecular group, the radius of the B-site metal ion, the effective radius of the X-site molecular group, and the effective height of the X-site molecular group, respectively [5]. The 3D perovskite lattice can be stably formed when the calculated TF is between 0.8 and 1.2. Otherwise, low-dimensional perovskites or non-perovskite structures would be generated. Divalent metal ions and halogen can easily form an MX$_6$ octahedron for the building block of perovskites, and the dimensionality of the formed HOIPs depends on the alignment of the MX$_6$ octahedron. When large organic cations separate 3D perovskite lattice in one, two, and three directions, the corresponding low-dimensional HOIPs are formed.

Due to the chemical diversity and structural tunability of HOIPs, their spatial symmetry can be modified easily by varying the organic components and corresponding inter-molecular interactions, which offers them tremendous opportunities for realizing physical properties that require specific crystal symmetry [6], such as ferroelectricity, ferroelasticity, and second-order non-linear optics. In this chapter, the ferroelastic properties of some representative HOIPs will be discussed in detail, from the perspectives of crystal structure, phase transition, the evolution of ferroelastic domains, to related physical properties, as well as their potential applications. In addition, a few other representative ferroelastic hybrid crystals are included.

9.2 Improper ferroelasticity

(Parts of this section have been reproduced with permission from [7]. Copyright 2020 John Wiley and Sons.)

Starting with a formate HOIP, [AZE][Mn(HCOO)$_3$] (AZE = (CH$_2$)$_3$NH$_2^+$) [8], which crystalizes in an orthorhombic space group *Pnma* at room temperature, has cell parameters of $a = 8.6939(3)$, $b = 12.3048(4)$, $c = 8.8768(3)$, $V = 949.61(6)$ Å^3. The MnO$_6$ octahedron is linked to six neighboring octahedra via HCOO$^-$ ligands,

(a)

(b)

(c)

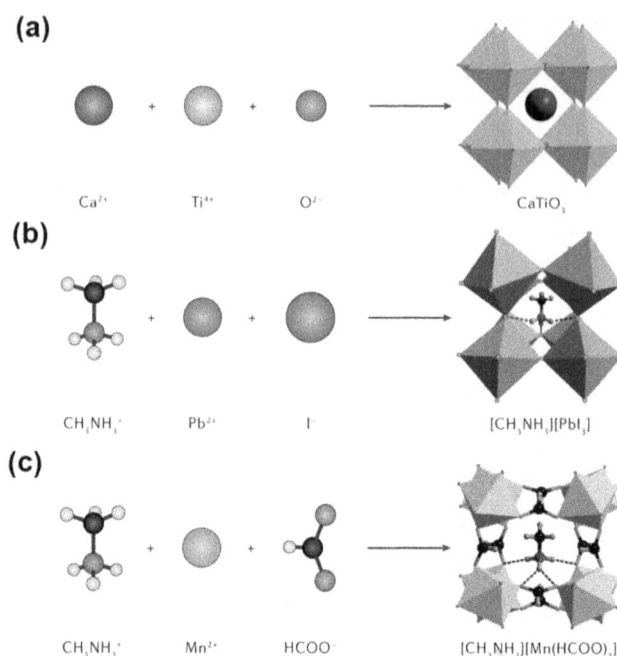

Figure 9.1. Evolution from perovskite oxides to hybrid organic–inorganic perovskites. (a) Perovskite oxide, $CaTiO_3$; (b) hybrid perovskite with the organic A-site, [MA][PbI$_3$]; and (c) hybrid perovskite with the organic A- and X-site, [MA][Mn(HCOO)$_3$]. Color schemes: C, black; N, blue; O, red; and H, gray. (Reproduced with permission from [7]. Copyright 2020 John Wiley and Sons.)

forming a 3D ReO_3-type anionic [Mn(HCOO)$_3$]$^-$ framework, and the AZE cations with two-fold disordered states are accommodated in the framework cavities. There are four N–H$\cdots$O hydrogen bonds in total within one pseudo-cubic perovskite unit, half of which are formed by the –NH$_2$ hydrogen atom of the AZE group and two oxygen atoms from one formate ligand normal to the –NH$_2$ plane with the perovskite framework (figure 9.2(a)). According to the Glazer notation, the octahedral tilting mode of the orthorhombic structure would be $a^-b^+c^-$. It should be noted that the AZE ring is dynamically disordered between two equal positions, leading to a flattened shape instead of a butterfly conformation at room temperature.

As demonstrated for several hybrid frameworks in which the disordered guest molecule can usually become ordered upon cooling, [AZE][Mn(HCOO)$_3$] also undergoes an order–disorder phase transition to a monoclinic $P2_1/n$ phase at about 265 K, with cell parameters of $a = 12.30956(6)$, $b = 12.25093(5)$, $c = 12.29019(6)$ Å, and $\beta = 92.8653(4)°$. In conventional perovskite oxides, the orthorhombic $Pnma$ could have a parent cubic structure with cell parameter a_c, and the cell parameters of $Pnma$ phase are $\sqrt{2}a_c \times 2a_c \times \sqrt{2}a_c$. Upon cooling to the monoclinic $P2_1/n$ phase, the cell parameters are transformed to $2a_c \times 2a_c \times \sqrt{2}a_c$ or $\sqrt{2}a_0 \times 2b_0 \times \sqrt{2}c_0$, where a_0, b_0 and c_0 are lattice parameters of the structure at room temperature. In addition, the AZE group is fully ordered in a butterfly conformation in the

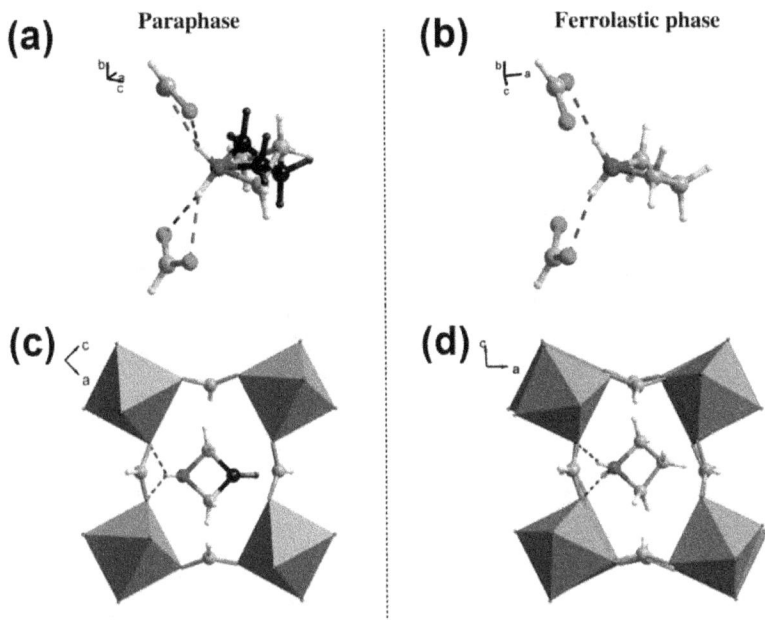

Figure 9.2. Crystal structure of HOIP [AZE][Mn(HCOO)$_3$]. (a), (b) The disordered and ordered AZE cation conformations in the orthorhombic and monoclinic phases, respectively. (c), (d) Para- and ferroelastic structures along the (010) and (101) orientations. Color scheme: Mn, green; O, red; N, blue; C, gray; and H, light gray. (Reproduced with permission from [8]. Copyright 2013 Elsevier.)

low-temperature phase, which leads to significant changes in hydrogen bonding. The two –NH$_2$ hydrogen atoms of the AZE group form two N–H$\cdots$O hydrogen bonds in total with the two oxygen atoms of two individual formate ligands along the diagonal direction of the perovskite framework (figure 9.2(b)).

As the orthorhombic to monoclinic phase transition is not common in conventional perovskite oxides, the phase transition that occurred in [AZE][Mn(HCOO)$_3$] presents a rare example. The detailed group theory analysis is conducted using the program ISPTROPY, which reveals that the critical point is the U-point ($k = 1/2, 0$, and $1/2$) of the Brillouin zone for the *Pnma* phase and the X-point ($k = 1/2, 0$, and 0) for the monoclinic phase. Therefore, this transition indicates that [AZE][Mn (HCOO)$_3$] is an improper ferroelastic, although the origin is different from conventional perovskite oxides, whose ferroelastic transitions are usually driven by tilting or Jahn–Teller distortions. For the setting of the *Pnma* structure, i.e. with Mn at the center of the cell (Wycoff 4b), the irreducible representations for octahedral tilting would be M_2^+ and R_5^-. The tilt system of the monoclinic structure would be $a^-b^+c^-$ in Glazer notation. Combining these with either irreducible M_3^+ or M_4^+, which represents the configurational ordering of the AZE cation, is sufficient to generate the $P2_1/n$ structure. In other words, the octahedral tilting and the ordering of the A-site AZE cation are both essential driving forces of the monoclinic phase.

Temperature-dependent single-crystal x-ray diffraction measurements further confirm the orthorhombic to monoclinic phase transition. There is a sudden jump in β angle with decreasing temperature, indicating the phase transition is first order and agrees with the group/subgroup relationship for the two structures. Differential scanning calorimetry (DSC) measurements show the thermal anomalies at ~273 K/ 272 K upon heating/cooling, respectively, displaying a first-order nature in thermal hysteresis. The entropy change extracted from the DSC result is 3.39 J mol^{-1} K^{-1}, comparable to the two-fold order–disorder model, whose entropy change is calculated as $R\ln 2 = 5.76$ J mol^{-1} K^{-1}.

The strain analysis of [AZE][Mn(HCOO)$_3$] is performed based on the above experimental lattice parameter data. For an orthorhombic to monoclinic phase transition, values of strains e_{1-} e_6 can be expressed as

$$e_1 = \frac{a/\sqrt{2}}{a_0} - 1 \tag{9.2}$$

$$e_2 = \frac{b/\sqrt{2}}{b_0} - 1 \tag{9.3}$$

$$e_3 = \frac{c/\sqrt{2}}{c_0} \sin \beta - 1 \tag{9.4}$$

$$e_5 = \frac{c/\sqrt{2}}{c_0} \cos \beta \approx \cos \beta \tag{9.5}$$

$$e_4 = 0 \tag{9.6}$$

$$e_6 = 0, \tag{9.7}$$

where a, b, c, and β are the lattice parameters of the monoclinic phase and a_0, b_0 and c_0 are the lattice parameters of the orthorhombic phase extrapolated into the stability field of the monoclinic phase. A baseline function is used to account for the saturation effect as $T \rightarrow 0$ K, which can be expressed by

$$a_0 = a_1 + a_2 \theta_s \coth \frac{\theta_s}{T}, \tag{9.8}$$

where a_0 is the non-linear baseline, a_1 and a_2 are constants, and θ_s is a temperature at which the variations of lattice parameters are flattened out at low temperatures. Figure 9.3(a) illustrates the fitted $\theta_s = 585$ K for b_0, and the same value is used for a_0 and c_0. Strain values calculated from the lattice parameters are shown in figure 9.3(b). The results indicate that the values of e_1, e_2, and e_3 are in the range of -2.0 to $+1.0\,\%$, indicating a total volume strain ($\approx e_1 + e_2 + e_3$) of about -1.5%. It should be noted that e_5 reaches a remarkable value of up to 5%, about five times the shear strains in improper ferroelastic perovskite oxides (usually less than 1%), such as (Ca, Sr)TiO$_3$. As e_5 is scaling with the square of the driving parameter, q^2, for an improper ferroelastic transition, which can be modeled by the first-order solution for a 246 Landau potential,

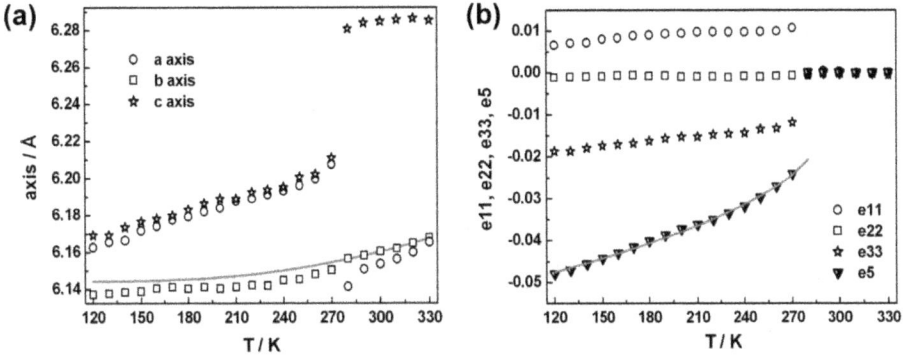

Figure 9.3. (a) Variations of unit cell parameters a, b, c, and β angle of [AZE][Mn(HCOO)$_3$] as a function of temperature. The a, b, and c of the para- and ferroelastic structures are adapted from the pseudo-cubic cell parameters. The red solid line represents the reference lattice parameter, b_0, obtained by fitting to data in the temperature interval 280 K–330 K. (b) Variation of spontaneous strains through the ferroelastic phase transition in [AZE][Mn(HCOO)$_3$]. The individual strain components appear to show a sharp discontinuity at ~270 K. The red solid line is the fit using the 246 potential (equation (9.9)) with $T_{tr} = 272$ K, $T_c = 239$ K, and $e_{5,0} = -2.4\%$. (Reproduced with permission from [8]. Copyright 2013 Elsevier.)

$$e_5 = \frac{2}{3}e_{5,\,0}\left\{1 + \left[1 - \frac{3}{4}\left(\frac{T}{T_{tr}} - \frac{T}{T_c}\right)\right]^{\frac{1}{2}}\right\}, \qquad (9.9)$$

where T_{tr} is the transition temperature; T_c is the critical temperature; $e_{5,0}$ is the magnitude of the discontinuity in e_5 at $T_{tr} = T_c$. The fitted T_c and $e_{5,0}$ are 239 K and -2.4%, respectively, when $T_{tr} = 272$ K. The difference between T_c and T_{tr} is only 33 K, which implies a weak first-order ferroelastic transition that is close to tricritical. Therefore, the large shear strain during the ferroelastic phase transition is ascribed to the configurational ordering of the AZE groups and variations of hydrogen bonding with the framework and is substantially greater than that of perovskite oxides, which is generally associated with octahedral tilting.

Resonant ultrasound spectroscopy (RUS) study is further conducted on a single-crystal sample of [AZE][Mn(HCOO)$_3$] to detect the dynamic response in the temperature range of 6 K–300 K upon heating. The square of vibrational frequency, f^2 (proportional to the elastic constants), and mechanical quality factor, Q^{-1} (a measure of acoustic dissipation or energy loss), are extracted from selected resonance peaks (figure 9.4). The f^2 values of all resonance peaks display a decrease when approaching the transition point (~270 K) from both above and below, indicating a significant elastic softening before and after the ferroelastic transition. The acoustic loss (Q^{-1}) exhibits a frequency-dependent feature below 270 K, ascribed to the movement of twin walls, as demonstrated in other ferroelastic materials at similar frequencies. The RUS measurements on SrTiO$_3$ and LaAlO$_3$ reveal that the second-order improper ferroelastic transition should have a step-like softening character on most single-crystal elastic constants below the transition point. Meanwhile, a sharp non-linear recovery below the transition point should

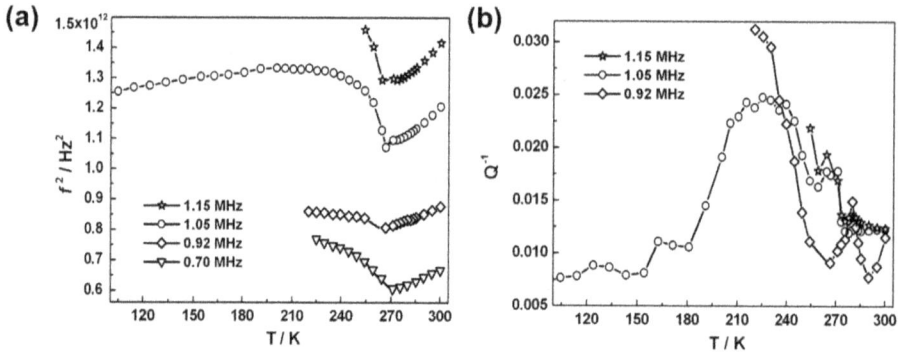

Figure 9.4. Temperature dependencies of (a) f^2 and (b) Q^{-1}, from fitting of selected resonance peaks detected by resonant ultrasound spectroscopy. (Reproduced with permissionfrom [8]. Copyright 2013 Elsevier.)

occur in weakly first-order or tricritical cases, such as $SrZrO_3$. In this case, all the single-crystal resonances show similar softening, indicating the mechanism might likely be coupling with dynamic effects. This is consistent with the observation that the AZE cations are ordered and dynamically disordered in the ferroelastic and paraelastic phases. The twin walls would be immobile under lower temperatures, and the influence of any coupling between the strain and the magnetic order parameter should become evident as changes in f^2 and Q^{-1}. However, very trivial variations in f^2 and Q^{-1} are observed at the temperature below 20 K, indicating the negligible magnetoelastic coupling in $[AZE][Mn(HCOO)_3]$.

9.3 Evolution of ferroelastic domains

Hydrogen/fluorine (H/F) substitution and halogen substitution strategies have been recognized as effective methods for synthesizing ferroic perovskite materials. For instance, a significant symmetry breaking could be achieved by substituting hydrogen atoms in organic cations with fluorine atoms, which have a high electronegativity. This symmetry breaking, in turn, could enhance the phase transition temperature, which is essential for the ideal ferroelastic materials. Chen *et al.* [9] successfully synthesized a series of 3D cyano-bridged perovskite ferroelastic materials $[C_3H_5FNH_2]_2[MFe(CN)_6]$ ($C_3H_5FNH_2$ = 3-fluoroazetidinium, M = K, Rb or Cs), using a H/F substitution strategy (figure 9.5). These materials feature a double perovskite structure. This framework is formed by connecting Fe^{3+} and M^+ ions through cyanide ions. The organic cations within the cages act as guests, providing structural flexibility.

All these materials experience a ferroelastic phase transition with the Aizu notation of $m\bar{3}mF\bar{1}$. Taking $[C_3H_5FNH_2]_2[RbFe(CN)_6]$ as an example, in the low-temperature phase (LTP), it crystallizes in the triclinic $P\bar{1}$ space group (figure 9.6(e) and (g)), with C − F − Rb coordination bonds between the ordered 3-fluoroazetidine cations and Rb^+ ions. Upon heating, these C − F − Rb coordination bonds break, and both the guest molecules and nitrogen atoms in the inorganic framework

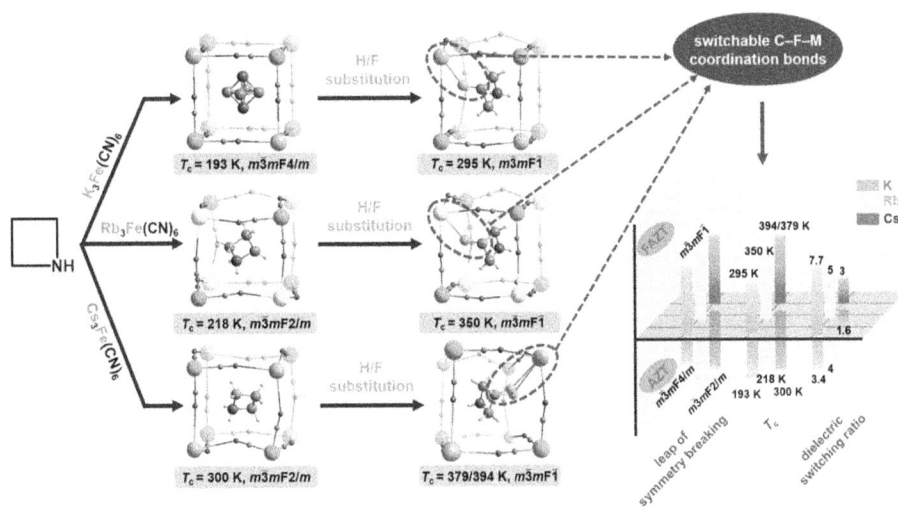

Figure 9.5. Schematic strategy of H/F substitution for designing compounds $[C_3H_5FNH_2]_2[KFe(CN)_6]$, $[C_3H_5FNH_2]_2[RbFe(CN)_6]$ and $[C_3H_5FNH_2]_2[CsFe(CN)_6]$. (Reproduced with permission from [9]. Copyright 2024 the Royal Society of Chemistry.)

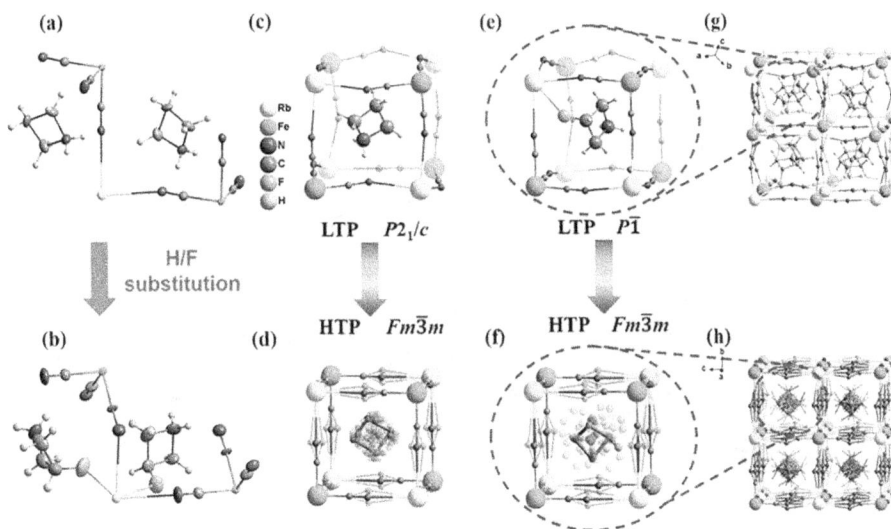

Figure 9.6. Crystal structures of $[C_3H_5FNH_2]_2[RbFe(CN)_6]$ and $[C_3H_6NH_2]_2[RbFe(CN)_6]$. The asymmetric unit in the LTP of $[C_3H_6NH_2]_2[RbFe(CN)_6]$ (a) and $[C_3H_5FNH_2]_2[RbFe(CN)_6]$ (b) is drawn as displacement ellipsoids with 50% probability. The basic unit of the framework in the LTP and HTP of (c), (d), $[C_3H_6NH_2]_2[RbFe(CN)_6]$ and (e), (f) $[C_3H_5FNH_2]_2[RbFe(CN)_6]$. Perspective view of the packing structure of $[C_3H_5FNH_2]_2[RbFe(CN)_6]$ in the (g) LTP and (h) HTP. For clarity, the hydrogen atoms in the HTP are omitted. (Reproduced with permission from [9]. Copyright 2024 Royal Society of Chemistry.)

become disordered (figure 9.6(f) and (h)). $[C_3H_5FNH_2]_2[RbFe(CN)_6]$ transitions into a high-temperature phase (HTP) with a high symmetry, crystallizing in the cubic $Fm\overline{3}m$ space group. DSC experiments confirm that the phase transition of

$[C_3H_5FNH_2]_2[KFe(CN)_6]$, $[C_3H_5FNH_2]_2[RbFe(CN)_6]$ and $[C_3H_5FNH_2]_2[CsFe(CN)_6]$ is a first-order phase transition, at approximately 295 K, 350 K, 379 K (T_{c1} of $[C_3H_5FNH_2]_2[CsFe(CN)_6]$), and 394 K ($T_{c2}$ of $[C_3H_5FNH_2]_2[CsbFe(CN)_6]$), respectively. Notably, $[C_3H_5FNH_2]_2[RbFe(CN)_6]$ shows a significant enhancement in T_c of 132 K compared to its parent compound $[C_3H_6NH_2]_2[MFe(CN)_6]$, referred to as $[C_3H_5FNH_2]_2[RbFe(CN)_6]$ (figure 9.6(c) and (d)). This enhancement is attributed to the H/F substitution (figure 9.6(a) and (b)), which introduces switchable C−F−Rb coordination bonds. These bonds significantly strengthen the interactions between the organic cations and the cyano-bridged framework, increasing the phase transition temperature (T_c).

The $m\bar{3}mF\bar{1}$ transition characteristics of $[C_3H_5FNH_2]_2[MFe(CN)_6]$ are identified as potential ferroelastics. The evolution of ferroelastic domains in $[C_3H_5FNH_2]_2[RbFe(CN)_6]$ was observed using polarized light microscopy (figure 9.7). Before reaching T_c, the crystals showed stable, interlaced stripe-like ferroelastic domains. Above T_c, these domains disappeared, indicating a transition to a high symmetry paraelastic phase, while with subsequent cooling below T_c, the striped domain patterns re-emerged, demonstrating the reversibility of the ferroelastic phase transition. Due to their inherent non-symmetry, these ferroelastic materials also exhibit excellent dielectric switching performance. Specifically, $[C_3H_5FNH_2]_2[RbFe(CN)_6]$ showed a significant ratio between the high and low dielectric states, reaching up to 5.

In the exploration of diacyanamide ferroelastic perovskites, Fu *et al.* [10] have successfully constructed a 3D HOIP ferroelastic $[Et_3P(CH_2)_2F][Cd(dca)_3]$ [dca = dicyanamide, $N(CN)_2$]) (**2**), through halogen substitution of the cation from $[Et_3P(CH_2)_2Cl]$ to $[Et_3P(CH_2)_2F]$. $[Et_3P(CH_2)_2F][Cd(dca)_3]$ features a perovskite structure that comprises $[Cd_8(N(CN)_2)_{12}]$ cages filled with phosphonium cation guests (figure 9.8(a)). Compared to $[Et_3P(CH_2)_2Cl][Cd(dca)_3]$, the variation in

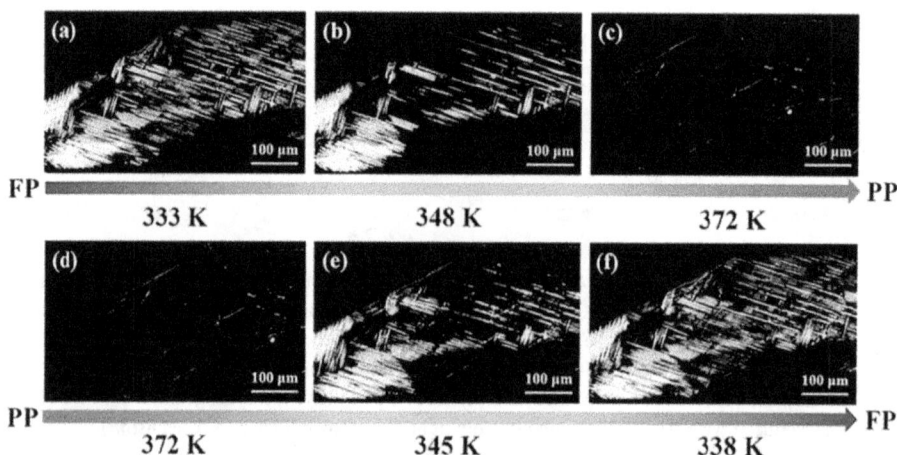

Figure 9.7. Evolution of ferroelastic domains for $[C_3H_5FNH_2]_2[RbFe(CN)_6]$ during the (a)–(c) heating process and (d)–(f) cooling process. (Reproduced with permission from [9]. Copyright 2024 the Royal Society of Chemistry.)

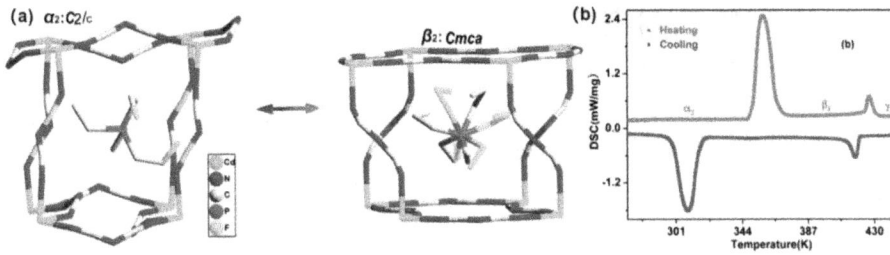

Figure 9.8. (a) Crystal structures of $[Et_3P(CH_2)_2F][Cd(dca)_3]$ in α_2 phase β_2 phases. Different configurations of organic cations are labeled in different colors in the β_2 phases. (b) DSC curves of $[Et_3P(CH_2)_2F][Cd(dca)_3]$ obtained in the heating and cooling cycles. (Reproduced with permission from [10]. Copyright 2018 the Royal Society of Chemistry.)

carbon–halogen bond length and the configuration of organic cations influences the arrangement of the host structures, consequently affecting the symmetry evolution of $[Et_3P(CH_2)_2Cl][Cd(dca)_3]$ and $[Et_3P(CH_2)_2F][Cd(dca)_3]$. DSC measurements (figure 9.8(b)) revealed that $[Et_3P(CH_2)_2F][Cd(dca)_3]$ undergoes two sequentially reversible high-temperature first-order phase transitions. Variable-temperature powder x-ray diffraction (PXRD) further confirmed the phase transitions of $[Et_3P (CH_2)_2F][Cd(dca)_3]$, which involve two ferroelastic to paraelastic phase transitions, from the centrosymmetric monoclinic $C2/c$ (α_2 phase) to orthorhombic $Cmca$ (β_2 phase), then to the simulated $P4_2/nnm$ (γ_2 phase) identified by PXRD.

$[Et_3P(CH_2)_2F][Cd(dca)_3]$ undergoes two phase transitions with Aizu notations of $4/mmmFmmm$ and $mmmF2/m$, which are ferroelastic phase transitions. The domain structures of $[Et_3P(CH_2)_2F][Cd(dca)_3]$ were observed (figure 9.9(a)–(d)), revealing a well-defined domain structure at 333 K with angles between domain walls slightly deviating from 90 degrees. Upon heating, the dominant domain structure transitions from 'A' to 'B', eventually leading to a sharp transition characterized by the disappearance of 'B'. This observation is consistent with the two-step ferroelastic phase transitions of $[Et_3P(CH_2)_2F][Cd(dca)_3]$. It is worth noting that $[Et_3P(CH_2)_2F]$ $[Cd(dca)_3]$ is the first above-room temperature 3D ferroelastic material with two ferroelastic phases, making it an ideal candidate for multifunctional ferroelastic device applications.

Furthermore, $[Et_3P(CH_2)_2F][Cd(dca)_3]$ exhibits temperature-induced bi-step switchable dielectric properties, along with pronounced anisotropy along different crystallographic axes, as shown in figures 9.9(e)–(f). These findings demonstrate the effectiveness of chemical substitutions, including H/F and Cl/F substitutions, in facilitating the development of multifunctional ferroelastic materials, paving new avenues for the engineering of innovative ferroics.

Another prototypical non-perovskite example is the R_3MX_6 type ferroelastic $(CH_3C(NH_2)_2)_3[BiBr_6]$, which was the first reported [11] with the coexistence of ferroelectric phase transition. The structure of the room temperature phase, the paraelastic phase, is tetragonal with a space group of $P4_2/n$, which is constituted of

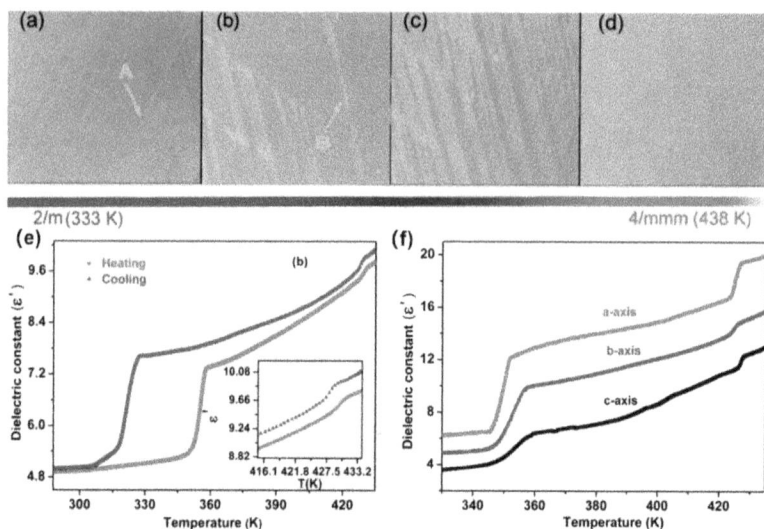

Figure 9.9. (a)–(d) Evolution of the domain structures of [Et$_3$P(CH$_2$)$_2$F][Cd(dca)$_3$] during the heating process, (e) temperature-dependent dielectric constant measured at 1 MHz of [Et$_3$P(CH$_2$)$_2$F][Cd(dca)$_3$], and (f) anisotropic dielectric permittivity of [Et$_3$P(CH$_2$)$_2$F][Cd(dca)$_3$] along the *a*-, *b*- and *c*-axes. (Reproduced with permission from [10]. Copyright the Royal Society of Chemistry.)

one isolated [BiBr$_6$]$^{3-}$ octahedron and four acetamidinium cations (figure 9.10(a)). In the paraelastic phase, two organic cations show a two-fold disorder where inorganic motifs have interacted with the organic cations via N–H...Br hydrogen bonds. Upon cooling, the cations become more ordered, resulting in symmetry decreasing from tetragonal to triclinic and the ferroelastic phase transition taking place at 277 K. The low-temperature ferroelastic phase has a triclinic $P1$ space group with an Aizu notation of 4/mF1. DSC curves also confirmed the ferroelastic phase transition with a well-shaped thermal anomaly. Based on the DSC data, the obtained value of the entropy change for this phase transition is 9.4 J mol^{-1} K^{-1}, indicating an order–disorder dynamic change of organic cations. During this phase transition, the Bi–Br bond lengths of the paraelastic phase lie in a narrow region of 2.8265(11)–2.8739(11) Å, resulting in small octahedral distortions and nonpolar structures. The bond lengths of the ferroelastic phase show great changes within a wide region of 2.772(4)–2.932(4) Å, leading to enhanced inorganic octahedral distortions and axial deformations. Such deformations of crystal structure would inevitably influence the microscopic strain of domains. The polarizing microscopic images revealed the domain structure evolutions and confirmed the ferroelastic phase transition in figure 9.10(b). In the paraelastic phase, the crystal shows a mono-domain morphology without any external strains. When cooling down, striated domains gradually emerged, and the domain walls intersected at angles of 180°, indicating the occurrence of the ferroelastic phase due to the octahedral distortions.

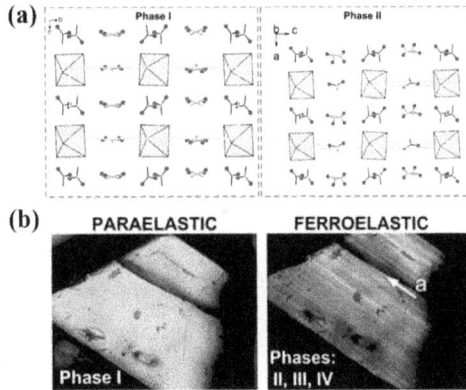

Figure 9.10. (a) The crystal packing of $(CH_3C(NH_2)_2)_3[BiBr_6]$ of the paraelastic phase (left) and the ferroelastic phase (right). (b) The ferroelastic domain structures of $(CH_3C(NH_2)_2)_3[BiBr_6]$. (Reproduced from [11]. CC BY 4.0.)

9.4 Ferroelasticity in low-dimensional HOIPs and related systems

Further work on the ferroelasticity of HOIPs was extended to the low-dimensional structures and some related systems [12]. Compared with 3D HOIPs, 2D HOIPs provide a wider platform to study lattice strain and ferroelasticity via tuning the number of inorganic sheets between the organic cations (n). In this study, three layered HOIPs, BA_2PbI_4 ($n = 1$, $BA = CH_3(CH_2)_3NH_3^+$), $BA_2MAPb_2I_7$ ($n = 2$, $MA = CH_3NH_3^+$), and $BA_2MA_2Pb_3I_{10}$ ($n = 3$), were synthesized to investigate the ferroelastic behavior. The as-grown crystals were fabricated onto a polydimethylsiloxane substrate to study the ferroelasticity and the domain wall motion under macroscopically applied stress. When the composite films were bent upward or downward, an obvious stripe pattern for $BA_2MAPb_2I_7$ was observed, and the domain walls showed obvious motion under different applied strains (figure 9.11), indicating the ferroelastic nature of $BA_2MAPb_2I_7$. Moreover, the different domain patterns did not show complete reversibility when the applied stress was released, which matched well with standard ferroelastic systems. However, no domain wall patterns for BA_2PbI_4 can be observed, suggesting the absence of natural ferroelasiticity. To understand the layer-number-dependent ferroelasticity mechanism, density functional theory (DFT) calculations were conducted. The geometric relaxation calculations of $BA_2MAPb_2I_7$ show a phase transition from the paraelastic $I4/mmm$ space group to the ferroelastic $Pnma$ space group, denoted in Aizu's notation as $4/mmmFmmm$. The ferroelasticity in $BA_2MAPb_2I_7$ is attributed to the rotation of aspherical MA molecules confirmed by an average angle of $33°$ between organic cations and the y-axis. Due to the absence of MA cations in BA_2PbI_4, it is reasonable that BA_2PbI_4 did not show the ferroelastic phase transition. For photoluminescence (PL) materials, the ferroelastic domains would influence the carrier recombination kinetics considering the lattice distortions and the symmetry breaking. A $BA_2MAPb_2I_7$ single-crystal sample was tested using polarized optical microscopy and showed clear ferroelastic domain patterns. By scanning the sample

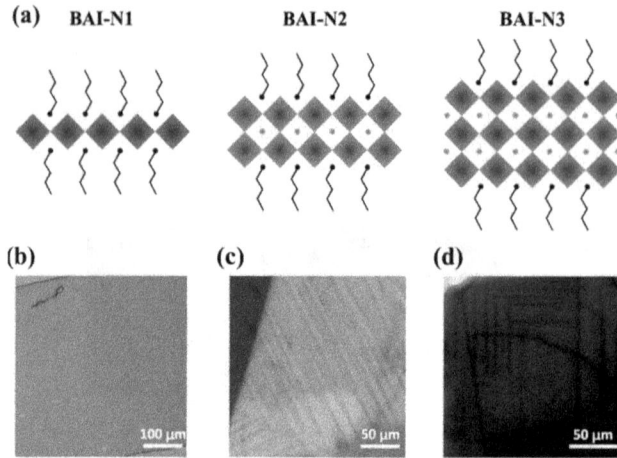

Figure 9.11. (a) Scheme of crystal structures for BA_2PbI_4, $BA_2MAPb_2I_7$, and $BA_2MA_2Pb_3I_{10}$; (b)–(d) corresponding polarized optical microscope images for BA_2PbI_4, $BA_2MAPb_2I_7$, and $BA_2MA_2Pb_3I_{10}$, respectively. (Reproduced from [12]. CC BY 4.0.)

along the X- and Y-axes, the mapping results of PL intensity and recombination lifetime were collected. However, both results show a uniform distribution over the mapping area in ferroelastic domains, indicating that the ferroelastic domain walls have little influence on photophysical properties, including both PL intensity and lifetime. Although the MA molecular rotation would disturb the crystal structures by introducing lattice distortions, the electronic structure is dominated by the inorganic Pb s–p and I p orbitals. This work gives insight into understanding strain-dependent material properties in layered HOIPs and paves an alternative pathway to engineering the ferroelastic domain.

Wang *et al.* [13] reported a 1D HOIP $C_6H_5N(CH_3)_3CdCl_3$, which exhibits a ferroelastic phase transition from monoclinic Cc space group to orthorhombic $Ama2$ space group that can be described by the Aizu notation of $mm2Fm$. In the low-temperature phase, 1D scaffold-typed edge-sharing cadmium halide (CdX_5^-) chains are connected with organic cations via hydrogen bonds between the methyl group and the halide anions. In the high-temperature phase, the longest Cd–X bond of the hexahedron was broken, and thus the coordination mode of Cd^{2+} transforms into a corner-sharing tetrahedral CdX_4^- (figure 9.12). Meanwhile, the organic cations display an order–disorder phase transition with two degenerate positions. Intriguingly, the shear strain measurements demonstrate that it undergoes a record-breaking shear strain of 21.5% during the phase transition process, which is two orders of magnitude greater than those of conventional polymers and oxides. It is facilitated by structural confinement of the large organic moieties, which hinders undesired 180° polarization reversal. In addition, it was found that the C_6H_5N $(CH_3)_3CdCl_3$ sample was poled into an almost single-domain state, while the back-switched ferroelastic domains are activated in the samples with Br substitution, which can be attributed to the chemical-substitution-induced bond softening as

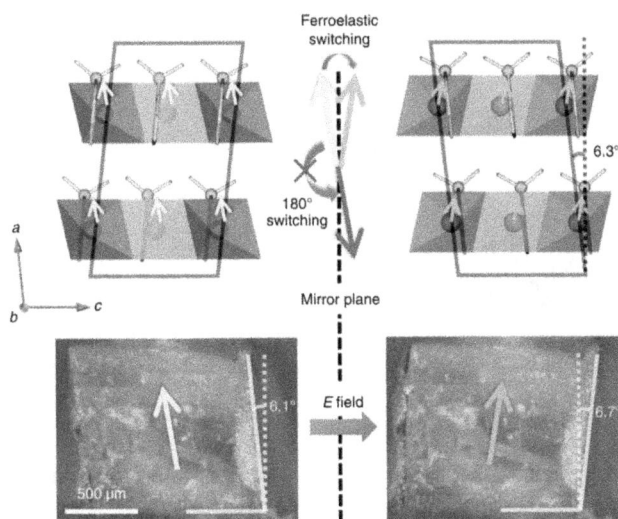

Figure 9.12. Top: schematic diagram of structural units of $C_6H_5N(CH_3)_3CdCl_3$ in the two ferroelastic states and structure confinement effect for ferroelectric/ferroelastic switching. The yellow and green arrows indicate the polarization directions. Bottom: photos of a bulk single-crystal poled to the two ferroelastic states by the electric field, showing a macroscopic shear strain in good correlation with the unit cell distortion. (Reproduced with permission from [13]. Copyright 2021 Springer.)

confirmed by DFT calculations. This not only enhances the extrinsic contribution of ferroelastic domain wall motion to the electromechanical response but also increases the intrinsic piezoelectricity for the Br-rich compound. This work suggests that by combining the structural confinement strategy with the diversity of organic molecules, it is possible to tailor the electromechanical outputs of organic–inorganic hybrids.

It is worth mentioning a 0D spin crossover compound $[Mn(3,5\text{-diCl-sal}_2(323))]$ BPh_4, $([Mn(3,5\text{-diCl-sal}_2(323))]^+ = [C_{22}H_{24}N_4O_2MnCl_4]^+$, $BPh_4^- = [C_{24}H_{20}B]^-)$, though it does not have a perovskite architecture. Morgen *et al.* [14] proved that the Jahn–Teller effect in the spin quintet form of Mn^{3+} systems would give rise to a strong and thermally switchable distortion and is favorable for developing the ferroelastics. Single-crystal x-ray diffraction revealed that this compound crystallizes in the polar monoclinic space group *Cc* at 250 K, and the asymmetric unit consists of one independent mononuclear $[Mn(3,5\text{-diCl-sal}_2(323))]^+$ complex cation and one disordered BPh_4^- counter anion (figure 9.13). Each Mn^{3+} ion connects with the *trans*-anionic phenolate oxygen donors and *cis*-amine and *cis*-imine nitrogen atoms to form a distorted octahedron. With decreasing temperature, this compound exhibits three distinct symmetry-breaking phase transitions, including $Cc \rightarrow Pc$, $Pc \rightarrow P1$, and $P1 \rightarrow P\bar{1}$ (figure 9.13), in which both triclinic phases are ferroelastic with a transition point of about 140 K. The ferroelastic transition of $[Mn(3,5\text{-diCl-sal}_2(323))]BPh_4$ from monoclinic to triclinic has been revealed by group–subgroup analysis and formal strain analysis. The results demonstrate that these ferroelastic transitions belong to the first order and the coupling between structural and spin state order parameters is

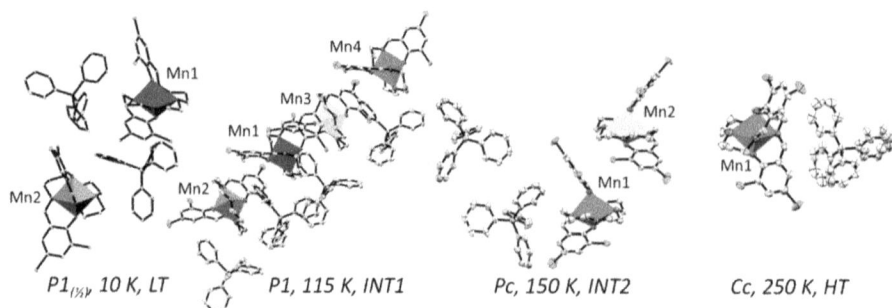

Figure 9.13. The asymmetric unit of [Mn(3,5-diCl-sal$_2$(323))]BPh$_4$ under different temperatures. (Reproduced from [14]. CC BY 4.0.)

important for the structure and properties of ferroelectric elastic domain walls. Meanwhile, the resonance peak changes of RUS from [Mn(3,5-diCl-sal$_2$(323))]BPh$_4$ indubitably confirmed the formation of ferroelastic domain walls below the ferroelastic transition point and movement under the influence of dynamic shear stress. When the temperature was below 40 K, due to the pinning/freezing effects, the domain wall mobility was reduced. In addition, the ferroelastic distortion has been proven to play a key role in the magnetoelectric coupling of this compound.

9.5 Ferroelasticity in MAPbI$_3$ (MA $=$ CH$_3$NH$_3{}^+$)

In the past decade, MAPbI$_3$ has drawn significant attention due to its remarkable optoelectronic features of highly efficient optical absorption and transition, long charge diffusion length and carrier lifetime, and especially high carrier mobility [15, 16]. Its excellent performance makes MAPbI$_3$ a promising candidate for applications in low-cost solar cells [17]. However, the knowledge about the fundamental properties of MAPbI$_3$ fall far behind that required for photoelectric converting efficiency improvement [18–20]. For instance, there is still no conclusive evidence confirming the existence of ferroelectricity in MAPbI$_3$ [21]. The work by Strelco's group demonstrated twin domain structures in MAPbI$_3$ during the cubic–tetragonal phase transition due to strain, thus indicating a ferroelastic rather than a ferroelectric behavior [22].

MAPbI$_3$ crystalizes in a typical 3D perovskite architecture; each Pb^{2+} metal cation is six coordinated via six I$^-$ ions and form a PbI$_6$ octahedron, and each PbI$_6$ polyhedron connects its neighboring counterparts in three dimensions in a corner-sharing mode. The organic amine cations are located in PbI$_6$ cavities. There are three crystallographic phases in MAPbI$_3$: orthorhombic, *Pnma*, below 162 K; tetragonal, *I4/mcm*, between 327 K and 162 K; and cubic, *Pm3̄m*, above 327 K [23]. The cubic–tetragonal transition can be categorized as purely ferroelastic based on the Aizu's classification [24, 25], because the symmetry of MAPbI$_3$ is lowered from the *m3m* to *4/mmm* centrosymmetric point group. This ferroelastic transition [16] can generate six types of domain orientations, which are referred to as ZY, Z–Y, ZX, Z–X, XY, and –XY [26]. Domain walls intersecting at 0°, 45°, or 90° in the principal axis system of the crystal are used to mark these domain orientations. A large (2 mm × 5 mm) and thin (about 20 μm) MAPbI$_3$ single crystal was synthesized and bent upward or downward to generate tensile or compressive strain within the crystal, respectively. Strelco *et al.* defined a stress-dependent area fraction (F, $F = \frac{\text{Area}_{\text{bright}} - \text{Area}_{\text{dark}}}{\text{Area}_{\text{bright}} + \text{Area}_{\text{dark}}}$) [22] to be a valid parameter to quantify the evolution of the domain architecture in the single crystal under strain. Based on this relevant descriptor, $F = 1$ (or −1) demonstrates that the field of view is filled with a bright (or dark) single domain, while $F = 0$ indicates an equal area of bright and dark domains. The data points in the complex domain area fraction (F) versus stress curve are started from state A, for which $F = 0$. Obviously, the system turned into a state dominated by bright domains under tensile stress with increasing F. More importantly, a new series of bright domains occurred with the new domain walls tilted 70° and 109° in contrast to the old walls in response to the application of a larger tensile stress. The formation of the newly observed domains is always accompanied by a minimum value of F at the stress of about 0.5 a.u.. Consequently, the phase transition generating 70° domains must be responsible for the previous increase of F maximum value by minimizing the energy of the system. Furthermore, on releasing stress, the value of F became negative, and the density of the new domains gradually decreased until they disappeared due to the decrease of the elastic energy storage. Then, with additional increase of compressive stress, the bright domains completely disappeared, and the value of F changed suddenly with the phase transition. Beyond this point, the single crystal of MAPbI$_3$ was cracked and thereby relieved part of the applied compressive strain. Thereafter, the domain pattern, which is similar to the pattern under zero stress, reappeared and the response of the system to external stress became significantly weak. It can be concluded that the bright and dark fields could be defined as Z and −Y domains, respectively. The non-linear and hysteretic pattern depends on the applied stress, which is a characteristic feature for ferroelastic materials. Moreover, Strelco's group [15] also provided solid evidence of ferroelastic properties and intergrain strain in MAPbI$_3$ polycrystalline films using piezoresponse force microscopy. As the ferroelastic domain boundaries have a great impact on the long-term stability of functional HOIP devices, the discovery of ferroelasticity in MAPbI$_3$ offers a new direction to modify and improve the stability of these functional materials.

Ferroelastic materials generally exhibit transformations in crystal morphology, crystal architecture, and mechanical behaviors when they are exposed to external mechanical stress [27]. For example, ferroelastic single crystals can show various plastic deformations upon loading shearing strain along a specific crystal orientation. Specifically, an inferior phase occurs at the bending domain, in which the orientation of atoms and molecules changes together, while the relative structural arrangement does not change; the single crystal thus goes through a twinning deformation without crystal structure transformation. Up to now, only a few studies reporting mechanical-stress-induced shape changes have been published for organic molecular crystals [28–32].

Another closely related metal–organic system is also discussed here. Recently, Seki *et al.* reported the first example of ferroelastic organometallic gold complexes that contain arylgold(I) (N-heterocyclic carbene) (NHC) (figure 9.14(a)) [33]. The crystal of arylgold(I) (N-heterocyclic carbene) exhibits distinct mechanical bending upon exposure to stress. This bent shape is maintained upon removing the applied stress which validates plastic bending. Repeated experimental progress demonstrated that the bending angle of arylgold(I) (N-heterocyclic carbene) is about $45°$. SEM characterizations for bent crystals of arylgold(I) (N-heterocyclic carbene) confirmed a continuous surface at the boundaries between the bent and unbent parts (figure 9.14(b)). Even a $45°$ bending for the local domain (about 5 μm) of the bent crystal edge of arylgold(I) (N-heterocyclic carbene) has been discovered when loading stress (figure 9.14(c)), indicating a high tendency of bending at this angle. Single-crystal X-ray diffraction (XRD) indicated that the compound arylgold(I) (N-heterocyclic carbene) crystallized in the $P\bar{1}$ space group at 20 °C [34]. In the independent molecule (figure 9.14(a)), the two aromatic rings include a dihedral angle of $4.7(3)°$, and the independent molecules stacked along the a-axis in a head-to-tail form with the π-stacking distance of about 3.415 Å. The face indexes of the bent and unbent parts of crystal arylgold(I) (N-heterocyclic carbene) demonstrated that the twining deformation was generated by an occurrence of the inferior phase α_1 from the original mother phase α_0 (figure 9.15(a)) by rotating the bent part by $180°$ along the [100] direction [35].

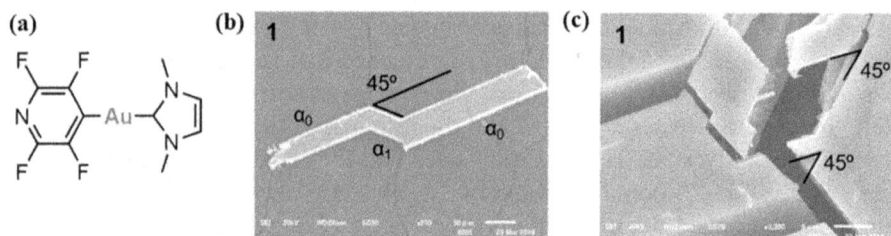

Figure 9.14. (a) Chemical structure of the ferroelastic organometallic gold complexes that contain arylgold(I) (N-heterocyclic carbene). (b), (c) SEM morphologies of twinning deformations of crystal arylgold(I) (N-heterocyclic carbene). (Reproduced with permission from [33]. Copyright 2020 John Wiley and Sons.)

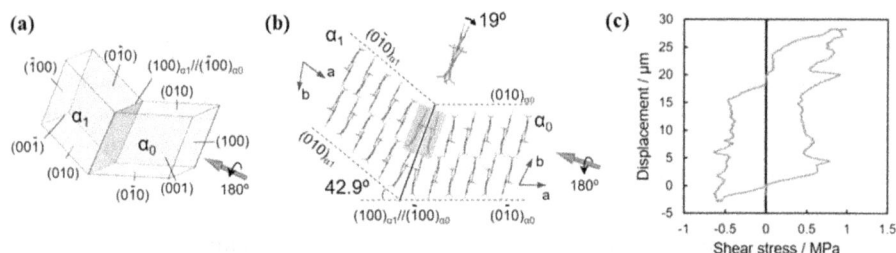

Figure 9.15. (a) The crystal face indexes and (b) the molecule packing arrangements for bent and unbent parts of crystal arylgold(I) (N-heterocyclic carbene). (c) Stress versus strain hysteresis loop for the ferroelastic bending of crystal arylgold(I) (N-heterocyclic carbene). (Reproduced with permission from [33]. Copyright 2020 John Wiley and Sons.)

Moreover, the (100) face of the α_1 phase and $(\bar{1}00)$ face of the α_0 phase formed a twining interface; both faces are perpendicular to the molecules' stacking direction (figure 9.15(b)). According to these discovered crystallographic features, the bending angle between the α_0 and α_1 phases can be calculated as 42.9°, which agrees well with the bending angle predicted via SEM (45°) (figure 9.14(b)). In order to achieve continuous boundaries between the bent and unbent parts in the twinning deformation, molecular rotations of about 19° (figure 9.15(c)) can be predicted. Furthermore, a feasible slippage of π-stacked columns also occurred with the overall column structure maintained. Subsequently, the stress versus strain plot, presented in figure 9.15(c), was determined by applying a shear force via a jig on the specific crystallographic face of crystal arylgold(I) (N-heterocyclic carbene) [30]. Crystal arylgold(I) (N-heterocyclic carbene) started to respond to the force almost immediately when the jig touched the crystal surface. The domain of the α_1 phase began to grow with an applied force of 0.61 MPa. Similar to previously reported ferroelastic crystals, the external shear stress is relatively constant during the generation progress of the α_1 phase. After removing the stress, a spontaneous build-up strain was observed. When loading a force of 0.49 MPa in the opposite direction further, a reverse deformation of crystal arylgold(I) (N-heterocyclic carbene) was detected to recover the original α_0 phase. Such hysteresis loop and the spontaneous build-up strain clearly proved the ferroelastic characteristic for arylgold(I) (N-heterocyclic carbene).

9.6 Summary

In this chapter, the recent advances of ferroelasticity in some representative HOIPs and related systems are summarized, including synthesis, crystal structures, phase transitions, domain evolution, and physical properties. The emergence of organic–inorganic hybrid crystals has expanded the realm of ferroelastic materials from purely inorganic to hybrid systems, hence opening an entirely new branch of this class of technologically important materials. Strikingly, the union and integration of both inorganic and organic components at a molecular level in these hybrid crystals give rise to enormous opportunities for introducing new properties and functionalities in addition to ferroelasticity, from the degree of structural and electronic dimensions. Moreover, the soft nature of hybrid crystals endows them with high sensitivity in response to external perturbations, such as temperature, pressure, force, and light, therefore enriching the avenue for the regulation of ferroelasticity and related properties.

Acknowledgments

We are grateful for the financial support from the National Key Research and Development Program of China (Grant no. 2022YFA1503301) and the National Science Foundation of China (Grant no. 22375105).

Bibliography

[1] Rose G 1839 *De novis quibusdam fossilibus quae in montibus Uraliis inveniuntur* (AG Schade) pp 3–5
[2] Wenk H and Bulakh A 2004 *Minerals: Their Constitution and Origin* (Cambridge University Press)

[3] Li W, Wang Z, Deschler F, Gao S, Friend R H and Cheetham A K 2017 Chemically diverse and multifunctional hybrid organic–inorganic perovskites *Nat. Rev. Mater.* **2** 16099

[4] Goldschmidt V M 1926 *Die gesetze der krystallochemie Naturwissenschaften* **14** 477–85

[5] Kieslich G, Sun S and Cheetham A K 2014 Solid-state principles applied to organic–inorganic perovskites: new tricks for an old dog *Chem. Sci.* **5** 4712–5

[6] Cheetham A K and Rao C N R 2007 *There's room in the middle* Science **318** 58–9

[7] Li W, Stroppa A, Wang Z-M and Gao S 2020 *Hybrid Organic–Inorganic Perovskites* (Weinheim: Wiley) pp 79–149

[8] Li W, Zhang Z, Bithell E G, Batsanov A S, Barton P T, Saines P J, Jain P, Howard C J, Carpenter M A and Cheetham A K 2013 Ferroelasticity in a metal–organic framework perovskite; towards a new class of multiferroics *Acta Mater.* **61** 4928–38

[9] Hu S-Q, Li M-Z, Chen Z-H, Zhou J-S, Ji L-Y, Ai Y and Chen X-G 2024 Switchable coordination bonds in 3D cyano-bridged perovskite ferroelastics: achieving the largest leap of symmetry breaking and enhanced dielectric switching performance *Inorg. Chem. Front.* **11** 4647–53

[10] Zhao M-M, Zhou L, Shi P-P, Zheng X, Chen X-G, Gao J-X, Geng F-J, Ye Q and Fu D-W 2018 Halogen substitution effects on optical and electrical properties in 3D molecular perovskites *Chem. Commun.* **54** 13275–78

[11] Mencel K, Kinzhybalo V, Jakubas R, Zaręba J K, Szklarz P, Durlak P, Drozd M and Piecha-Bisiorek A 2021 0D bismuth(III)-based hybrid ferroelectric: tris(acetamidinium) hexabromobismuthate(III) *Chem. Mater.* **33** 8591–601

[12] Xiao X, Zhou J, Song K, Zhao J, Zhou Y, Rudd P N, Han Y, Li J and Huang J 2021 Layer number dependent ferroelasticity in 2D Ruddlesden–Popper organic-inorganic hybrid perovskites *Nat. Commun.* **12** 1332

[13] Jakobsen V B *et al* 2022 Domain wall dynamics in a ferroelastic spin crossover complex with giant magnetoelectric coupling *J. Am. Chem. Soc.* **144** 195–211

[14] Hu Y *et al* 2021 Ferroelastic-switching-driven large shear strain and piezoelectricity in a hybrid ferroelectric *Nat. Mater.* **20** 612–17

[15] Li J, Zhu Y, Huang P-Z, Fu D-W, Jia Q-Q and Lu H-F 2022 Ferroelasticity in organic–inorganic hybrid perovskites *Chem.—Eur. J.* **28** e202201005

[16] Fu D-W, Gao J-X, He W-H, Huang X-Q, Liu Y-H and Ai Y 2020 High-T_c enantiomeric ferroelectrics based on homochiral Dabco-derivatives (Dabco=1,4-Diazabicyclo[2.2.2] octane) *Angew. Chem. Int. Ed.* **59** 17477–81

[17] Chen B, Zheng X, Yang M, Zhou Y, Kundu S, Shi J, Zhu K and Priya S 2015 Interface band structure engineering by ferroelectric polarization in perovskite solar cells *Nano Energy* **13** 582–91

[18] Dazzi A, Glotin F and Carminati R 2010 Theory of infrared nanospectroscopy by photothermal induced resonance *J. Appl. Phys.* **107** 124519

[19] Deng Y, Peng E, Shao Y, Xiao Z, Dong Q and Huang J 2015 Scalable fabrication of efficient organolead trihalide perovskite solar cells with doctor-bladed active layers *Energy Environ. Sci.* **8** 1544–50

[20] Dong Q, Fang Y, Shao Y, Mulligan P, Qiu J, Cao L and Huang J 2015 Electron-hole diffusion lengths > 175 μm in solution-grown $CH_3NH_3PbI_3$ single crystals *Science* **347** 967–70

[21] Frost J M, Butler K T, Brivio F, Hendon C H, van Schilfgaarde M and Walsh A 2014 Atomistic origins of high-performance in hybrid halide perovskite solar cells *Nano Lett.* **14** 2584–90

[22] Strelcov E, Dong Q, Li T, Chae J, Shao Y, Deng Y, Gruverman A, Huang J and Centrone A 2017 $CH_3NH_3PbI_3$ perovskites: ferroelasticity revealed *Sci. Adv.* **3** e1602165

[23] Frost J M, Butler K T and Walsh A 2014 Molecular ferroelectric contributions to anomalous hysteresis in hybrid perovskite solar cells *APL Mater.* **2** 081506

[24] Honig M, Sulpizio J A, Drori J, Joshua A, Zeldov E and Ilani S 2013 Local electrostatic imaging of striped domain order in $LaAlO_3/SrTiO_3$ *Nat. Mater.* **12** 1112–18

[25] Jacobsson T J, Schwan L J, Ottosson M, Hagfeldt A and Edvinsson T 2015 Determination of thermal expansion coefficients and locating the temperature-induced phase transition in methylammonium lead perovskites using x-ray diffraction *Inorg. Chem.* **54** 10678–85

[26] Hermes I M *et al* 2016 Ferroelastic fingerprints in methylammonium lead iodide perovskite *J. Phys. Chem.* C **120** 5724–31

[27] Baranyai P, Marsi G, Jobbágy C, Domján A, Oláh L and Deák A 2015 Mechano-induced reversible colour and luminescence switching of a gold(I)–diphosphine complex *Dalton Trans.* **44** 13455–59

[28] Crudden C M *et al* 2014 Ultra stable self-assembled monolayers of N-heterocyclic carbenes on gold *Nat. Chem.* **6** 409–14

[29] Fiebig M, Lottermoser T, Meier D and Trassin M 2016 The evolution of multiferroics *Nat. Rev. Mater.* **1** 16046

[30] Ito H, Muromoto M, Kurenuma S, Ishizaka S, Kitamura N, Sato H and Seki T 2013 Mechanical stimulation and solid seeding trigger single-crystal-to-single-crystal molecular domino transformations *Nat. Commun.* **4** 2009

[31] Jin M, Sumitani T, Sato H, Seki T and Ito H 2018 Mechanical-stimulation-triggered and solvent-vapor-induced reverse single-crystal-to-single-crystal phase transitions with alterations of the luminescence color *J. Am. Chem. Soc.* **140** 2875–79

[32] Jin M, Seki T and Ito H 2017 Mechano-responsive luminescence via crystal-to-crystal phase transitions between chiral and non-chiral space groups *J. Am. Chem. Soc.* **139** 7452–55

[33] Seki T, Feng C, Kashiyama K, Sakamoto S, Takasaki Y, Sasaki T, Takamizawa S and Ito H 2020 Photoluminescent ferroelastic molecular crystals *Angew. Chem. Int. Ed.* **59** 8839–43

[34] Penney A A, Sizov V V, Grachova E V, Krupenya D V, Gurzhiy V V, Starova G L and Tunik S P 2016 Aurophilicity in action: fine-tuning the gold(I)–gold(I) distance in the excited state to modulate the emission in a series of dinuclear homoleptic gold(I)–NHC complexes *Inorg. Chem.* **55** 4720–32

[35] Osawa M, Kawata I, Igawa S, Hoshino M, Fukunaga T and Hashizume D 2010 Vapochromic and mechanochromic tetrahedral gold(I) complexes based on the 1,2-bis (diphenylphosphino)benzene ligand *Chem.—Eur. J.* **16** 12114–26

IOP Publishing

Ferroelastic Materials

Guillaume F Nataf, Blai Casals and Ekhard K H Salje

Chapter 10

Applications of ferroelastic materials: past, present and future

Tadej Rojac, Etienne Lemaire and Dwight Viehland

10.1 Introduction

There are many types of materials that exhibit ferroelastic-like effects. They are identified by a unique microstructure whose formation and physical attributes are governed by the minimization of the elastic strain energy [1–5]. The term shape-memory alloy (SMA) was first used in describing martensitic and/or ferroelastic materials that develop a spontaneous elastic strain ($\varepsilon_{\text{spont}}$) on cooling below a symmetry reducing phase transition. The term SMA refers to the ability of a low temperature deformed elastic state to recover its originally shape on heating above the phase transition [6].

The unique microstructure is a polydomain state that forms on cooling from a high-temperature prototypic state through a symmetry reducing displacive phase transformation [7]. Bonds are not broken, rather the crystal lattice parameters simply elastically distort. On cooling through the phase transformation, a poly-domain state forms which consists of multiple symmetry equivalent variants or directions along which structural distortions are oriented. An example of a ferroelastic polydomain state is given in figure 10.1. The symmetry lost at the phase transformation is the symmetry operation(s) that relates the equivalent variants to each other. Each individual domain has a uniform spontaneous strain, where the thermodynamics of the transformation can be described by Landau (LD) phenomenology expanded in terms of an order parameter $\varepsilon_{\text{spont}}$ [3]. Neighboring domains are separated by twin boundaries that consists of strain gradients that can be described by adding Ginzburg terms to the phenomenology. The gradient terms represent an excess (positive) elastic energy, whose elimination would reduce the free energy of the ensemble of domains, yielding a uniformly and coherently strained material throughout the body. The shape-memory effect, in this case, is the recovery of the polydomain ensemble state on heating from an *a priori* elastically deformed

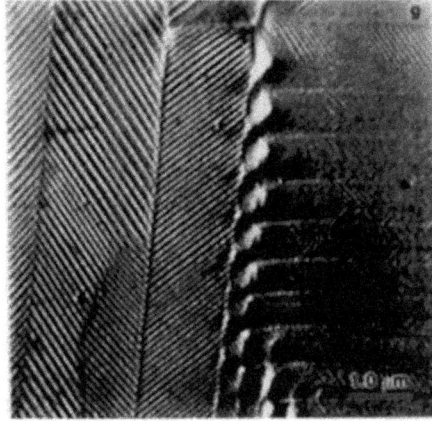

Figure 10.1. Typical domain structure of the ferroelastic/martensitic phase. (Reproduced with permission from [7]. Copyright 2003 AIP Publishing.)

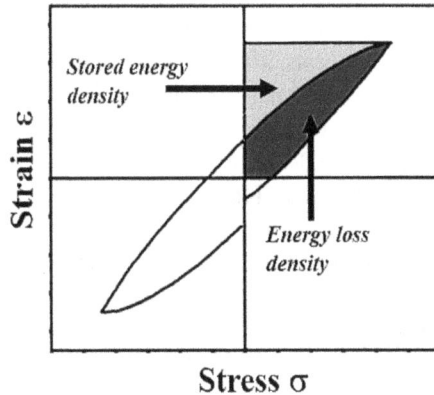

Figure 10.2. Stress–strain hysteresis curve for ferroelastic/martensite. The stored energy is the area shown in light gray, and the energy loss is that shown in dark gray.

condition. There are also SMA effects that can sometimes occur between two low temperature deformed conditions [8], without cycling through the higher temperature phase transition: such memory effects can occur in highly defective or substituted alloys, where internal nonuniform stress fields can assist in memory effects between two low temperature strained conditions.

Application of stress σ to a ferroelastic/martensitic material results in a nonlinear hysteretic ε–σ curve, where the domain orientation is switched between symmetry equivalent variants by σ. Energy is both stored and dissipated on cycling the ε–σ hysteresis curve. As shown in figure 10.2, the dissipated energy is the area of the hysteresis curve, and the stored mechanical energy is the area above that curve over the σ range between the remanent strain ε_r and the saturation strain ε_{sat}. The area of the hysteresis curve represents a material's ability to dissipate energy from impact, yielding a robustness to, for example, martensitic steels.

10.2 The material can be the machine: transduction of energy and efficient power transfer

The term SMA has come to describe the behavior of many other types of materials other than just ferroelastic and/or martensite materials which have an order parameter of ε_{spont}. These other types include (i) ferroelectrics which have significant strains, such as Pb-based perovskite ferroelectrics such as $Pb(Zr,Ti)O_3$ ceramics and $Pb(Mg_{1/3},Nb_{2/3})O_3$–$PbTiO_3$ single crystals [9]. In this case, the order parameter is a spontaneous polarization (P_s), but because of electrostriction, the shape of the unit cell changes with the insertion of polarization into the lattice [10]. Thus, a ferroelectric polydomain state can have excess elastic energy, due to the presence of domain boundaries between symmetry equivalent variants [7]. In this case, the domain distribution can be changed both by the applied electric field E and stress σ. In fact, poled ferroelectric materials that are strongly piezoelectric can be completely de-poled by σ [11]. Accordingly, a poled ferroelectric material that is elastically deformed at low temperatures can also recover its original shape on heating above the ferroelectric phase transformation. This is a shape-memory effect, via a proper ferroelectric that is also an improper ferroelastic [12].

Furthermore, the term SMA has also been used to describe (ii) ferromagnetic shape-memory alloys, which undergo an elastically displacive martensitic transformation, and which are also ferromagnetic, such as Ni–Mn–Ga [12–16]. Application of either a magnetic field H or stress σ can alter the domain distribution, inducing significant shape changes. Heating above the structural phase transformation temperature again results in recovery of the original shape. The inclusion of other types of material into the SMA classification has also been expanded by some to include (iii) shape-memory polymers [17, 18]. These are polymers that establish a residual elastic strain state by the application of stress during a processing step. This sets up a memory condition that can be recovered after application of a secondary mechanical stress at lower temperatures.

The various applications of ferroelastic and other co-elastic materials are due to their unique ability to recover their shape on temperature cycling. They can also dissipate energy via the hysteresis curve, and store and/or transduce mechanical energy.

At the heart of the energy transduction capability is the concept popularized by Bhattacharya and James [19], i.e. 'the material is the machine'. There are no gears, rotators or other moving parts typical of a machine. Rather, the coordination of atomic displacements and the elastic accommodation between domain regions does all the work. Following the original geometric/crystallographic theory by Wechsler–Lieberman–Read (WLR) from 1955 [1] for elastic compatibility conditions of martensitic phase transformations, the ferroelastic microstructure consists of poly-domain plates: a structurally heterogeneous condition. Each plate is formed by alternating layers of twin-related domains, as shown in figure 10.3 [7]. A macro-scopic invariance of the habit plane is achieved by adjusting the relative thicknesses of the domain layers [1, 20], eliminating long-range stress fields generated by crystal lattice misfits. A crystallographic pane exists that has no distortion or rotation,

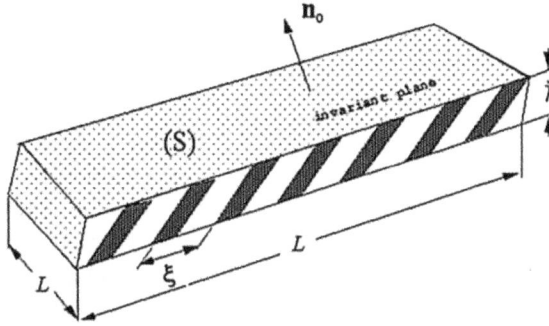

Figure 10.3. Schematic illustrating the polydomain structure formed by alternating twin-related domains in stress-free martensite. (Adapted with permission from [7]. Copyright 2003 AIP Publishing.)

called the invariant plane. This special plane achieves elastic accommodation with long-range stress fields eliminated—a structurally heterogeneous state with zero net stress. In ferroelectric materials with large strain, a similar invariant plane can be achieved, with the additional requirement of head to tail patterns of neighboring domains.

Bhattacharya and James then formulated their concept using a mathematical extension (Γ convergence) applied to thin films for the WLR theory of the elastic compatibility conditions in bulk. They found that special eigenvalue solutions exist for the matrix equation for elastically compatible interfaces of thin films, which reduces dissipation by restricting the crystallographic pathway along which domain redistribution occurs under an applied field. The interfacial condition offers an approach to dramatically reduce dissipation due to special boundary conditions of the thin-film geometry.

There is also another recently understood special condition [21, 22] that can minimize the energy of domain redistribution. First, we need to understand the balance between interfacial and elastic strain energies, which in turn governs the size of structural domains (λ_0) [22–24]. Minimization of the sum of these two energies yields a well-known relationship between λ_0, the polydomain plate thickness D, and the domain wall energy γ. This relationship is given as

$$\lambda_0 = \beta \sqrt{\frac{\gamma}{\mu \varepsilon_0^2} D}, \tag{10.1}$$

where μ is the shear modulus, ε_0 is the twinning strain, and β is a dimensionless constant. Large domain wall energies γ result in large domains that are energetically expensive to move and dissipative. This is the case in most martensitic polydomain structures.

However, there are special situations where $\gamma \rightarrow 0$ [7, 21, 22, 25]. This is the adaptive martensite and/or ferroelectric state, which is a structurally heterogenous state, where the scale of structural heterogeneity is very small. This is the case of pre-martensite [26] and of poled $(1 - x)$at%Pb$(Mg_{1/3}Nb_{2/3})O_3 - x$at%PbTiO$_3$ type piezoelectric crystal [27–31]. The domain size in both these materials is on the nanometer scale, which is below the coherence length for the x-rays and neutrons used in diffraction studies for structural determination [32, 33]. These materials have been shown to form a domain hierarchy which results in apparent symmetry reduction due to averaging over multiple nanodomain orientations [7, 25]. The nanodomains self-assemble into patterns that reduce the excess elastic energy. In this case, for example, an ensemble of tetragonal nanodomains appears monoclinic by x-ray diffraction (XRD) and neutron diffraction. For the PMN-PT type piezo-electric single crystals in the poled condition, the readjustment of the ensemble of polar nanodomains is not dissipative. The low energy domain walls readjust easily under an applied field. Dissipation is only incurred by re-nucleation under strong reverse bias. The low domain wall energy condition, along with the compatibility conditions, is a special condition that has importance to the reduction of energy dissipation required for the redistribution of domain distributions under ordering fields.

Both the special cases of thin layers and the reduction of domain wall energy offer important approaches to highly efficient transduction of energy. They offer unique ways by which the material 'can act like a machine'. It has important applications in actuators, energy transfer and transduction.

10.3 Multifunctionality can be enabled

The stored mechanical energy imparts an ability to do reversible work, and/or to transduce mechanical energy to other forms (electrical, magnetic, etc). This can be achieved by coupling to other co-order parameters, or to other elastically bonded layers in a hybrid material. Salje was the first to study how coupled order parameters are important to the thermodynamic and physical properties of minerals and other co-elastic solids, introducing biquadratic coupling coefficients [3].

Ferroelastic and other co-elastic type materials can couple strain, polarization, and/or magnetization—this enables multifunctional properties. For example, ferro-electrics are a type of co-elastic material that is similar to ferroelastics. They offer the ability to transduce electrical to mechanical energies, and vice versa, at high coupling and good efficiency. These materials are the basis for electromechanical coupling and transduction, used in sonar and ultrasonic applications. Most ferromagnetic materials do not have a significant coupling between magnetization and lattice distortion. However, there are special classes of magnetostrictive materials (example, Terfenol-D [34], Galfenol [35], and Metglas [36]) that can transduce magnetic and elastic energies at good efficiency and coupling. They are the basis of magneto-mechanical transduction and can be used in sonar and ultrasonic applications [37]. There are also special cases of ferromagnetic shape-memory alloys, where an input magnetic energy results in ferroelastic domain redistribution.

These electromechanical and magneto-mechanical materials inherently involve the transfer of energy between mechanical and electric forms, and vice versa, and mechanical and magnetic forms, and vice versa. Some of them can transduce energy at high coupling coefficients and high efficiency. Piezoelectric PMN-PT crystals can have longitudinal electromechanical coupling coefficients greater than 90% [9], and magnetostrictive Metglas and Terfonel-D [34–36] can have similarly high magneto-mechanical coupling coefficients and efficiencies. These energy transduction capabilities underlies the unique behavior enabling multifunctionality.

Furthermore, multiferroic and magnetoelectric hybrid materials have been developed using a layered architecture [38, 39]. For example, magnetostrictive layers can be bonded to piezoelectric and/or ferroelastic ones [40–42]. Magnetic energy can then be transduced to electrical energy, via an elastic strain. This makes possible the strain tuning of the magnetic, electrical and optical properties of hybrid materials. It forms the basis of many types of emerging multifunctional properties in future device applications [43]. These hybrid materials also can have layers that can remember prior elastic strain states.

Symmetry is broken in the twin walls [25, 44, 45], localizing specialized functional properties with in the twin walls. Ferroelectric, ferroelastic, and ferromagnetic twin walls all offer an approach to domain boundary engineering. The topology of the twin walls, their interaction with dopants/defects, and the loss of translation symmetry at the walls can result in properties not found in the bulk of the material.

Superconducting, chiral, magnetoelectric, ionic conducting, chemical transport, heat transport, and other unique properties can be engineered in the confinement of twin boundaries that are not inherent to the bulk. The functional interfaces offer the potential to be optimized by decreasing the twin wall energy, and consequently increasing the twin boundary density. This approach offers new functional and multifunctional properties to materials for numerous important emerging applications.

In summary, the special energy transduction capabilities of ferroelastic and co-elastic materials impart to them a wide range of present and future applications. Some of these are summarized in figure 10.4 [46]. In the following sections, we will

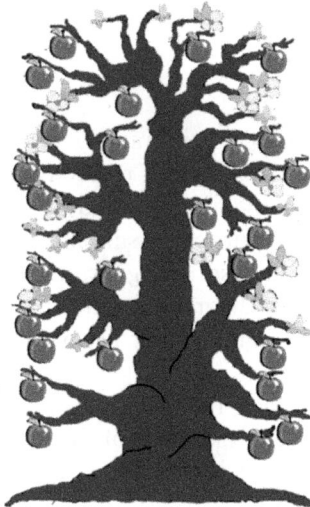

Ferroelastic/SMA
• Memory wires for dental, medical and bioengineering
• Vibration dampening of structures
• Actuators in aerospace platforms
• Retrofitting structures for seismic activity

Ferroelectric
• Piezoelectric transducers and actuators
• Medical ultrasound
• Acoustic sensors
• Transformers
• Acousto-optic modulators
• Servo-mechanisms
• Memory elements
• Memristors

Magnetostriction and magnetic SMA
• Magneto-elastic transducers and actuators
• Transformers
• Phase shifters
• Sensors

Magnetoelectric
• Magnetic sensors
• Stress tunable filters and gyrators
• Stress tunable electronics
• E-field writable magnetic memory
• Low frequency mechanical antenna

Applications of ferroelastic and co-elastic materials

Figure 10.4. Summary of the various applications of ferroelastic and related co-elastic materials. (Reproduced from [46]. CC BY 4.0.)

talk in more detail about some of these classes of materials and their special properties that enable device applications, starting in the section with ferroelectric ceramic oxides in electromechanical applications.

10.4 Tailoring the electromechanical response of ferroelectric/ ferroelastic perovskites for piezoelectric applications

Historically, ferroelectric/ferroelastic materials based on oxide perovskites have been the primary choice for applications in electromechanical devices. Soon after the first studies in 1950s [47, 48], Pb(Zr,Ti)O$_3$ (PZT), a solid solution between antiferro-electric PbZrO$_3$ and ferroelectric PbTiO$_3$, became the commercially dominant material for a range of sensor, actuator and ultrasound devices. Despite the significant progress in the search for lead-free alternatives, to date, PZT cannot be replaced on a general basis [49, 50]. Interestingly, the reason for the uniqueness of PZT is not so much its (otherwise excellent) piezoelectric properties, but the ability to tailor them using chemical modifiers, such as dopants. These concepts are called hardening and softening [51] and will be presented in more detail in the following paragraphs. The mechanism lying behind these property-engineering approaches is related to the microscopic interactions between the ferroelectric/ferroelastic non-90° domain walls (present in poled ceramic samples) and charged point defects, such as ionic vacancies combined with the dopant lattice sites [52–55]. The contribution of ferroelectric/non-ferroelastic 180° domain walls (e.g. to the dielectric response) in poled samples is usually disregarded, first because 180° domain walls lack electro-mechanical coupling [56] (although they can contribute to the electromechanical response in special circumstances [57]), and second due to the expected low concentration of such interfaces after the switching process occurs during the poling procedure of ceramics or films [58].

10.4.1 Hardening and softening in PZT

Hardening and softening concepts are intimately related to the ordering and disordering of pinning centers in the material, respectively. An illustration of these concepts using PZT as an example is presented in figure 10.5. It should be emphasized that the schematic is purposely simplified but it captures the main characteristics of the hardening and softening principles. The images were con-structed based on a large number of theoretical and experimental studies performed in the past several decades on PZT and similar perovskites (for details consult, for example, the review papers [52, 59]).

In the case of the hard PZT variant, the material is doped with an acceptor, i.e. a foreign ion of a lower oxidation state than that of the host ion. An example is Fe^{3+} ions replacing B-site ions in PZT of higher charge, i.e. Zr^{4+} and Ti^{4+}, creating negatively charged Fe$'_{Ti}$ defects in the lattice (given in standard Vink–Kröger notation where the Zr host site is omitted for simplicity). The compensating defects are thus positively charge and often assumed or shown to be oxygen vacancies, $V_O^{\bullet\bullet}$ (figure 10.5(a), hard). First-principles calculations [60–63] supported by a large

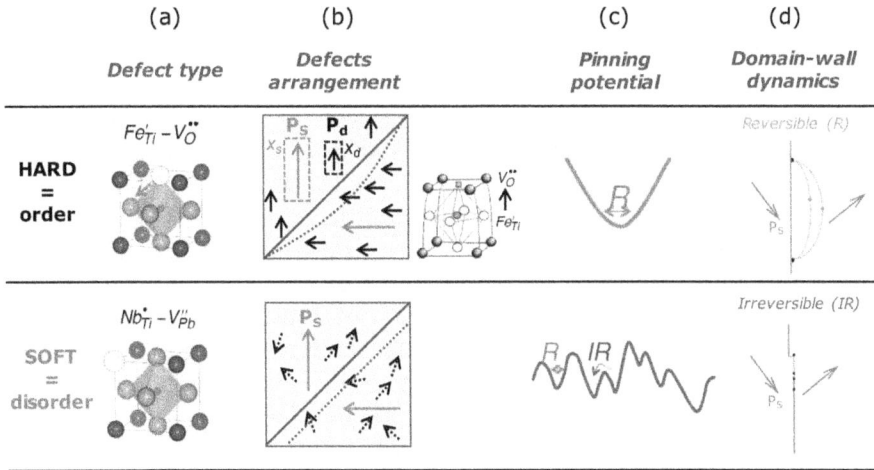

Figure 10.5. Schematic illustration of the hardening and softening concepts using PZT as an example. (a) Type of charged point defects associated with the pinning centers, (b) defect arrangements inside non-180° ferroelectric/ferroelastic domains and (c) the resulting pinning potential together with (d) the effect on domain wall dynamics in hard and soft PZT. Hard PZT is represented by Fe doping, while soft PZT by Nb doping, creating Fe'_{Ti} and $Nb^{\bullet}_{Ti}$ defects, respectively (defects are represented in Vink–Kröger notation; for simplicity, only Ti host sites are included). In panel (a), $V^{\bullet\bullet}_O$ and V''_{Pb} denote an oxygen vacancy and lead vacancy as compensating defects, respectively, drawn as white circles in the respective perovskite cells. In panel (b), P_s and x_s denote spontaneous polarization and spontaneous strain; in analogy, P_d and x_d denote the polarization and strain associated with the defect complexes of the bound $Fe'_{Ti}-V^{\bullet\bullet}_O$ defects in hard PZT (see also the schematic of the perovskite-cell deformation associated with the $Fe'_{Ti}-V^{\bullet\bullet}_O$ defect complex for the case of hard PZT). The smaller black arrows in the case of hard PZT depict the electric dipoles associated with the $Fe'_{Ti}-V^{\bullet\bullet}_O$ complexes. The smaller dashed black arrows in the case of soft PZT denote the $Nb^{\bullet}_{Ti}$ and V''_{Pb} defects, which are mostly disordered (since such defects do not have a tendency to form complexes [60]). The straight and dashed lines in the schematic show a domain wall at rest (before field application) and after application of the electric (or stress) field, respectively. In panel (c) and (d), R and IR denote reversible and irreversible domain wall displacements, respectively. Parts of this figure are courtesy of Professor Dragan Damjanovic.

amount of experimental evidence, particularly using electron-paramagnetic resonance (EPR) spectroscopy [64–71], confirm the tendency of such defect pairs in perovskite to bind into defect complexes. Theoretical examples include a large variety of acceptor-doped $PbTiO_3$ with the dopants including K, Ca, Sc, Mn, Fe, Co, Ni, Cu and Zn [60–63]. On the other hand, experimental support for the existence of defect complexes have been provided in various acceptor-doped lead-based ($PbTiO_3$, $PbZrO_3$, PZT) [64–68] and lead-free ($BaTiO_3$, KNN, BNT) [64, 69–71] perovskites (in most cases, the acceptor dopants are Cu, Fe or Mn).

It is important to note that the defect complexes do not only form their own electric dipole due to the opposite charges of the constituent defects (P_d in figure 10.5(b), hard), but that they are also associated with an elastic deformation of the lattice (x_d in figure 10.5(b), hard; [53]). The latter is induced due to the mismatch between the ionic radius of the dopant and the host ion as well as due to the chemical strain associated with the compensating $V^{\bullet\bullet}_O$ (see also the deformed perovskite-cell schematic in figure 10.5(b), hard) [72]. Owing to the electric and elastic coupling between the defect complexes (P_d, x_d) and ferroelectric distortion (P_s, x_s), the most favorable electrostatic and elastic-energy state will be achieved when the complexes are aligned parallel to P_s and x_s [53, 73]. This equilibrium defect state creates strong electric and elastic pinning effects, described phenomenologically as a parabolic-like pinning potential [53] (figure 10.5(c), hard), so that under applied electric or stress fields the domain wall can only move in a reversible fashion, ideally returning to its stable positions upon field removal (figure 10.5(d), hard) [73]. The deep pinning potential for domain walls in hard PZT is thus a consequence of the aligned defects [58]. This alignment will start naturally as P_s and x_s develop in the material, that is, after the sintering process when the sample is cooled just below the Curie temperature (T_c) [52, 74]. Both the tendency of defects to bind together and the ability of such complexes to align via $V_O^{\cdot\cdot}$ site-hopping creates the conditions for ordering, which lies at the origin of hardening. It has to be noted, however, that such ordering is not necessarily realized via defect-complex alignment inside domains, as drawn in figure 10.5(b) (hard), but can also be achieved by segregation of defects in grain-boundary and/or domain wall regions [55, 75–77].

In contrast to hard PZT, donor-doped soft PZT is created by doping with donors, meaning foreign ions with higher oxidation states than that of the replacing host ion. An example is Nb^{5+} dopant in Zr^{4+}/Ti^{4+} B sites, creating positively charge $Nb_{Ti}^{\cdot}$ defects (figure 10.5(a), soft). It is often assumed that the donor-compensating defects are negatively charged lead vacancies, $V_{Pb}^{''}$ [78]. While softening mechanisms are still not entirely clear [60], what prevents the donor-doped material from hardening is the expected reduced concentration of $V_O^{\cdot\cdot}$, counterbalancing the introduction of $Nb_{Ti}^{\cdot}$ sites with same positive charge as $V_O^{\cdot\cdot}$. Nevertheless, relative to the hard PZT variant, the situation in soft PZT is fundamentally different in two aspects: (i) as suggested by recent first-principles calculations [60], which are well supported by EPR analysis (using Gd as a donor dopant [78]), the introduced defect pairs, $Nb_{Ti}^{\cdot}$ and $V_{Pb}^{''}$, do not have the tendency to form complexes (this is represented by the dashed arrows in figure 10.5(b), soft), and (ii) none of those defects are sufficiently mobile below the typical T_c of PZT ($\sim$370 °C for the morphotropic composition [48]). The absence of defect-complex formation and the inability of defects to align will result in a disordered defect state. The resulting disordered pinning potential (figure 10.5(c), soft) is in principle equivalent to the potential assumed by Rayleigh equations. In fact, Rayleigh law describes the weak-field nonlinearity and hysteresis arising from irreversible domain wall displacements (figure 10.5(d), soft) in a hypothetical material with perfectly disordered pinning centers [79, 80]. This idea is supported by experiments, i.e. the weak-field piezoelectric and dielectric responses of soft PZT obey satisfactorily those predicted by Rayleigh equations, while hard PZT does not [59, 81].

The different pinning potential shaped by the different types and distributions of charged point defects in hard and soft PZT have profound implications in both the high-field (super-coercive) and weak-field (sub-coercive) electrical and electromechanical properties of PZT. The consequence of the strong electric and elastic pinning effects of acceptor–oxygen-vacancy complexes is the difficulty of domains to switch to a permanent state under application of the electric field, reflected macroscopically in pinched (or double) hysteresis loops [54, 55, 74] (figure 10.6(a), black

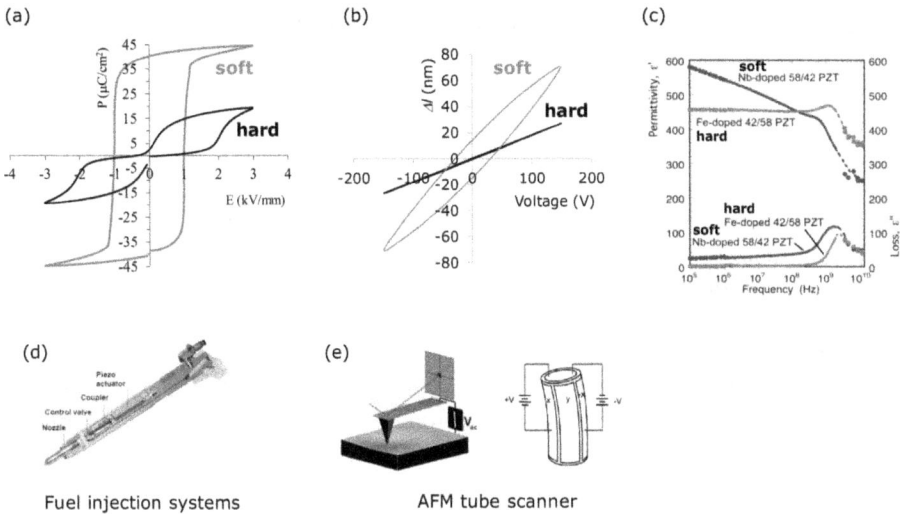

(a)

(b)

(c)

(d)

Fuel injection systems

(e)

AFM tube scanner

Figure 10.6. High-field (super-coercive) and weak-field (sub-coercive) electrical and electromechanical responses of hard and soft PZT and their implications in piezoelectric applications. (a) High-field polarization–electric-field (P–E) hysteresis loops, (b) weak-field piezoelectric hysteresis loops and (c) frequency-dependent dielectric permittivity for soft and hard PZT. (d) A fuel injection system for cars and (e) an atomic force microscopy (AFM) tube scanner as representatives of applications where soft PZT (fuel injection) or hard PZT (tube scanner) are preferred. ((c) Reproduced with permission from [96]. Copyright 2010 World Scientific Publishing.)

loop). This particular hysteresis shape, contrasting the more common open-like P–E hysteresis loop observed in soft PZT (figure 10.6(a), red loop), arises as a result of a double switching event during each half of the electric field cycle whereby the first switching event upon field increase is followed by a back-switching event (due to pinning) during the field release [73].

10.4.2 Pinning-dependent electrical and electromechanical responses and their role in applications

Following the pinning-induced reversibility of the response in hard PZT, Ren [73] proposed using this effect for generating large reversible electromechanical strain responses. The concept was show for hard $BaTiO_3$ single crystals in which the aging process was designed to obtain ordered defect complexes (as shown in figure 10.5(b) for hard PZT), leading to large recoverable strains of up to 0.8% [73]. This concept has been recently renewed by a series of publications on large bipolar electromechanical strains in so-called defect-engineered lead-free perovskites, including (K, Na)NbO_3-based (KNN) [82, 83] and (Na,Bi)TiO_3–$BaTiO_3$-based (NBT–BT) solid solutions [84–86]. Strains in excess of 1% have been reported; for a quantitative comparison, a typical strain response achieved in PZT is on the level of 0.5% [87], however, newer studies suggest that it may depend on the sample thickness and details of the experimental set-up for strain measurement [88, 89]. Interestingly, the large electromechanical strains in lead-free ceramic samples were reported to be accompanied by an unusual asymmetry in the strain–electric-field hysteresis loops, i.e. a large difference between strains measured at positive and negative field cycles. Different scenarios have been proposed to explain the peculiar strain response, perhaps the most interesting is the presumable direct electromechanical effect of the oxygen-vacancy-related defect complexes [90]. Interestingly, concurrent studies suggested that the hysteresis-loop asymmetry may originate from sample bending during measurements [88, 89, 91], apparently related to uneven distribution of oxygen vacancies at the two sample surfaces close to the electrodes [88]. While the bending effect is an issue regularly encountered during electromechanical measurements of thin films [92], it appears relevant also during measurements of thin ceramic bulk samples (with thicknesses <0.5 mm [88]). The exact role of all these phenomena is yet to be investigated, however, the bending effect linked to the inhomogeneous defect distribution across sample thickness may be beneficial, rather than a problem. For example, the effect may be used in the development of actuators alternative to monomorphic actuators where asymmetry between sample surfaces is created through complicated processes, such as (i) the design of Schottky barriers between the piezoelectric element and electrodes [93], (ii) the creation of an electric-conductivity gradient across the thickness of the beam-actuator [94] or (iii) the local chemical reduction of one of the surfaces of a beam, such as in so-called Rainbow actuators [95].

Similar to the case of the properties measured under switching conditions, the sub-switching behavior in hard and soft PZT is also affected by the domain wall pinning potential set by the different types of pinning centers in the two variants [81]. At the microscopic level, the key characteristic in hard PZT is the blockage of extensive irreversible motion of non-180° domain walls. Since such movements lead to non-linearity and hysteresis in the macroscopic response (see e.g. physical interpretations of the Rayleigh law [79]). Hard PZT will show a quasi-linear piezoelectric response with reduced hysteresis (figure 10.6(b), black loop) and weak frequency dependence of dielectric and piezoelectric properties (an example of dielectric permittivity is shown in figure 10.5(c), red curves [96]). Conversely, in soft PZT where the pinning potential is more random (figure 10.5(c), soft) and thus overall flatter than the parabolic-like potential in hard PZT (figure 10.5(c), hard), and domain walls may move more easily, overcoming pinning barriers and thus leading to a strongly nonlinear and hysteretic response (figure 10.6(b), red) arising from local (irreversible) switching events (figure 10.5(d), soft). This same reasoning applies for the large frequency dependence of weak-field properties in soft PZT (see the log-normal increase of permittivity in soft PZT as the frequency is decreased from the GHz region toward the MHz region in figure 10.6(c), blue curve [96] and compare it to the flat behavior of hard PZT, red curve). Obviously, the dielectric and piezoelectric coefficients of soft PZT are superior but this is at the expense of a more frequency-dependent response and more hysteretic, lossy behavior (see the larger dielectric losses of soft PZT compared to hard PZT in figure 10.6(c) in the whole measured frequency range) [51].

An enhanced response of soft materials is relevant in applications requiring, for example, large strokes, such as the multilayer actuators used in fuel injection systems in automobiles (figure 10.6(d)), an application area representing a huge market world-wide [49]. Here, the figure-of-merit (FOM) is primarily the longitudinal driving-field-normalized maximum strain response, S_{max}/E_{max} [97]. A parameter that is often underrated despite playing an important role in such applications is the self-heating of actuators during operation [98]. The most simplistic approach to the problem is to look at the generalized dissipated power density of a leaky capacitor, which is proportional to the driving frequency (w), the square of the driving electric field amplitude (E_0) and the imaginary permittivity coefficient (e'') (see classical textbooks, such as [99]). The latter is essentially correlated with the amplitude of the out-of-phase component of electric-field-induced polarization, responsible for P–E hysteresis. In fact, the hysteresis area of a generalized P–E loop is the measure of the dielectric loss and thus dissipated electrical energy [100]. The practical consequence is that the hysteresis generated due to irreversible domain wall motion in soft PZT will lead to self-heating of the material (although this same motion beneficially enhances the piezoelectric coefficients). In their pioneering studies, Uchino *et al.* [101] elaborated upon this problem, showing the excessive heating of a soft PZT ceramic at off-resonance conditions, reflected in increased temperatures of the material up 150 °C when driven at 3 kV mm^{-1} and 300 Hz of frequency. The self-heating is strongly enhanced with increased driving-field amplitude, frequency and reduced surface-to-volume ratio of the piezoelectric element (due to the smaller area through which the self-generated heat can be dissipated into the environment). In that study it was also shown that most of the heat generated in a soft PZT under stress-free condition originates from electric loss, not mechanical loss, meaning that a good approximation of the self-heating effects may be obtained by analysing the P–E hysteresis area. Obviously, in real applications the piezoelectric materials are subjected to more complex mechanical boundary conditions as they must be pre-stressed, mostly to avoid cracking. Consequently, blocking force becomes an important factor as the elements must provide strain under a certain amount of static stress [102]. In addition to the herein-discussed strain magnitude, hysteresis and blocking force, other important parameters are reliability, cycling stability, thermal stability against depoling and cost. All of these requirements and the high market competitiveness in the area of actuators make the transition from PZT to lead-free materials sluggish [49].

A completely different subset of actuators is that used in high-precision positioning. An example is the piezoelectric tube scanner used in probe scanning systems, such as in atomic force microscopes (AFMs; figure 10.6(e)), which typically consists of a radially poled piezoelectric tube with outer sectioned electrodes and a continuous inner electrode that allow both longitudinal and bending displacements [103]. Hysteresis, nonlinearity and creep (time-dependent strain upon application of constant electric field) of the piezoelectric element are all undesired. For example, in the case of linear piezoelectrics it is possible to determine the strain x from the constant piezoelectric coefficient d and the applied field E, using the linear relation $x = d \cdot E$. Nonlinearity, meaning the dependence of the piezoelectric coefficient on the field, $d(E)$, coupled to hysteresis, makes the strain response a complex multi-valued function of the driving field, necessitating expensive control drive systems [104]. Of importance is also a stable frequency (or time) response and its temperature stability, both of which can be reduced by hardening PZT [96, 105].

10.4.3 Conformal domain miniaturization and reduced hysteresis in PMN-PT

A alternative method to control material properties, rather than the defect engineering presented in the previous section, is the control of domain structure. Studies have shown that conformal miniaturization of domains is another common mechanism in numerous high-performance transduction materials [13, 26, 32, 106], and achieves enhanced coupling coefficients and power transfer efficiency. The conformal miniaturized domains is characterized by what has commonly been called tweed-like structures, as illustrated in figure 10.7. They were first reported in pre-martensite states of Ni–Al [26], and subsequently in piezoelectric PMN-PT as an intermediate bridging phase near a morphotropic phase boundary between tetragonal and rhombohedral ferroelectric phases [31, 32], and later in magnetic shape-memory Co–Ni–Ga alloys [13, 106]. The stable states in these cases are in fact structurally heterogenous states, where the length scale of the heterogeneity is notably smaller than the coherence length of diffraction; rather than a conventional long-range ordered state which a homogeneous order parameter and crystal structure with translational invariance.

Thus, the local symmetry on the atomic scale is different than the average symmetry determined by diffraction in these special materials. Monoclinic structures have been reported in pre-martensite alloys [26], poled PMN-PT ceramics and crystals [107–109], and magnetic SMAs—all of which are sandwiched between high-temperature cubic and low tetragonal/rhombohedral structures in unique ways. The apparent monoclinic structures result from an averaging over numerous tetragonal or rhombohedral nanodomain ensembles [7, 25], rather than having a local symmetry that is monoclinic. This concept is in distinction to *ab initio* approaches [110, 111] which have predicted a monoclinic unit cell, where the intermediate monoclinic phase is structurally homogeneous.

Following the adaptive phase theory for structurally heterogeneous states, under applied fields, the distribution of miniaturized domains changes between equivalent variants, within the geometric constraints imposed by the domain platelets. This is in distinction to the predictions by *ab initio* approaches where the changes induced under applied field are due to a rotation of the order parameter direction within each unit cell.

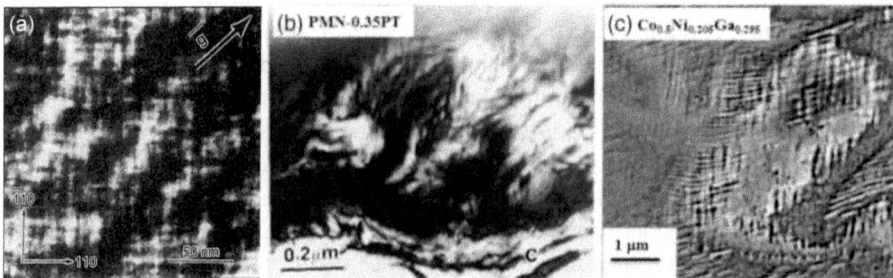

Figure 10.7. Bright field images illustrating tweed-like structures in (a) Ni–Al pre-martensite, (b) PMN-PT and (c) Co–Ni–Ga magnetic shape-memory alloy. ((a) Reprinted from [181], Copyright (1990), with permission from Elsevier. (b) Reproduced with permission from [31]. Copyright 1995 AIP Publishing. (c) Reproduced with permission from [13]. Copyright 2003 Microscopy Society of America.)

In figure 10.8, we illustrate the changes in the miniaturized domain distribution within macrodomain plates induced by changes in applied electric field history [112, 113]. Inspection of this figure will show significant changes between the zero-field-cooled (ZFC) and field-cooled (FC) conditions. In both cases, the macrodomain plates are of the same size and orientation. However, the miniaturized domains within the macrodomain plates are notably different for the ZFC and FC conditions. In the ZFC state they are nearly colinear with an orientation along the [110] direction, whereas in the FC condition they are zig-zagged between numerous neighboring regions with different (110) type orientations. The apparent crystal structure in the ZFC condition is cubic, whereas that in the FC condition is monoclinic. Changes in the miniaturized domain distribution have resulted in changes in the apparent symmetry by diffraction. This is unconventional behavior, which is not expected for a traditional uniform structure.

However, the difference in apparent structure due to changes in the distribution of miniaturized domains within the domain platelets can readily be explained by the adaptive phase theory of structurally heterogeneous states [7, 25]. Conformal domain miniaturization requires low twin wall energy and elastic strain accommodation. As a consequence, the lattice parameters themselves follow the WLR theory of martensite. Multiple twin variants are averaged over during diffraction, resulting in a symmetry reduction following special geometric invariant conditions. This results in the apparent lattice parameters of the adaptive state (a_{ad}, b_{ad}, c_{ad}) having an inter-relationship with those of a tetragonal phase (a_t, b_t, c_t) and cubic (a_c), as follows:

$$a_{ad} = a_c; \ b_{ad} = a_t; \ c_{ad} = c_t + a_t - a_c. \tag{10.2}$$

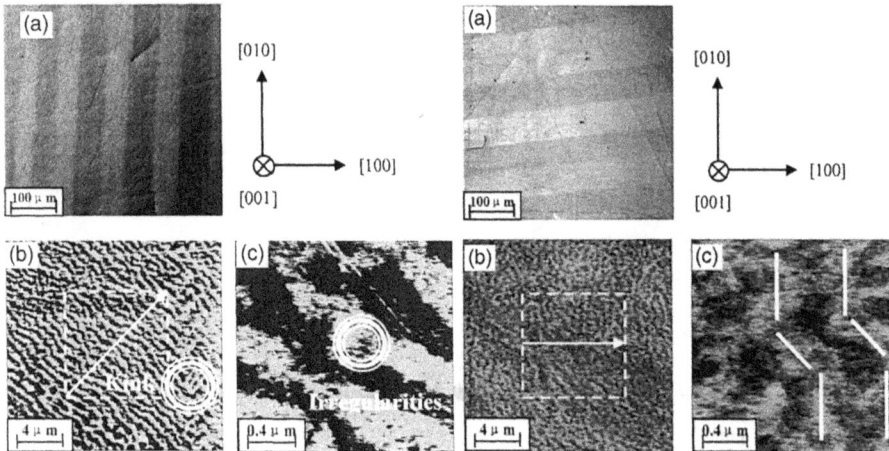

Figure 10.8. Polarized light microscopy images of PMN-PT in (a) the zero-field-cooled (ZFC) condition and (b) the field-cooled condition. ((a) Reproduced with permission from [112]. Copyright 2004 AIP Publishing. (b) Adapted with permission from [113]. Copyright 2005 AIP Publishing.)

This is a special geometric invariant condition that restricts the lattice parameter of the adaptive phase.

Investigations have shown that the lattice parameter changes of the monoclinic phases of PMN-PT crystals for compositions near the MPB follows this restriction inter-relating the adaptive and tetragonal/cubic phases. Detailed investigations of the lattice parameters of poled PMN-PT crystals have been performed for various compositions as a function of temperature and applied electric field [107, 108]. In fact, the changes in the lattice parameters have been shown to be invariant with composition, temperature and electric field, as shown in figure 10.9 [7, 25]. Similar invariance of the lattice parameters has been reported for Fe-30.1at%Pd [7] in the pre-martensite state, and in ferromagnetic shape-memory thin films of Ni–Mn–Ga on (100) MgO substrates [114]. These findings show that the average lattice parameters of monoclinic adaptive ferroelectric states result from the redistribution of miniaturized domains within the geometric invariant conditions determined by complete elastic strain relaxation.

The unique structural heterogeneous nature of the adaptive state is important to achieving its excellent transduction abilities. Key is the reduction of hysteresis loss, which can limit the energy coupling coefficient and power transfer efficiency. In these materials, as long as the domain platelets are not disturbed, the redistribution of the miniaturized domains is nearly non-dissipative. Figure 10.10 shows unipolar and bipolar ε–E curves [11]. The unipolar curves can be seen to have slim hysteresis loops, revealing little to no energy dissipation. However, under bipolar drive, strong hysteresis is found. It is the very low hysteresis in the ε–E curves which enables the ability of PMN-PT crystals to achieve longitudinal piezoelectric coupling coefficients of $k_{33} = 0.90$–0.95, and ultra-high weak-field longitudinal piezoelectric coefficients of $d_{33} = 2000$ pC/N. This redistribution process underlies the unique electromechanical characteristic of PMN-PT crystals.

Single-crystal PMN-PT offers notably lower hysteretic losses than untextured polycrystalline ones. This is because the single-crystal form offers the ability of the

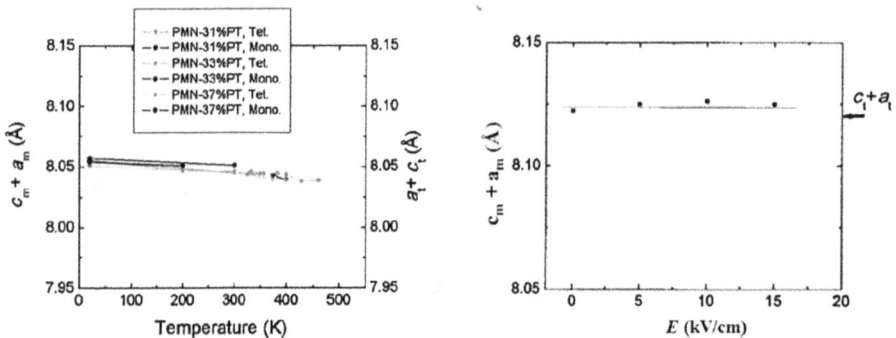

Figure 10.9. Invariance of the lattice parameters with respect to the geometrical conditions of the adaptive phase theory for (a) various PMN-PT compositions around the MPB as a function of temperature and (b) PZN-PT as a function of electric field E. (Reproduced with permission from [7]. Copyright 2003 AIP Publishing.)

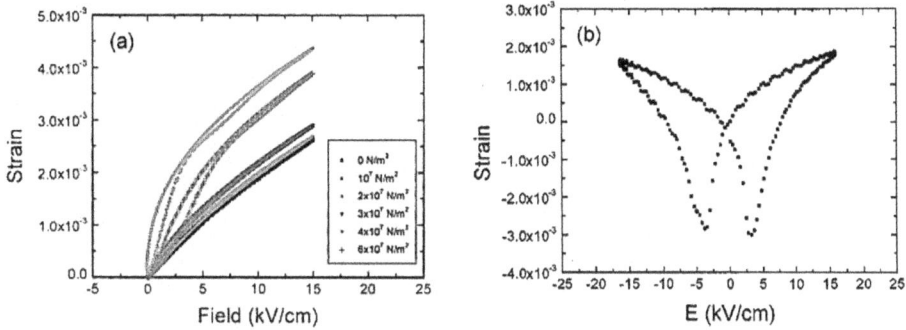

Figure 10.10. Electric field induced strain for PMN-PT under (a) unipolar drive taken at different uniaxial stresses and (b) bipolar drive. (Reproduced with permission from [11]. Copyright 2001 AIP Publishing.)

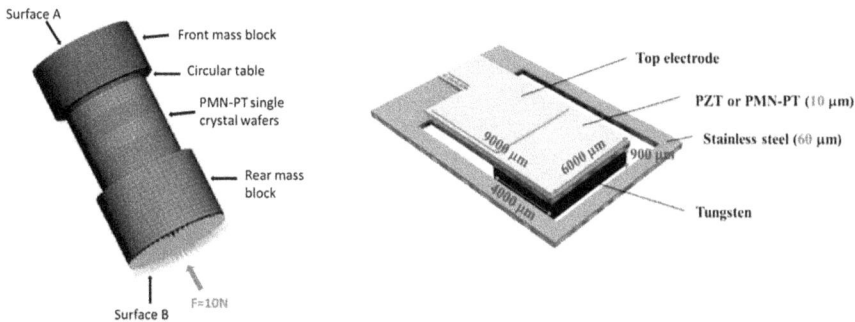

Figure 10.11. Schematics of PMN-PT acoustic-device applications of (a) underwater sound and (b) energy harvesting. ((a) Reproduced with permission from [115]. Copyright 2021 Elsevier. (b) Reproduced from [116]. CC BY 4.0.)

entire crystal to achieve the elastically relaxed state. In a polycrystalline material, this is not feasible.

Thus, PMN-PT crystals offer superior properties to conventional piezoelectric PZT ceramics. They offer lower losses in the poled condition, and higher values of k_{33} and d_{33}. These properties allow for enhancements in the power density and bandwidth of acoustic and ultrasonic transducers, offering performance characteristics which are not achievable in systems based on conventional piezoelectric ceramics. PMN-PT crystal technology has already penetrated the market for medical ultrasonic devices. The improved properties of PMN-PT crystals also offer advancements in actuators [115], energy harvesting devices [116] and acoustic sensors. Designs for acoustic devices for transducers and energy harvesting devices are shown in figure 10.11.

10.4.4 Towards lead-free piezoelectric applications

In order to provide as broad applicability as that characteristic for PZT or PMN-PT, hardening and softening concepts have been of paramount importance in the development of lead-free piezoelectric ceramics [52]. Since the launch of EU environmental legislations in 2006 [97], the research on lead-free piezoceramics has been all but simple. One of the many reasons is that in lead-free materials microscopic pinning effects might be more complex and different than those encountered in PZT. An excellent example is $BiFeO_3$, a perovskite that has recently been subject of continuous studies due to its high Curie temperature ($\sim$830 °C), making it useful in high-temperature applications [117]. After more than a decade of speculation around the defect chemistry of BFO and the key role of reduced Fe^{2+} defects (Fe'_{Fe}) and its n-type conductivity, the complete scenario has been revealed by confirming the tendency of Fe to oxidize into Fe^{4+} state ($Fe^{\bullet}_{Fe}$) when annealed in air, giving rise to significant p-type conductive behavior (figure 10.12(a) [118–120]). This is contrast to the defect chemistry of classical perovskites, such as $BaTiO_3$ and PZT, which show weak p-type conduction when annealed in an environment with high partial pressure of O_2, $p(O_2)$, and are thus often labeled as 'p-type insulators' [121].

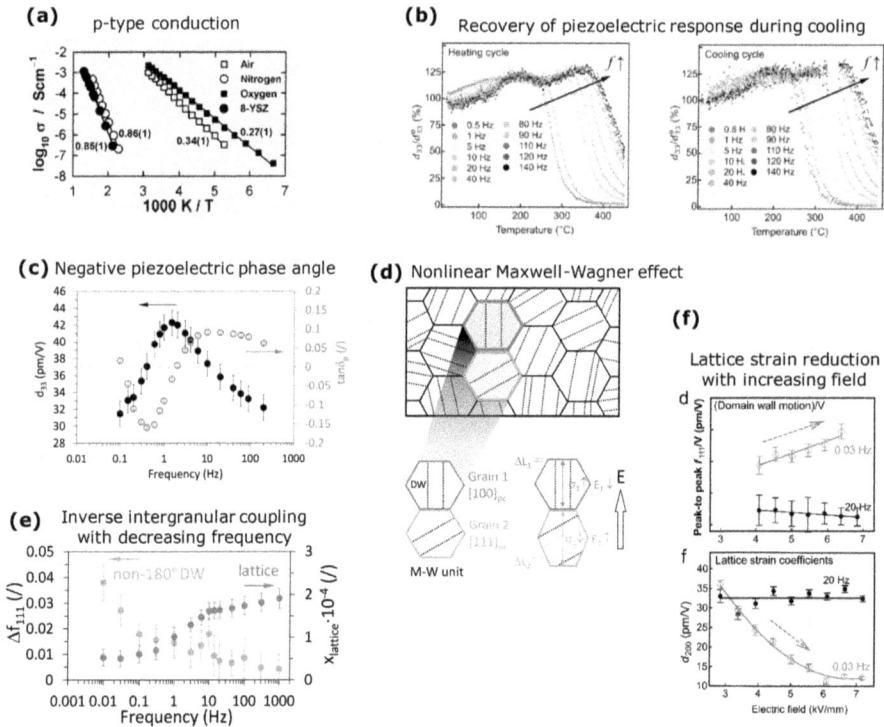

Figure 10.12. Peculiarities of the piezoelectric response of $BiFeO_3$ (BFO) ceramics. (a) One of the first publications showing the dominance of p-type conduction in air-annealed BFO [118]. The plot shows the electrical conductivity versus inverse of temperature for Ca-doped BFO ceramics annealed in air, O_2 and N_2. The data on yttria stabilized zirconia (YSZ), a known ionic conductor, are shown for comparison. The authors report on a p-type conduction in Ca-doped BFO when equilibrated in air and O_2, which can be reduced by annealing in N_2, leading to predominant ionic conductivity, similar to that in YSZ. (b) The direct piezoelectric response of BFO ceramics, shown as the d_{33} coefficient measured at different frequencies and normalized to its room-temperature value, upon heating (left) and cooling (right). The results reveal a frequency-dependent

disappearance of the piezoelectric response upon heating, followed by its recovery upon cooling [122]. (c) Frequency dependence piezoelectric coefficient and tangent of the piezoelectric phase angle at room temperature revealing a complex behavior with the appearance of negative piezoelectric phase angle, an indication of Maxwell–Wagner (M–W) piezoelectric relaxation arising due to local variations in the electrical conductivity [75, 123]. (d) Schematic of the nonlinear M–W effect arising from conductive domain walls in BFO [59]. The effect originates from the different orientation of conductive domain walls in neighboring grains with respect to the direction of the applied field, leading to a redistribution of internal fields in individual grains. In this particular case, due to the higher conductivity of grain 1 (related to the vertically oriented conductive domain walls), the internal field will be reduced in this grain and, vice versa, it will be enhanced in grain 2 (in which the conductivity is lower due to non-vertical orientation of conductive domain walls). The net result of the internal field redistribution at low frequencies is an enhanced domain wall contribution in grain 2 ($[111]_{pc}$ oriented) and reduced lattice strain in grain 1 ($[001]_{pc}$ oriented) [59, 123]. (e) Experimental validation of this inverse intergranular coupling phenomenon as observed using *in situ* XRD diffraction analysis; an increase in the non-180° domain texture in $[111]_{pc}$ grains was identified with decreasing frequency (blue), coupled to a decrease of lattice strain (red) [124]. (f) A peculiar reduction of the lattice-strain coefficient with increasing electric field amplitude at 0.03 Hz (lower plot, see the red dashed arrow) occurring alongside an increase of the equivalent coefficient extracted for non-180° domain wall contribution (upper plot, see the red dashed arrow), as observed from *in situ* XRD analysis [125]. The apparent strain decoupling at low frequencies was tentatively explained using a hybrid model of nonlinear M–W effect and transverse elastic intergranular coupling. ((a) Reproduced with permission from [118]. Copyright 2012 ACS Publications. (b) Reproduced with permission from [122]. Copyright 2024 John Wiley and Sons. (c) Reproduced from [126]. CC BY 3.0. (d) Reproduced from [59]. CC BY 4.0. (e) Reproduced from [124]. CC BY 4.0. (f) Reproduced with permission from [125]. Copyright 2022 The Authors. Published by Elsevier Ltd on behalf of Acta Materialia Inc. CC BY-NC-ND 4.0.)

The particular conductivity of BFO results in emerging phenomena. One such example is shown in figure 10.12(b) displaying the direct piezoelectric response of BFO ceramics as a function of temperature for various driving-stress frequencies [122]. A clear frequency-dependent decay of the coefficient is observed above 300 °C, which gradually decreases toward zero (see the heating cycle in figure 10.6(b)). It is commonly assumed that such behavior is due to depoling of the sample. The cooling data (figure 10.12(b)), however, surprisingly reveal an almost complete recovery of the piezoelectric response, which is obviously not consistent with permanent depoling. In fact, additional *ex situ* XRD and piezo-response force microscopy (PFM) analyses confirmed the stability of the poling-induced domain texture up to 750 °C [122], which is not far from the T_c of BFO (~830 °C). Therefore, the unusual frequency-dependent disappearance and recovery of the piezoelectric response is likely related to charge migration and elevated electrical conductivity of BFO whereby the domain texture and poling state are very stable.

The first discovered signature of the piezoelectric response of BFO that is not typical was the negative piezoelectric phase angle (figure 10.12(c)) [123]. Such a response is common in, for example, Aurilivius piezoelectrics with strong anisotropic conductivity [127], and indicates the so-called Maxwell–Wagner (M–W) relaxational effects arising from variation in the electrical conductivity at the local scale [58]. In the field of dielectrics, M–W relaxation is often associated with insulating grain boundaries and/or variations in the conductivity at the electrode–sample interface due to Schottky-type barriers [128]. In the case of BFO, however, the M–W-like relaxation was shown to be likely dominated by conductive domain walls, first discovered in BFO thin films [129], followed by BFO ceramics [123]. Since the local p-type conductivity is confined to the domain wall [130], which can move under field leading to nonlinearity (as discussed for soft PZT in figure 10.5(c), (d) and figure 10.6(b)), the case of BFO is the first example of the so-called *nonlinear* M–W piezoelectric effect [59, 123].

A schematic of the nonlinear M–W effect arising from the conductivity at the walls is shown in figure 10.12(d) [59]. The idea that was modeled analytically [59, 124] is that the local conductivity of each grain depends on the orientation of the conductive domain walls inside the grains. When applying low driving-field frequencies, which allow charges to move along the domain walls, the externally applied electric field will be redistributed internally such as to exhibit (i) a decreased field in grains with high conductivity (grain 1 with conductive domain walls aligned vertically, thus parallel with the applied field, will exhibit leakage and thus drop of internal field; figure 10.12(d)) and, vice versa, (ii) an increased field in other grains with unfavorable wall orientation (grain 2 with lower conductivity because domain walls are oriented under an angle with respect to the field vector; figure 10.12(d)). The net result is a larger response from domain wall motion in the field-enhanced grain 2 and the concurrent reduction of lattice strain in grain 1 in which the field is reduced. This counterintuitive phenomenon was indeed confirmed experimentally as a function of field frequency using *in situ* XRD diffraction (figure 10.12(e)) [124]. The inverse relation between domain wall and lattice strains in differently oriented grains under applied field was initially counterintuitive for the sole reason that for PZT it is well known that microstrain contributions are coupled, meaning that surrounding grains should respond coherently and in synergy to the applied field (e.g. increased domain wall motion in a given grain is elastically transferred to surrounding grains increasing their microstrain lattice response, not decreasing it as observed in BFO [131]). Another emerging effect discovered soon after the inverse intergranular coupling was the reduction of the lattice strain in individual grains with increasing field amplitude, occurring despite the domain wall switching fraction in other grains being increased (figure 10.12(f), see red dashed arrow) [125]. This effect could not be explained either by using the nonlinear M–W effect, or classical transverse elastic cross-interactions between grains (as used to explain microstrain coupling for PZT [131]). It turned out that a hybrid model using both mechanisms was capable of explaining the field-inverse lattice-strain behavior.

The example in figure 10.12 of BFO illustrates how microstrain mechanisms, defects, conductivity and, consequently, hardening effects may be entirely different in lead-free piezoelectrics when compared to those known for PZT (figure 10.5). Detangling the microscopic relationships between charged defects and ferroelectric/ferroelastic domain walls is not an easy task but will have to be realized to bring new materials to the next readiness level toward full replacement of PZT. In parallel, new approaches are currently being evaluated, such as for example dislocation [132] and precipitation hardening [133]. On the other hand, despite decades of use of soft PZT in applications, softening is still not well understood [51, 60]. Promising novel approaches are also emerging, such as the so-called 'relaxor softening' in which domain wall mobility is enhanced through lattice disorder induced by multiple cations with variable valence states and ionic radii [134].

10.5 Piezo- or ferroelectric micromachined ultrasonic transducer (PMUT or FMUT)

Ultrasonic transducers are known for their applications in the medical field of ultrasound imaging [135]. The standard version of a transducer consists of a *bulk wave resonator* made of piezoelectric ceramics that emits and receives waves with frequencies above 20 kHz, which conventionally represents the human ear's sound perception threshold (ultrasonic waves). With the advent of microfabrication techniques with smart materials, it has been possible to construct micro-electromechanical systems (MEMS) and in particular micromachined ultrasonic transducers (MUTs) [136] by including a thin film of piezoelectric material that can perform the same task [137]. It is also clear that established microfabrication techniques allow for far greater precision in manufacturing processes than is characteristic of bulk ceramic elements. However, the operation of the thin-film microsystem turns out to be based on a different principle to that characterizing the standard transducer. The thin film is deposited between two electrodes on a laminated plate in contact with the fluid in which the ultrasonic waves propagate. Therefore, the structure that results from a non-piezoelectric laminated plate and the piezoelectric layer is also called a heteromorphic structure. The piston-like motion of the system that resonates at the fundamental resonant frequency couples well with a medium low acoustic characteristic impedance such as air or water. As the medium carrying the signal varies, the operating frequencies are different, from a few MHz to 30 MHz for water applications [138] and from 40 to 400 kHz and above for air applications [139, 140]. Early work concerning such layered structures involves microsonars, filters, peristaltic micropumps, microstators, and micromotors mainly made with PZT films [141]. Parts of the operational principles are explained below using the dynamics of a circular piezoelectric plate based on the full description and model initially made by Muralt *et al.* [142]. Initially these type of devices were called 'micromachined high frequency ferroelectric sonar transducers' by some authors [143], but later the term piezoelectric micromachined ultrasonic transducer (PMUT) was adopted by the community [144]. Note that the term FMUT (with the F-stranding for ferroelectric) arose recently [145] and could be adopted, in particular if the ferroelectricity provides additional features compared to the piezoelectricity, such as tunable properties.

The elastic energy in a circular laminated plate is determined by the inflection of the plate. The bending moment in a piezoelectric thin-film system is due to the electromechanical coupling caused by the electric field in the piezoelectric material layer via the piezoelectric effect of the same name. Taking into consideration the fundamental resonance frequency of a linearly elastic homogeneous circular plate embedded on the edge this results in

$$f_r = \frac{(3.19)^2}{2\pi a^2}\left(\frac{D}{\mu}\right)^{1/2} = \frac{(3.19)^2}{2\pi a^2}\left(\frac{Et^2}{12(1-\vartheta^2)\rho}\right) \tag{10.3}$$

$$D = \frac{Et^3}{12(1-\vartheta^2)}, \quad \mu = \rho t, \tag{10.4}$$

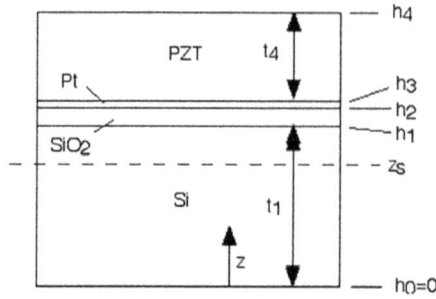

Figure 10.13. Schematic representation of the cross section of a laminated plate. (Reproduced with permission from [142]. Copyright 2005 IEEE.)

where a and t are the radius and thickness of the plate, respectively, E, v, and ρ are the Young's modulus, Poisson's coefficient, and density of the constituent material, D is the plate flexural stiffness, and μ is the mass per unit area [142].

In the case of a laminated plate, the previous expressions must be modified to include the various constants related to the different materials. For this purpose, the position of the neutral axis must be determined (figure 10.13).

Here the neutral axis is

$$z_s = \frac{1}{2}\frac{\dfrac{h_n^2 - h_{n-1}^2}{\mathrm{se}(n)}}{\dfrac{t_n}{\mathrm{se}(n)}}. \tag{10.5}$$

The term $\mathrm{se}(n)$ depends on the position within the plate. Expressing by $s_{11}^{(n)}$ the 11th term of the elastic yielding matrix of the nth layer, the flexural stiffness and mass per unit area of the laminated system are determined as follows:

$$D = \frac{1}{3}\sum\frac{(h_n - z_s)^3 - (h_{n-1} - z_s)^3}{s_{11}^{(n)}\left(1 - \left(v_{12}^{(n)}\right)^2\right)} \tag{10.6}$$

$$\mu = \sum \rho_n t_n. \tag{10.7}$$

The fundamental frequency of the laminated plate can be calculated using the initial expression with the coefficients for the layered system. The following figure 10.14 shows how the plate diameter varies as a function of the fundamental frequency [142].

The diameter of a transducer should be neither too large nor too small in relation to λ. Efficient transmission of the signal generated by membrane oscillation occurs for diameter values of half wavelength. Interestingly, the *matching zone* shifts from fundamental frequencies of 100 kHz for air to 10 MHz for water for typical thicknesses of these objects. This represents a necessary system design requirement for standard applications. From 100 kHz–1 MHz the frequency–diameter relationship fits well with the pattern followed by $\lambda/2$ determining a good frequency range for short-range airborne sensors. Each PMUT can function as a transmitter and receiver. As a transmitter, the

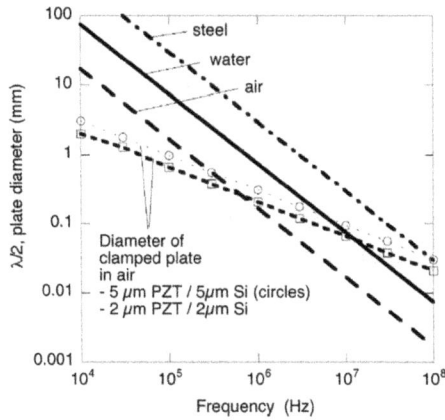

Figure 10.14. Frequency–diameter versus $\lambda/2$ relationship where $\lambda = vs/f$ is the wavelength of the signal emitted in air, water and steel. (Reproduced with permission from [142]. Copyright 2005 IEEE.)

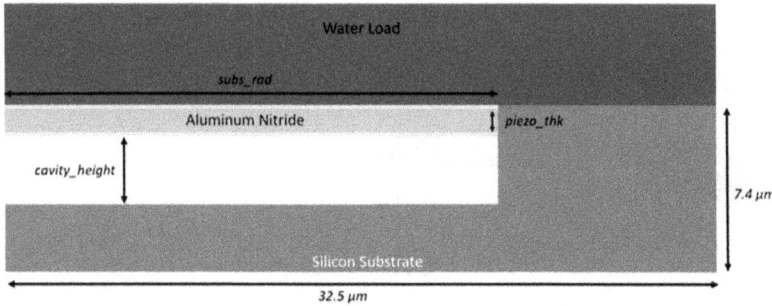

Figure 10.15. Cross section view of a single PMUT cell having electrodes fully covering the thin 1 μm thick piezoelectric layer (piezo_thk), with a cavity height of 3 μm and a cell radius (subs_rad) of 22.5 μm. (Reproduced with permission from [147].)

electric field causes a state of stress (strain) in the piezoelectric layer between the two electrodes (inverse piezoelectric effect). The stress generated causes a bending moment that forces the membrane to flex out of plane. Through the application of alternating current to each oscillation of the system an acoustic wave is transmitted to the fluid in contact. As a receiver, an incident acoustic pressure wave causes flexural deformation of the membrane resulting in a state of stress transverse to the cross-sectional area of the piezoelectric layer, which results in a creation of surface charge at the electrodes (direct piezoelectric effect) whereby there is an output voltage at the ends.

A single PMUT cell needs to be designed accordingly. Once the vertical stack dimensions are fixed, the 2D design of the PMUT cell and in particular the electrode design will have a strong impact on PMUT performance [146]. To study design variations, an option is to use FEM simulation. As an example, the dynamic response of a single PMUT membrane has been simulated with electrodes covering the full surface of both sides of a piezoelectric membrane made of aluminum nitride (AlN) as represented in figure 10.15. This model of a simple PMUT single cell with an AlN active layer with gold electrodes residing on a silicon substrate is operated in water.

The outputs of such FEM simulations could give the fast Fourier transform (FFT) of the Y-displacement at the central surface of the piezoelectric membrane (figure 10.16). On the spectrum we could identify that the second mode (at 27 MHz) has a higher amplitude than the first mode (at 5 MHz). This is due to electrode design. The simulated example with a full covering electrode on both sides is not optimal for first mode output displacement, but it enhances second mode out-of-plane displacement.

Thus, the proposed design is not well suited for PMUT cells that operate at 5 MHz but it should be possible to use this simple fully metal covered design by exploiting it at 27 MHz.

The fine discretization of the top electrode has already been tested in the literature through simulation [148]. If high grade discrete electrodes on the membrane could be useful for the exploitation of high range modes, the optimal design for the first mode is quite simple. For example, with a square PMUT membrane, having a reduced square as the upper electrode with a centered surface area of approximately 70% of that of the membrane is a valid choice (figure 10.17(a)). Additionally, a frame-like electrode (figure 10.17(b)) can be considered for better sensitivity [148].

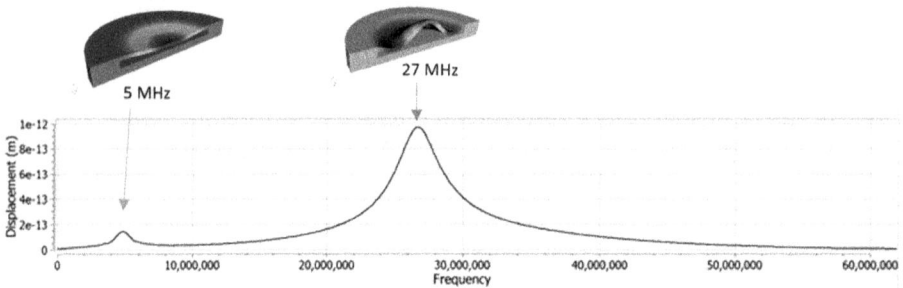

Figure 10.16. FFT of the Y-displacement at the central surface of the membrane obtained with OnScale. (Reproduced with permission from [147].)

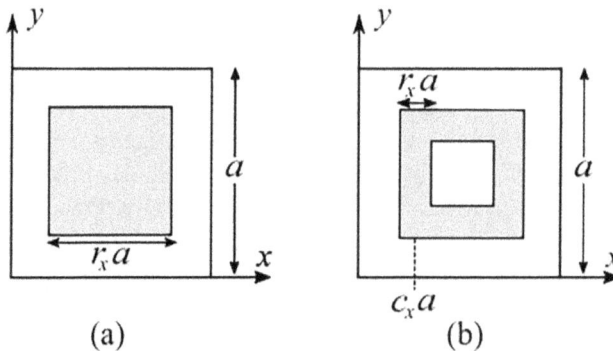

Figure 10.17. Top electrode configurations for square PMUT cell: (a) single continuous and (b) frame-like. (Reproduced from [148]. CC BY 4.0.)

With these examples we have highlighted the fact that the optimization of electrode design on the membrane for a PMUT cell must be done according to the vibration modes exploited in practice for the considered application.

Considering the PMUT elementary cell as already presented, 'which is the best thin-film piezoelectric material?' This this the exact question Muralt *et al.* answered in 2017 [149]. Several materials can be considered, such as the AlN already mentioned or scandium doped AlN that could have enhanced properties [150]. Zinc oxide (ZnO) can also be a candidate. However, the historically winning technology remains the PZT group that has been recently challenged by potassium sodium niobate (KNN), which has the double advantage of reaching material properties close to PZT but also being lead-free. See table 10.1 for a comparison of these materials.

The material properties of the piezoelectric layer of the PMUT cell will have an impact on the performance of the device, as shown in figure 10.18. For the same design, the acoustic output pressure with respect to frequency is compared for three

Table 10.1. Comparison of piezoelectric materials most commonly used for PMUTs (partly from [154]).

Materials	$e_{31,f}$ (C.m^{-2})	ε_{33}	E (GPa)
AlN [151, 152]	-1	10	340
PZT (2 μm) [142]	-13.1	854	76
KNN (2 μm) [153]	-12	1000	65
Single-crystal PZT [154]	$-16 \sim -24$	308	85
20% ScAlN [149, 155]	-1.6	12	200

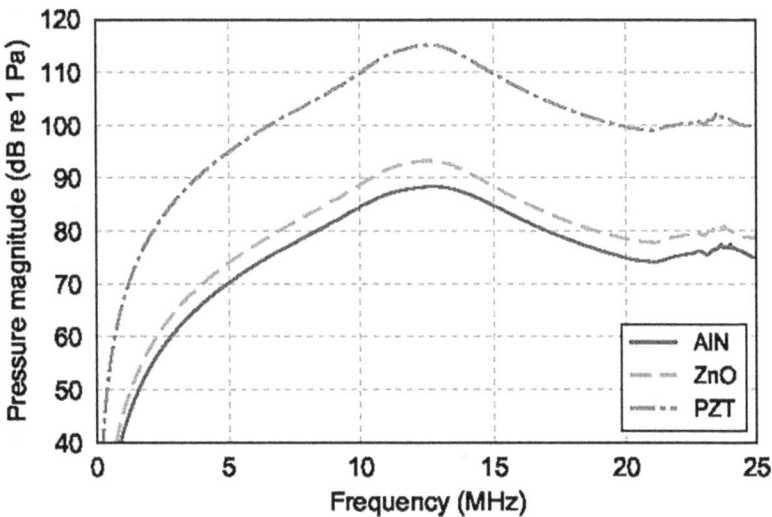

Figure 10.18. The frequency response for AlN, ZnO, and PZT on Si PMUTs. (Reproduced with permission from [156]. Copyright 2017 IEEE.)

active materials (AlN, ZnO, and PZT) for PMUT membranes [156]. Up to 30 dB could distinguish the 'technological gold standard' (PZT) from the others. This highlights the fact that unless the effort is made to build a lead-free technological word, PZT remains in a good position [157].

It is one thing is to design and predict the PMUT elementary cell behavior with a specific material, and another thing to manufacture it. Thus, the problem of inserting ferroelectric thin films between metal electrodes on the silicon layer is one of the major challenges in PMUT microfabrication. One of the problems to be addressed is the deposition of the piezoelectric thin layer and the improvement of the quality of the final product by optimizing, for example, the residual stresses due to the manufacturing processes [158]. There are three main approaches for making PMUTs: (i) deposition of the piezoelectric thin film on a silicon substrate with appropriate insulating and conductive layers followed by surface or bulk micromachining (the additive approach), (ii) the direct bulk micromachining process of single or piezoelectric polycrystals (the subtractive approach) and (iii) the integration of the silicon structure by (wafer-)bonding techniques (the integrative approach).

The additive process consists of the deposition of passive and active layers on the silicon substrate, after which the structure is defined in the 'wet' or 'dry' etching step and results in a suspended piezoelectric sensor. The difficulty is in controlling the residual stresses when releasing the membranes. Nevertheless, PMUTs result many times in direct micromachining and subtractive manufacturing. The integrative process, on the other hand, involves bonding precise silicon microstructures onto piezoelectric substrates. The latter process requires low temperatures and complex processes. The additive process, which is widely used, allows thin films of wurtzite and perovskite crystalline structure materials to be deposited on the silicon substrate [159]. The integration of these smart materials occurs in two basic steps: (i) deposition of amorphous material on the substrate and (ii) crystallization before after or deposition. This mechanism allows the nucleation of piezoelectric grains to be controlled, and through the energy balance required by growth and nucleation, the orientation of the lattice structure can be defined. The latter depends, therefore, on the thermodynamics of the process.

In general, a PMUT cell is not alone as a matrix of transducers is manufactured and packaged. For ultrasound imaging, the matrix configuration allows beam-forming to be performed [160]. Other types of imaging such as thermos or photo-acoustic imaging can be also done with PMUT arrays [161]. Most of the time, an impedance matching circuit combined with dedicated electric driving and receiving circuits also need to be connected or integrated with the PMUT array [162]. As illustration of a complete system, a schematic representation of an integrated PMUT-based imager is shown in figure 10.19. This PMUTs-on-CMOS array is designed for catheter ultrasound imaging [163]. The system on chip (SoC) contains 7×7 AlScN PMUT cells, CMOS-based high-voltage (HV)-pulsers and a low-noise amplifier (LNA). To image the medium, it uses ultrasound pulses at 7.7 MHz with a normalized pressure of about 2 kPa.mm V^{-1} and it has a receiving sensitivity of 3.3 V MPa^{-1}. It was able to image sub-wavelength objects with a diameter of 25 μm. This is what is also achieved by advanced ultrasound equipment [164], which is

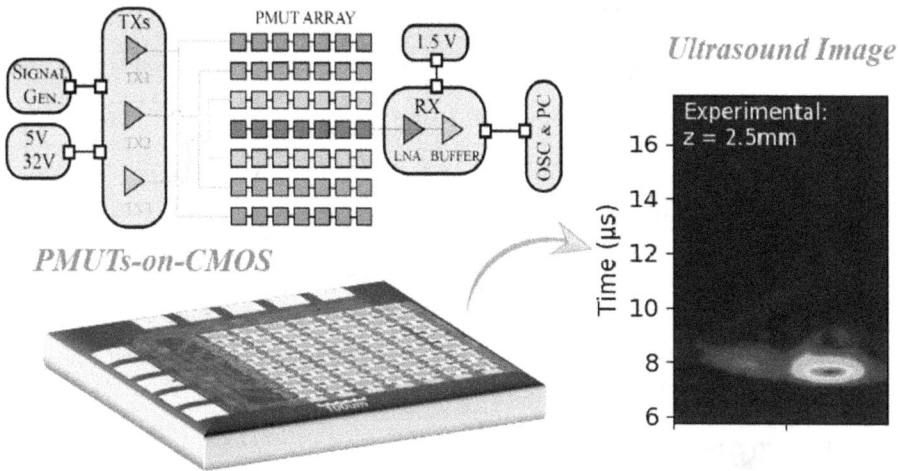

Figure 10.19. Fully integrated PMUT-on-CMOS-based ultrasound imaging device. (Reproduced from [163]. CC BY 4.0.)

much less integrated compared to this one. This PMUT-based embedded system might be one of the most convincing integrated ultrasound imaging devices presented in the literature. In conclusion, this ultrasound imaging SoC highlights concretely the potential of PMUT-based technology for the future.

10.6 Co-elastic marriage of piezoelectric and magnetostrictive layers in bonded composites

The device applications are all related to the ability of piezoelectric crystals to convert electrical energy into mechanical, and vice versa. However, the applications of piezoelectric crystals, in particular PMN-PT, are not limited to just traditional electromechanical transduction. Rather, the co-elastic nature of the crystal allows the elastic strain to be coupled to bonded and/or physically deposited magneto-strictive, magnetic shape memory, and ferroelastic layers.

E-field tunability of magnetic shape anisotropy in ferrite layers epitaxially grown on PMN-PT. Heteroepitaxial layers of ferrites have been grown on (100) and (110) PMN-PT crystals [165, 166]. These heterostructures have shown large changes in the magnetic shape anisotropy on application of electric fields, as illustrated in figure 10.20 [165]. These results demonstrate that large magnetoelectric effects can be obtained in epitaxial heterostructures, offering an approach to integrated sensing.

The physical properties of these ME heterostructures have been well studied. However, magnetic sensing units have not yet been developed based on these heterostructures, nor has their ability for magnetic sensing yet been demonstrated.

Vertically integrated two-phase layers for multi-state magnetic memory. Cobalt ferrite (CFO) and barium titanate (BTO) two-phase films have been deposited on SRO/PMN-PT substrates [167, 168]. Deposition on (100) PMN-PT results in a two-phase morphology with a quasi-regular pattern of CFO nanopillars within a BTO

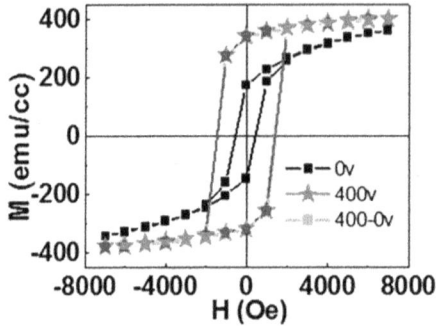

Figure 10.20. *M–H* curve of an epitaxial $CoFe_2O_4$ layer on PMN-PT crystal under different voltages. (Reproduced with permission from [165]. Copyright 2013 AIP Publishing.)

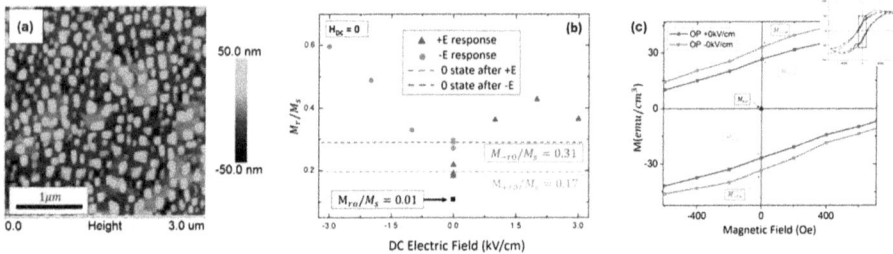

Figure 10.21. Vertically integrated two-phase epitaxial layer of CFO and BTO on PMN-PT substrate. (a) AFM image of two-phase distribution, (b) history dependence of remanent magnetization, and (c) blow-up of the *M–H* curve illustrating four different remanent values. (Reproduced from [180]. CC BY 4.0.)

dielectric matrix, as shown in figure 10.21(a) [169]. The CFO vertical nanopillars are columns with facets that minimize the surface energy. The two-phase film develops this unique self-assembling microstructure in order to minimize the elastic free energy.

The magnetization *M–H* curves of these vertically integrated heterostructures have been studied as a function of the electric field. Multiple values can be found for both the spin up $(+M_r)$ and spin-down $(-M_r)$ remanent magnetization states [169]. At least four different remanent magnetic states can be identified.

These vertically integrated two-phase heterostructures thus have a multi-state ($N \geqslant 4$) memory [169]. The co-elastic nature of the two-phase layer results in shape-memory-like effects that depend upon the film's history. This potentially offers an approach to neuromorphic-like memories and logic.

Microwave filters and isolators that can be tuned by electric fields. Ferrite materials have been used for microwave circulators and filters [170, 171]. They have non-reciprocal properties for microwave EM wave transmission. Rotation of the plane of polarization under application of DC magnetic bias occurs, enabling port circulator devices with isolation. Transmission occurs when the frequency of the incident EM wave is equal to that of the ferromagnetic resonance (FMR) frequency. If the polarization orientation of the device and that of the EM wave match, then transmission occurs; however, if the polarization orientation is 180° phase shifted from the EM wave, then the EM wave has zero transmission through the ferrite material.

One of the difficulties is that the conventional ferrimagnetic microwave circulators are large, as the size is determined by the wavelength of the microwave radiation. Circulators are important devices; for example, every qubit in quantum computers needs numerous circulators in order to prevent backaction and the introduction of noise from the classical side of a port. Innovation has been needed to reduce the size, increase the ease of tunability, and to reduce attenuation and losses.

An innovative approach has been studied involving piezoelectric–ferrite heterostructures [172, 173]. For example, yttrium iron garnet (YIG) has been epitaxially deposited on (100) PMN-PT. The interaction between the ferrite layer and piezoelectric substrate is co-elastic, and thus the size of the device has the potential to be significantly decreased. Application of an electric field to the piezoelectric substrate then results in a shift of the FMR frequency, as illustrated in figure 10.22. The property characteristics of piezoelectric–ferrite heterostructures have demonstrated potential for *E*-field tunable circulators, filters and delay lines.

E-field tunability of emerging materials grown on PMN-PT. Numerous other epitaxial thin-film layers have been grown on PMN-PT substrates whose properties can be tuned by an electric field. Tuning of the band-gap under *E* has been shown in numerous epitaxial layers, such as NNO [174]. Such band-gap tuning offers an approach to control the electron and hole concentrations and effective mass, amongst other important semiconducting properties.

The superconducting phase transition temperature of Fe–Se–Te grown on PMN-PT has been demonstrated to be tunable by an electric field [175–177], as shown in figure 10.23(a). The phase transition can be switched on/off at a set temperature by *E*, resulting in dramatic changes in resistivity and flux exclusion. One could modulate the transition temperature in qubits, turning them on and off, and/or effect the coherence of the state using the applied field. It is an emerging field of materials science with potential special applications in quantum information science (QIS).

Furthermore, 2D materials have also been grown on PMN-PT. Application of an electric field to the PMN-PT substrate has been shown to transfer stress across the

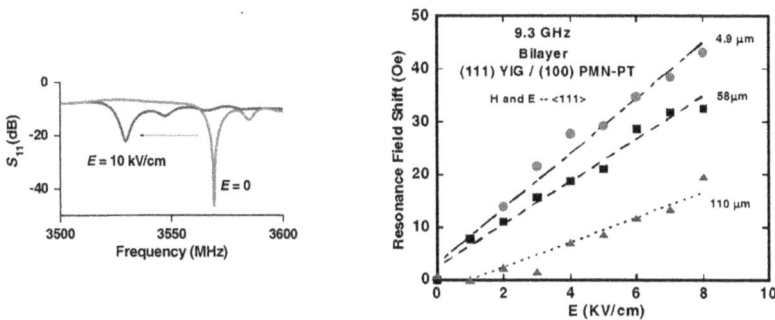

Figure 10.22. Tunable microwave dielectric properties of an epitaxial YIG layer on a PMN-PT substrate. (Reproduced with permission from [38]. Copyright 2008 AIP Publishing.)

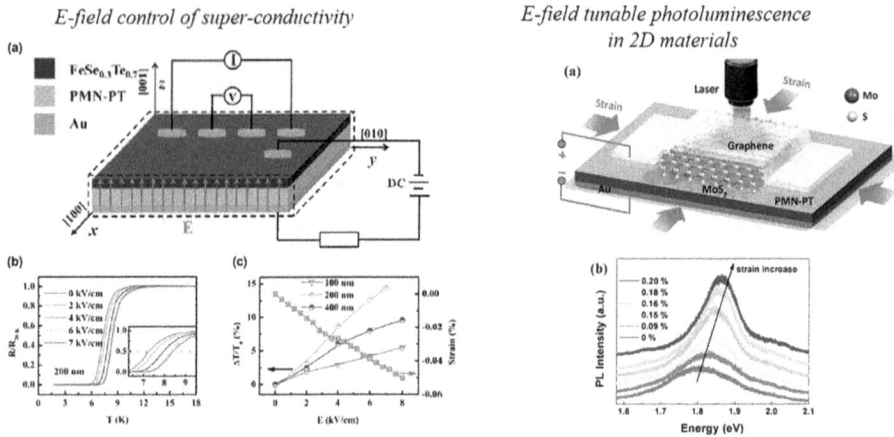

Figure 10.23. Electric field tunability of 2D emerging materials grown on PMN-PT substrates. (Left panel reproduced from [177]. CC BY 4.0. Right panel reproduced from [178]. CC BY 4.0.)

van der Waals interaction to the 2D layer. For example, in MoS$_2$, this results in an *E*-field tunability of the photoluminescence [178, 179]. Electric field tuning of 2D materials on PMN-PT substrates may offer promise for the future of QIS.

These are but a few of the many possibilities that the co-elastic marriage of functional thin films to single-crystal PMN-PT substrates offers. Many opportunities await discovery and innovative applications.

10.7 Summary

Ferroelastic and co-elastic materials, such as shape-memory alloys (SMAs), ferroelectrics, and magnetostrictive materials, exhibit unique properties that enable a wide range of applications. These materials can recover their shape upon temperature cycling, dissipate energy through hysteresis, and efficiently transduce mechanical energy. The concept that 'the material is the machine' underscores their potential to perform mechanical work without traditional moving parts, relying instead on coordinated atomic displacements and elastic accommodations.

These materials can couple mechanical strain with other order parameters, such as polarization and magnetization, enabling a wide range of applications in actuators, sensors, and energy harvesting devices. Recent advancements involve integrating these materials with other functional layers, such as magnetostrictive or superconducting films, to create heterostructures with tunable properties. These developments open new avenues for applications in quantum information science, neuromorphic computing, and advanced sensing technologies. Examples include epitaxial layers of ferrites on PMN-PT substrates, which show significant changes in magnetic shape anisotropy under electric fields, and vertically integrated two-phase heterostructures that offer multi-state memory capabilities. Adaptive and hierarchical domain structures, characterized by tweed-like formations, enhance coupling coefficients and power transfer efficiency, particularly in materials such as PMN-PT and magnetic shape-memory alloys. These conformal miniaturized domains result in efficient energy transduction and reduced hysteresis loss, making them suitable for high-performance applications. The development of lead-free piezoelectric ceramics is crucial due to environmental regulations. However, challenges remain in fully replacing traditional PZT ceramics with lead-free alternatives. Micromachined ultrasonic transducers (MUTs), which use thin films of piezoelectric materials, offer greater precision in manufacturing and are used in applications such as medical ultrasound imaging. The optimization of electrode design and selection of piezo-electric materials, such as AlN, ZnO, and PZT, are critical for enhancing MUT performance.

These materials finally show promise in energy harvesting devices, acoustic sensors, and advanced applications, highlighting their versatility and potential in various technological fields. Continued research and innovation in this field will likely lead to even more sophisticated applications, further bridging the gap between materials science and practical engineering solutions.

Acknowledgments

D.D.V.: the work on piezoelectric crystals was previously supported by the Office of Naval Research (Program manager, the late Jan F Lindberg). T.R. would like to thank the Slovenian Research and Innovation Agency for funding (research core funding P2–0105).

Bibliography

[1] Lieberman D S, Schmerling M A and Karz R W 1975 Ferroelastic 'memory' and mechanical properties in gold-cadmium *Shape Memory Effects in Alloys* (Boston, MA: Springer) pp 203–44

[2] Sapriel J 1975 Domain-wall orientations in ferroelastics *Phys. Rev.* B **12** 5128

[3] Salje E K 1991 *Phase Transitions in Ferroelastic and Co-Elastic Crystals* (Cambridge: Cambridge University Press)

[4] Bhattacharya K 2003 *Microstructure of Martensite* (Oxford: Oxford University Press)

[5] 2008 *Shape Memory Alloys* ed D C Lagoudas (Boston, MA: Springer)

[6] Wayman C M 1964 *Introduction to the Crystallography of Martensitic Transformations* (New York: Macmillan)

[7] Jin Y M, Wang Y U, Khachaturyan A G, Li J F and Viehland D 2003 Adaptive ferroelectric states in systems with low domain wall energy: tetragonal microdomains *J. Appl. Phys.* **94** 3629

[8] Wayman C M and Duerig T W 1990 An introduction to martensite and shape memory *Engineering Aspects of Shape Memory Alloys* (Amsterdam: Elsevier) pp 3–20

[9] Park S-E and Shrout T R 1997 Ultrahigh strain and piezoelectric behavior in relaxor based ferroelectric single crystals *J. Appl. Phys.* **82** 1804

[10] Haun M J 1988 *No Title* (University Park, PA: The Pennsylvania State University)

[11] Viehland D and Powers J 2001 Effect of uniaxial stress on the electromechanical properties of $0.7Pb(Mg_{1/3}Nb_{2/3})O_3$–$0.3PbTiO_3$ crystals and ceramics *J. Appl. Phys.* **89** 1820

[12] Ren X and Zhang L X 2006 Electro-shape-memory effect in ferroelectric martensite *Mater. Sci. Eng.* A **438–440** 1071

[13] De Graef M, Venkateswaran S, Kishi Y, Lograsso T A, Viehland D and Wuttig M 2003 Magnetic tweed contrast in ferromagnetic shape memory alloys *Microsc. Microanal.* **9** 584

[14] Wilson S A *et al* 2007 New materials for micro-scale sensors and actuators *Mater. Sci. Eng.* R **56** 1

[15] Sozinov A, Lanska N, Soroka A and Zou W 2013 12% magnetic field-induced strain in Ni–Mn–Ga-based non-modulated martensite *Appl. Phys. Lett.* **102** 021902

[16] Pagounis E, Szczerba M J, Chulist R and Laufenberg M 2015 Large magnetic field-induced work output in a NiMnGa seven-layered modulated martensite *Appl. Phys. Lett.* **107** 152407

[17] Hager M D, Bode S, Weber C and Schubert U S 2015 Shape memory polymers: past, present and future developments *Prog. Polym. Sci.* **49** 3

[18] Lu L, Wu S and Zhao R R 2024 Mechanics of magnetic-shape memory polymers *J. Mech. Phys. Solids.* **190** 105742

[19] Bhattacharya K and James R D 2005 The material is the machine *Science* **307** 53

[20] Bowles J and Mackenzie J 1954 The crystallography of martensite transformations I *Acta Metall.* **2** 129

[21] Khachaturyan A G, Shapiro S M and Semenovskaya S 1991 Adaptive phase formation in martensitic transformation *Phys. Rev.* B **43** 10832

[22] Viehland D 2000 Symmetry-adaptive ferroelectric mesostates in oriented $Pb(BI_{1/3}BII_{2/3})O_3$–$PbTiO_3$ crystals *J. Appl. Phys.* **88** 4794

[23] Khachaturyan A G and Shatalov G A 1969 Theory of macroscopic periodicity for a phase transition in the solid state *Zh. Eksp. Teor. Fiz.* **29** 1037

[24] Roytburd A L 1969 Elastic interactions in crystals and the formation of structure during a solid-state phase transition *Sov. Phys. Solid State* **11** 1191–8; *Fizika Tverdogo Tela* 11 (6)

[25] Viehland D D and Salje E K H 2014 Domain boundary-dominated systems: adaptive structures and functional twin boundaries *Adv. Phys.* **63** 267

[26] Robertson I M and Wayman C M 1983 Tweed microstructures I. Characterization in β-NiAl *Philos. Mag.* A **48** 421

[27] Randall C A, Barber D J and Whatmore R W 1987 *In situ* TEM experiments on perovskite-structured ferroelectric relaxor materials *J. Microsc.* **145** 275

[28] Dai X, Xu Z and Viehland D 1994 The spontaneous relaxor to normal ferroelectric transformation in La-modified lead zirconate titanate *Philos. Mag.* B **70** 33

[29] Cross L E 1987 Relaxor ferroelectrics *Ferroelectrics* **76** 241

[30] Xu Z, Kim M-C, Li J-F and Viehland D 1996 Observation of a sequence of domain-like states with increasing disorder in ferroelectrics *Philos. Mag.* A **74** 395

[31] Viehland D, Kim M-C, Xu Z and Li J-F 1995 Long-time present tweedlike precursors and paraelectric clusters in ferroelectrics containing strong quenched randomness *Appl. Phys. Lett.* **67** 2471

[32] Bhattacharyya S, Jinschek J R, Cao H, Wang Y U, Li J and Viehland D 2008 Direct high-resolution transmission electron microscopy observation of tetragonal nanotwins within the monoclinic MC phase of $Pb(Mg_{1/3}Nb_{2/3})O_3$-$0.35PbTiO_3$ crystals *Appl. Phys. Lett.* **92** 142904

[33] Cao H, Bai F, Li J, Viehland D, Xu G, Hiraka H and Shirane G 2005 Structural phase transformation and phase boundary/stability studies of field-cooled $Pb(Mg_{1/3}Nb_{2/3}O_3)$–32% $PbTiO_3$ crystals *J. Appl. Phys.* **97** 094101

[34] Clark A E 1993 High power rare earth magnetostrictive materials *J. Intell. Mater. Syst. Struct.* **4** 70

[35] Clark A E, Restorff J B, Wun-Fogle M, Lograsso T A and Schlagel D L 2000 Magnetostrictive properties of body-centered cubic Fe-Ga and Fe-Ga-Al alloys *IEEE Trans. Magn.* **36** 3238

[36] Malmhäll R, Bäckström G, Rao K V, Bhagat S M, Meichle M and Salamon M B 1978 Metglas 2826B—transport, magnetic and thermal properties *J. Appl. Phys.* **49** 1727

[37] Butler S C, Lindberg J F and Clark A E 1996 Hybrid magnetostrictive/piezoelectric tonpilz transducer *Ferroelectrics* **187** 163

[38] Nan C-W, Bichurin M I, Dong S, Viehland D and Srinivasan G 2008 Multiferroic magnetoelectric composites: historical perspective, status, and future directions *J. Appl. Phys.* **103** 031101

[39] Bichurin M and Viehland D 2012 *Magnetoelectricity in Composites* (Pan Standford Publishing)

[40] Wang Y, Gray D, Berry D, Gao J, Li M, Li J and Viehland D 2011 An extremely low equivalent magnetic noise magnetoelectric sensor *Adv. Mater.* **23** 4111

[41] Viehland D, Wuttig M, McCord J and Quandt E 2018 Magnetoelectric magnetic field sensors *MRS Bull.* **43** 834

[42] Wang Y J, Gao J Q, Li M H, Shen Y, Hasanyan D, Li J F and Viehland D 2014 A review on equivalent magnetic noise of magnetoelectric laminate sensors *Philos. Trans. R. Soc.* A **372** 20120455

[43] Leung C M, Li J, Viehland D and Zhuang X 2018 A review on applications of magneto-electric composites: from heterostructural uncooled magnetic sensors, energy harvesters to highly efficient power converters *J. Phys. D: Appl. Phys.* **51** 263002

[44] Aktas O and Salje E K H 2014 Functional twin boundaries and tweed microstructures: a comparison between minerals and device materials *Mineral. Mag.* **78** 1725

[45] Zykova-Timan T and Salje E K H 2014 Highly mobile vortex structures inside polar twin boundaries in $SrTiO_3$ *Appl. Phys. Lett.* **104** 082907

[46] Zygmunt S and Dariusz B 2008 Multiferroic materials for sensors, transducers and memory devices *Arch. Acoust.* **33** 243

[47] Haertling G H 1999 Ferroelectric ceramics: history and technology *J. Am. Ceram. Soc.* **82** 797

[48] Jaffe H 1958 Piezoelectric ceramics *J. Am. Ceram. Soc.* **41** 494

[49] Rödel J, Webber K G, Dittmer R, Jo W, Kimura M and Damjanovic D 2015 Transferring lead-free piezoelectric ceramics into application *J. Eur. Ceram. Soc.* **35** 1659

[50] Koruza J, Bell A J, Frömling T, Webber K G, Wang K and Rödel J 2018 Requirements for the transfer of lead-free piezoceramics into application *J. Mater.* **4** 13

[51] Eric Cross L 1993 Ferroelectric ceramics: tailoring properties for specific applications *Ferroelectric Ceramics* (Basel: Birkhäuser Basel) pp 1–85

[52] Genenko Y A, Glaum J, Hoffmann M J and Albe K 2015 Mechanisms of aging and fatigue in ferroelectrics *Mater. Sci. Eng.* B **192** 52

[53] Robels U and Arlt G 1993 Domain wall clamping in ferroelectrics by orientation of defects *J. Appl. Phys.* **73** 3454

[54] Carl K and Hardtl K H 1977 Electrical after-effects in $Pb(Ti,Zr)O_3$ ceramics *Ferroelectrics* **17** 473

[55] Lambeck P V and Jonker G H 1986 The nature of domain stabilization in ferroelectric perovskites *J. Phys. Chem. Solids* **47** 453

[56] Fancher C M, Brewer S, Chung C C, Röhrig S, Rojac T, Esteves G, Deluca M, Bassiri-Gharb N and Jones J L 2017 The contribution of 180° domain wall motion to dielectric properties quantified from *in situ* x-ray diffraction *Acta Mater.* **126** 36

[57] Trolier-McKinstry S, Bassiri Gharb N and Damjanovic D 2006 Piezoelectric nonlinearity due to motion of 180° domain walls in ferroelectric materials at subcoercive fields: a dynamic poling model *Appl. Phys. Lett.* **88** 202901

[58] Damjanovic D 2006 Hysteresis in piezoelectric and ferroelectric materials *The Science of Hysteresis* vol 3 (Amsterdam: Elsevier) pp 337–465

[59] Otonicar M, Dragomir M and Rojac T 2022 Dynamics of domain walls in ferroelectrics and relaxors *J. Am. Ceram. Soc.* **105** 6479

[60] Chandrasekaran A, Damjanovic D, Setter N and Marzari N 2013 Defect ordering and defect–domain-wall interactions in $PbTiO_3$: a first-principles study *Phys. Rev.* B **88** 214116

[61] Erhart P, Eichel R-A, Träskelin P and Albe K 2007 Association of oxygen vacancies with impurity metal ions in lead titanate *Phys. Rev.* B **76** 174116

[62] Erhart P, Träskelin P and Albe K 2013 Formation and switching of defect dipoles in acceptor-doped lead titanate: a kinetic model based on first-principles calculations *Phys. Rev.* B **88** 024107

[63] Erhart P and Albe K 2015 Dopants and dopant–vacancy complexes in tetragonal lead titanate: a systematic first principles study *Comput. Mater. Sci.* **103** 224

[64] Eichel R-A 2011 Structural and dynamic properties of oxygen vacancies in perovskite oxides—analysis of defect chemistry by modern multi-frequency and pulsed EPR techniques *Phys. Chem. Chem. Phys.* **13** 368

[65] Eichel R-A, Erhart P, Träskelin P, Albe K, Kungl H and Hoffmann M J 2008 Defect-dipole formation in copper-doped $PbTiO_3$ ferroelectrics *Phys. Rev. Lett.* **100** 095504

[66] Meštrić H, Eichel R-A, Dinse K-P, Ozarowski A, van Tol J and Brunel L C 2004 High-frequency electron paramagnetic resonance investigation of the Fe^{3+} impurity center in polycrystalline $PbTiO_3$ in its ferroelectric phase *J. Appl. Phys.* **96** 7440

[67] Meštrić H, Eichel R-A, Kloss T, Dinse K-P, Laubach S, Laubach S, Schmidt P C, Schönau K A, Knapp M and Ehrenberg H 2005 Iron-oxygen vacancy defect centers in $PbTiO_3$: Newman superposition model analysis and density functional calculations *Phys. Rev.* B **71** 134109

[68] Meštrić H *et al* 2006 Iron-oxygen vacancy defect association in polycrystalline iron-modified $PbZrO_3$ antiferroelectrics: multifrequency electron paramagnetic resonance and newman superposition model analysis *Phys. Rev.* B **73** 184105

[69] Zhang L, Erdem E, Ren X and Eichel R-A 2008 Reorientation of $(Mn_{Ti}''-V_O^{\bullet\bullet})^\times$ defect dipoles in acceptor-modified $BaTiO_3$ single crystals: an electron paramagnetic resonance study *Appl. Phys. Lett.* **93** 202901

[70] Aksel E, Erdem E, Jakes P, Jones J L and Eichel R-A 2010 Defect structure and materials 'hardening' in Fe_2O_3-doped $[Bi_{0.5}Na_{0.5}]TiO_3$ ferroelectrics *Appl. Phys. Lett.* **97** 2

[71] Eichel R-A, Erünal E, Jakes P, Körbel S, Elsässer C, Kungl H, Acker J and Hoffmann M J 2013 Interactions of defect complexes and domain walls in CuO-doped ferroelectric (K,Na) NbO_3 *Appl. Phys. Lett.* **102** 242908

[72] Tsvetkov D S, Sereda V V, Malyshkin D A, Ivanov I L and Zuev A Y 2022 Chemical lattice strain in nonstoichiometric oxides: an overview *J. Mater. Chem.* A **10** 6351

[73] Ren X 2004 Large electric-field-induced strain in ferroelectric crystals by point-defect-mediated reversible domain switching *Nat. Mater.* **3** 91

[74] Morozov M I and Damjanovic D 2008 Hardening-softening transition in Fe-doped Pb(Zr, Ti)O_3 ceramics and evolution of the third harmonic of the polarization response *J. Appl. Phys.* **104** 034107

[75] Cheng T and Li J (ed) 2022 *Piezoelectric Actuators* (London: IntechOpen)

[76] Genenko Y A, Glaum J, Hirsch O, Kungl H, Hoffmann M J and Granzow T 2009 Aging of poled ferroelectric ceramics due to relaxation of random depolarization fields by space-charge accumulation near grain boundaries *Phys. Rev.* B **80** 224109

[77] Postnikov V S, Pavlov V S and Turkov S K 1970 Internal friction in ferroelectrics due to interaction of domain boundaries and point defects *J. Phys. Chem. Solids* **31** 1785

[78] Eichel R-A, Meštrić H, Kungl H and Hoffmann M J 2006 Multifrequency electron paramagnetic resonance analysis of polycrystalline gadolinium-doped $PbTiO_3$—charge compensation and site of incorporation *Appl. Phys. Lett.* **88** 3

[79] Damjanovic D 1997 Stress and frequency dependence of the direct piezoelectric effect in ferroelectric ceramics *J. Appl. Phys.* **82** 1788

[80] Damjanovic D and Demartin M 1996 The Rayleigh law in piezoelectric ceramics *J. Phys. D: Appl. Phys.* **29** 2057

[81] Morozov M, Damjanovic D and Setter N 2005 The nonlinearity and subswitching hysteresis in hard and soft PZT *J. Eur. Ceram. Soc.* **25** 2483

[82] Zhao Z, Lv Y, Dai Y and Zhang S 2020 Ultrahigh electro-strain in acceptor-doped KNN lead-free piezoelectric ceramics via defect engineering *Acta Mater.* **200** 35

[83] Huangfu G, Zeng K, Wang B, Wang J, Fu Z, Xu F, Zhang S, Luo H, Viehland D and Guo Y 2022 Giant electric field-induced strain in lead-free piezoceramics *Science* **378** 1125

[84] Jia Y, Fan H, Zhang A, Wang H, Lei L, Quan Q, Dong G, Wang W and Li Q 2023 Giant electro-induced strain in lead-free relaxor ferroelectrics via defect engineering *J. Eur. Ceram. Soc.* **43** 947

[85] Feng W, Luo B, Bian S, Tian E, Zhang Z, Kursumovic A, MacManus-Driscoll J L, Wang X and Li L 2022 Heterostrain-enabled ultrahigh electrostrain in lead-free piezoelectric *Nat. Commun.* **13** 5086

[86] Luo H *et al* 2023 Achieving giant electrostrain of above 1% in $(Bi,Na)TiO_3$-based lead-free piezoelectrics via introducing oxygen-defect composition *Sci. Adv.* **9** eade7078

[87] Hoffmann M, Hammer M, Endriss A and Lupascu D 2001 Correlation between micro-structure, strain behavior, and acoustic emission of soft PZT ceramics *Acta Mater.* **49** 1301

[88] Das Adhikary G *et al* 2025 Longitudinal strain enhancement and bending deformations in piezoceramics *Nature* **637** 333

[89] He X *et al* 2024 Ultra-large electromechanical deformation in lead-free piezoceramics at reduced thickness *Mater. Horizons* **11** 1079

[90] Tian S, Wang B, Li B, Guo Y, Zhang S and Dai Y 2024 Defect dipole stretching enables ultrahigh electrostrain *Sci. Adv.* **10** eadn2829

[91] Das Adhikary G, Singh D N, Tina G A, Muleta G J and Ranjan R 2023 Ultrahigh electrostrain > 1% in lead-free piezoceramics: role of disk dimension *J. Appl. Phys.* **134** 054101

[92] Kholkin A L, Wütchrich C, Taylor D V and Setter N 1996 Interferometric measurements of electric field-induced displacements in piezoelectric thin films *Rev. Sci. Instrum.* **67** 1935

[93] Uchino K, Yoshizaki M, Kasai K, Yamamura H, Sakai N and Asakura H 1987 Monomorph actuators' using semiconductive ferroelectrics *Jpn. J. Appl. Phys.* **26** 1046

[94] Wu C C M, Kahn M and Moy W 1996 Piezoelectric ceramics with functional gradients: a new application in material design *J. Am. Ceram. Soc.* **79** 809

[95] Haertling G H 1997 Rainbow actuators and sensors: a new smart technology *Proc. SPIE* **3040** 81–92

[96] Damjanovic D, Klein N, Li J and Porokoonskyy V 2010 What can be expected from lead-free piezoelectric materials? *Funct. Mater. Lett.* **03** 5

[97] Rödel J, Jo W, Seifert K T P, Anton E, Granzow T and Damjanovic D 2009 Perspective on the development of lead-free piezoceramics *J. Am. Ceram. Soc.* **92** 1153

[98] Senousy M S, Rajapakse R K N D, Mumford D and Gadala M S 2009 Self-heat generation in piezoelectric stack actuators used in fuel injectors *Smart Mater. Struct.* **18** 045008

[99] Moulson A J and Herbert J M 2003 *Electroceramics* (New York: Wiley)

[100] Uchino K and Hirose S 2001 Loss mechanisms in piezoelectrics: how to measure different losses separately *IEEE Trans. Ultrason. Ferroelectr. Freq. Control.* **48** 307

[101] Zheng J, Takahashi S, Yoshikawa S, Uchino K and de Vries J W C 1996 Heat generation in multilayer piezoelectric actuators *J. Am. Ceram. Soc.* **79** 3193

[102] Webber K G, Aulbach E and Rödel J 2010 High temperature blocking force measurements of soft lead zirconate titanate *J. Phys. D: Appl. Phys.* **43** 365401

[103] Binnig G and Smith D P E 1986 Single-tube three-dimensional scanner for scanning tunneling microscopy *Rev. Sci. Instrum.* **57** 1688

[104] Ikeda H and Morita T 2011 High-precision positioning using a self-sensing piezoelectric actuator control with a differential detection method *Sensors Actuat. A* **170** 147

[105] Zhang X L, Chen Z X, Cross L E and Schulze W A 1983 Dielectric and piezoelectric properties of modified lead titanate zirconate ceramics from 4.2 to 300 K *J. Mater. Sci.* **18** 968

[106] Morii Y and Iizumi M 1985 Lattice instability in cubic $Cu_{69.2}Al_{25.4}Ni_{5.4}$ related to martensitic phase transition *J. Phys. Soc. Japan* **54** 2948

[107] Ye Z-G, Noheda B, Dong M, Cox D and Shirane G 2001 Monoclinic phase in the relaxor-based piezoelectric/ferroelectric $P(Mg_{1/3}Nb_{2/3})O_3$-$PbTiO_3$ system *Phys. Rev. B* **64** 184114

[108] Noheda B, Gonzalo J A, Cross L E, Guo R, Park S-E, Cox D E and Shirane G 2000 Tetragonal-to-monoclinic phase transition in a ferroelectric perovskite: the structure of $PbZr_{0.52}$ *Phys. Rev. B* **61** 8687

[109] Bai F, Wang N, Li J, Viehland D, Gehring P M, Xu G and Shirane G 2004 X-ray and neutron diffraction investigations of the structural phase transformation sequence under electric field in $0.7Pb(Mg_{1/3}Nb_{2/3})$–$0.3PbTiO_3$ crystal *J. Appl. Phys.* **96** 1620

[110] Fu H and Cohen R E 2000 Polarization rotation mechanism for ultrahigh electro-mechanical response in single-crystal piezoelectrics *Nature* **403** 281

[111] Bellaiche L, García A and Vanderbilt D 2000 Finite-temperature properties of $Pb(Zr_{1-x}Ti)$ *Phys. Rev. Lett.* **84** 5427

[112] Bai F, Li J and Viehland D 2004 Domain hierarchy in annealed (001)-oriented $Pb(Mg_{1/3}Nb_{2/3})O_3$-$X\%PbTiO_3$ single crystals *Appl. Phys. Lett.* **85** 2313

[113] Bai F, Li J and Viehland D 2005 Domain engineered states over various length scales in (001)-oriented $Pb(Mg_{1/3}Nb_{2/3})O_3$-$X\%PbTiO_3$ crystals: electrical history dependence of hierarchal domains *J. Appl. Phys.* **97** 054103

[114] Kaufmann S, Rößler U K, Heczko O, Wuttig M, Buschbeck J, Schultz L and Fähler S 2010 Adaptive modulations of martensites *Phys. Rev. Lett.* **104** 145702

[115] Tian F, Liu Y, Ma R, Li F, Xu Z and Yang Y 2021 Properties of PMN-PT single crystal piezoelectric material and its application in underwater acoustic transducer *Appl. Acoust.* **175** 107827

[116] Chen C-T, Lin S-C, Trstenjak U, Spreitzer M and Wu W-J 2021 Comparison of metal-based PZT and PMN–PT energy harvesters fabricated by aerosol deposition method *Sensors* **21** 4747

[117] Catalan G and Scott J F 2009 Physics and applications of bismuth ferrite *Adv. Mater.* **21** 2463

[118] Masó N and West A R 2012 Electrical properties of Ca-doped $BiFeO_3$ ceramics: from p-type semiconduction to oxide-ion conduction *Chem. Mater.* **24** 2127

[119] Schrade M, Masó N, Perejón A, Pérez-Maqueda L A and West A R 2017 Defect chemistry and electrical properties of $BiFeO_3$ *J. Mater. Chem. C* **5** 10077

[120] Wefring E T, Einarsrud M-A and Grande T 2015 Electrical conductivity and thermopower of $(1 - x)BiFeO_3 - xBi_{0.5}K_{0.5}TiO_3$ ($x = 0.1, 0.2$) ceramics near the ferroelectric to paraelectric phase transition *Phys. Chem. Chem. Phys.* **17** 9420

[121] Raymond M V and Smyth D M 1996 Defects and charge transport in perovskite ferroelectrics *J. Phys. Chem. Solids* **57** 1507

[122] Liu L *et al* 2024 Piezoelectric properties of $BiFeO_3$ exposed to high temperatures *Adv. Funct. Mater.* **34** 2314807

[123] Rojac T, Ursic H, Bencan A, Malic B and Damjanovic D 2015 Mobile domain walls as a bridge between nanoscale conductivity and macroscopic electromechanical response *Adv. Funct. Mater.* **25** 2099

[124] Liu L, Rojac T, Damjanovic D, Michiel M D and Daniels J 2018 Frequency-dependent decoupling of domain-wall motion and lattice strain in bismuth ferrite *Nat. Commun.* **9** 4928

[125] Liu L, Rojac T, Damjanovic D, Li J-F, Michiel M D and Daniels J 2022 Reduction of the lattice strain with increasing field amplitude in polycrystalline $BiFeO_3$ *Acta Mater.* **240** 118319

[126] Rojac T 2022 Piezoelectric nonlinearity and hysteresis arising from dynamics of electrically conducting domain walls *Piezoelectric Actuators* (IntechOpen)

[127] Damjanovic D, Demartin Maeder M, Duran Martin P, Voisard C and Setter N 2001 Maxwell–Wagner piezoelectric relaxation in ferroelectric heterostructures *J. Appl. Phys.* **90** 5708

[128] Lunkenheimer P, Bobnar V, Pronin A V, Ritus A I, Volkov A A and Loidl A 2002 Origin of apparent colossal dielectric constants *Phys. Rev. B* **66** 052105

[129] Seidel J *et al* 2009 Conduction at domain walls in oxide multiferroics *Nat. Mater.* **8** 229

[130] Rojac T *et al* 2017 Domain-wall conduction in ferroelectric $BiFeO_3$ controlled by accumulation of charged defects *Nat. Mater.* **16** 322

[131] Pramanick A, Damjanovic D, Daniels J E, Nino J C and Jones J L 2011 Origins of electro-mechanical coupling in polycrystalline ferroelectrics during subcoercive electrical loading *J. Am. Ceram. Soc.* **94** 293

[132] Höfling M *et al* 2021 Control of polarization in bulk ferroelectrics by mechanical dislocation imprint *Science* **372** 961

[133] Zhao C *et al* 2021 Precipitation hardening in ferroelectric ceramics *Adv. Mater.* **33** 2102421

[134] Otoničar M *et al* 2020 Connecting the multiscale structure with macroscopic response of relaxor ferroelectrics *Adv. Funct. Mater.* **30** 2006823

[135] Weng L 2006 Ultrasound transducers for imaging and therapy *J. Acoust. Soc. Am.* **120** 3456

[136] Eccardt P C, Niederer K and Fischer B n.d. Micromachined transducers for ultrasound applications *1997 IEEE Ultrasonics Symp. Proc. An Int. Symp.* vol 2 (Piscataway, NJ: IEEE) pp 1609–18

[137] Trolier-McKinstry S and Muralt P 2004 Thin film piezoelectrics for MEMS *J. Electroceramics* **12** 7

[138] Haider S T, Shah M A, Lee D-G and Hur S 2023 A review of the recent applications of aluminum nitride-based piezoelectric devices *IEEE Access* **11** 58779

[139] Mura M L, Giusti D, Prelini C L, Ferrera M and Savoia A S 2024 Air-coupled piezoelectric micromachined ultrasonic transducers (PMUTs) for high-resolution distance sensing *2024 IEEE Sensors* (Piscataway, NJ: IEEE) pp 1–4

[140] Baborowski J, Ledermann N and Muralt P 2003 Piezoelectric micromachined transducers (PMUT's) based on PZT thin films *2002 IEEE Ultrasonics Symp. Proc.* vol 2 (Piscataway, NJ: IEEE) pp 1051–4

[141] Muralt P 2000 PZT thin films for microsensors and actuators: where do we stand? *IEEE Trans. Ultrason. Ferroelectr. Freq. Control.* **47** 903

[142] Muralt P, Ledermann N, Paborowski J, Barzegar A, Gentil S, Belgacem B, Petitgrand S, Bosseboeuf A and Setter N 2005 Piezoelectric micromachined ultrasonic transducers based on PZT thin films *IEEE Trans. Ultrason. Ferroelectr. Freq. Control.* **52** 2276

[143] Bernstein J J, Finberg S L, Houston K, Niles L C, Chen H D, Cross L E, Li K K and Udayakumar K 1997 Micromachined high frequency ferroelectric sonar transducers *IEEE Trans. Ultrason. Ferroelectr. Freq. Control.* **44** 960

[144] Moisello E, Novaresi L, Sarkar E, Malcovati P, Costa T L and Bonizzoni E 2024 PMUT and CMUT devices for biomedical applications: a review *IEEE Access* **12** 18640

[145] Herrera B, Pirro M, Giribaldi G, Colombo L and Rinaldi M 2021 AlScN programmable ferroelectric micromachined ultrasonic transducer (FMUT) *2021 21st Int. Conf. on Solid-State Sensors, Actuators and Microsystems (Transducers)* (Piscataway, NJ: IEEE) pp 38–41

[146] Sammoura F, Smyth K and Kim S-G 2013 Optimizing the electrode size of circular bimorph plates with different boundary conditions for maximum deflection of piezoelectric micromachined ultrasonic transducers *Ultrasonics* **53** 328

[147] Allison C 2002 PMUT 2D *onscale* https://support.onscale.com/hc/en-us/articles/360006551932-PMUT-2D

[148] Mansoori A, Hoff L, Salmani H and Halvorsen E 2023 Electrode design based on strain mode shapes for configurable PMUTs *IEEE Open J. Ultrason. Ferroelectr. Freq. Control.* **3** 88

[149] Muralt P 2017 Which is the best thin film piezoelectric material? *International Ultrasonics Symposium* (Piscataway, NJ: IEEE) pp 1

[150] Ledesma E, Zamora I, Uranga A and Barniol N 2021 9.5% scandium doped ALN PMUT compatible with pre-processed CMOS substrates *34th Int. Conf. on Micro Electro Mechanical Systems (MEMS)* (Piscataway, NJ: IEEE) pp 887–90

[151] Luo G-L, Kusano Y and Horsley D 2018 Immersion PMUTs fabricated with a low thermal-budget surface micromachining process *2018 IEEE Int. Ultrasonics Symp. (IUS)* (Piscataway, NJ: IEEE) pp 1–4

[152] Horsley D, Lu Y and Rozen O 2017 Flexural piezoelectric resonators *Piezoelectric MEMS Resonators: Microsystems and Nanosystems* ed H Bhugra and G Piazza (Cham: Springer) pp 153–73

[153] Zhang S, Zhou Z, Luo J and Li J 2019 Potassium-sodium-niobate-based thin films: lead free for micro-piezoelectrics *Ann. Phys.* **531** 1800525

[154] Luo G-L, Kusano Y and Horsley D A 2021 Airborne piezoelectric micromachined ultrasonic transducers for long-range detection *J. Microelectromech. Syst.* **30** 81

[155] Wang Q, Lu Y, Fung S, Jiang X, Mishin S, Oshmyansky Y and Horsley D A 2016 Scandium doped aluminum nitride based piezoelectric micromachined ultrasound

transducers *2016 Solid-State, Actuators, and Microsystems Workshop Technical Digest* vol 1 (San Diego, CA: Transducer Research Foundation) pp 436–9

[156] Shieh B, Sabra K G and Degertekin F L 2018 A hybrid boundary element model for simulation and optimization of large piezoelectric micromachined ultrasonic transducer arrays *IEEE Trans. Ultrason. Ferroelectr. Freq. Control.* **65** 50

[157] Waqar M, Wu H, Chen J, Yao K and Wang J 2022 Evolution from lead-based to lead-free piezoelectrics: engineering of lattices, domains, boundaries, and defects leading to giant response *Adv. Mater.* **34** 1

[158] Jung J, Bastien J-C, Lefevre A, Benedetto K, Dejaeger R, Blard F and Fain B 2020 Wafer-level experimental study of residual stress in AlN-based bimorph iezoelectric micro-machined ultrasonic transducer *Eng. Res. Express* **2** 045013

[159] Francombe M H and Krishnaswamy S V 1990 Growth and properties of piezoelectric and ferroelectric films *J. Vac. Sci. Technol.* A **8** 1382

[160] Zhang Y and Demosthenous A 2021 Integrated circuits for medical ultrasound applications: imaging and beyond *IEEE Trans. Biomed. Circuits Syst.* **15** 838

[161] Wong S J Z, Roy K, Lee C and Zhu Y 2024 Thin-film piezoelectric micromachined ultrasound transducers in biomedical applications: a review *IEEE Trans. Ultrason. Ferroelectr. Freq. Control.* **71** 622

[162] Rong Z, Zhang M, Ning Y and Pang W 2022 An ultrasound-induced wireless power supply based on AlN piezoelectric micromachined ultrasonic transducers *Sci. Rep.* **12** 16174

[163] Ledesma E, Uranga A, Torres F and Barniol N 2024 Fully integrated pitch-matched AlScN PMUT-on-CMOS array for high-resolution ultrasound images *IEEE Sens. J.* **24** 15954

[164] Christensen-Jeffries K *et al* 2020 Super-resolution ultrasound imaging *Ultrasound Med. Biol.* **46** 865

[165] Wang Z, Yang Y, Viswan R, Li J and Viehland D 2011 Giant electric field controlled magnetic anisotropy in epitaxial $BiFeO_3$–$CoFe_2O_4$ thin film heterostructures on single crystal $Pb(Mg_{1/3}Nb_{2/3})_{0.7}Ti_{0.3}O_3$ substrate *Appl. Phys. Lett.* **99** 132909

[166] Yan L, Wang Y, Li J, Pyatakov A and Viehland D 2008 Nanogrowth twins and abnormal magnetic behavior in $CoFe_2O_4$ epitaxial thin films *J. Appl. Phys.* **104** 123910

[167] Zheng H *et al* 2004 Multiferroic $BaTiO_3$-$CoFe_2O_4$ nanostructures *Science* **303** 661

[168] Gao M, Yang Y, Rao W-F and Viehland D 2021 Magnetoelectricity in vertically aligned nanocomposites: past, present, and future *MRS Bull.* **46** 123

[169] Tang X, Gao M, Leung C M, Luo H, Li J and Viehland D 2019 Non-volatility using materials with only volatile properties: vertically integrated magnetoelectric heterostructures and their potential for multi-level-cell devices *Appl. Phys. Lett.* **114** 242903

[170] Harris V G 2012 Modern microwave ferrites *IEEE Trans. Magn.* **48** 1075

[171] Sebastian M T, Ubic R and Jantunen H 2017 *Microwave Materials and Applications* (New York: Wiley and Sons)

[172] Bichurin M I, Petrov R V and Kiliba Y V 1997 Magnetoelectric microwave phase shifters *Ferroelectrics* **204** 311

[173] Fetisov Y K and Srinivasan G 2006 Electric field tuning characteristics of a ferrite-piezoelectric microwave resonator *Appl. Phys. Lett.* **88** 5

[174] Yan J-M *et al* 2019 Manipulation of the electronic transport properties of charge-transfer oxide thin films of $NdNiO_3$ using static and electric-field-controllable dynamic lattice strain *Phys. Rev. Appl.* **11** 034037

[175] Chen C, Das P, Aytan E, Zhou W, Horowitz J, Satpati B, Balandin A A, Lake R K and Wei P 2020 Strain-controlled superconductivity in few-layer $NbSe_2$ *ACS Appl. Mater. Interfaces.* **12** 38744

[176] Lin Z *et al* 2015 Quasi-two-dimensional superconductivity in $FeSe_{0.3}Te_{0.7}$ thin films and electric-field modulation of superconducting transition *Sci. Rep.* **5** 14133

[177] Qi Y, Sadi M A, Hu D, Zheng M, Wu Z, Jiang Y and Chen Y P 2023 Recent progress in strain engineering on van der Waals 2D materials: tunable electrical, electrochemical, magnetic, and optical properties *Adv. Mater.* **35** e2205714

[178] Jin L *et al* 2022 The rise of 2D materials/ferroelectrics for next generation photonics and optoelectronics devices *APL Mater.* **10** 060903

[179] Hui Y Y, Liu X, Jie W, Chan N Y, Hao J, Hsu Y-T, Li L-J, Guo W and Lau S P 2013 Exceptional tunability of band energy in a compressively strained trilayer MoS_2 sheet *ACS Nano* **7** 7126

[180] Tang X *et al* 2018 Nanopillars with *E*-field accessible multi-state ($N \geq 4$) magnetization having giant magnetization changes in self-assembled $BiFeO_3$-$CoFe_2O_4$/Pb($Mg_{1/3}Nb_{2/3}$)-38at%PbTiO$_3$ heterostructures *Sci. Rep.* **8** 1628

[181] Tanner L E, Schryvers D and Shapiro S M 1990 Electron microscopy and neutron scattering studies of premartensitic behavior in ordered Ni Al β_2 phase *Mater. Sci. Eng.* **127** 205–13

www.ingramcontent.com/pod-product-compliance
Lightning Source LLC
Chambersburg PA
CBHW082137210326
41599CB00031B/6016